Vertebrate Ecology

in the NORTHERN NEOTROPICS

JOHN F. EISENBERG, *Editor*

Results from a
Team Research Effort
Sponsored by the
National Zoological Park,
Smithsonian Institution,
March 1973 to August 1978
and a
Workshop held at the
National Zoological Park
on December 14, 1978

Smithsonian Institution Press
Washington, D.C., 1979

Results from a Team Research Effort
Sponsored by the National Zoological Park,
Smithsonian Institution,
March 1973 to August 1978
and a
Workshop held at the
National Zoological Park
on December 14, 1978

Sponsored by the Friends of the National Zoo (FONZ)

Library of Congress Cataloging in Publication Data
Main entry under title:
Vertebrate ecology in the northern neotropics.
"Results from a team research effort sponsored by the National Zoological Park, Smithsonian Institution, March 1973 to August 1978 and a workshop held at the National Zoological Park on December 14, 1978."
1. Vertebrates—South America—Ecology. 2. Vertebrates—Panama—Ecology. I. Eisenberg, John Frederick. II. Washington, D.C. National Zoological Park. III. Series: Washington, D.C. National Zoological Park. Symposia of the National Zoological Park.
QL606.54A1V47 596'.09'8 79-9436
ISBN 0-87474-410-5
ISBN 0-87474-409-1 pbk.

Contents

JOHN F. EISENBERG
National Zoological Park
Washington, D.C. 20008

Preface

This volume brings together contributions from investigators who have been associated with Smithsonian Institution projects in Panama and Venezuela. Although much of our field research in the neotropics has been carried out in Panama, we have been actively developing long-term projects in other areas.

Venezuela became a focus for our efforts and we were joined by many Venezuelan colleagues who encouraged us and shared with us many of the joys and minor hardships of field work. We are especially grateful to Professor Edgardo Mondolfi who served as the Venezuelan coordinator for those research efforts sponsored by the National Zoological Park. Sr. Tomás Blohm made his ranch Fundo Pequario Masaguaral available to us as a study area. This working cattle ranch has been protected for over twenty-five years from fire and the wildlife has not been hunted since the ownership of the ranch came to him. I know of no other place in the northern neotropics where direct observation of such species as the crab-eating fox, the howler monkey, and the giant anteater can be carried out with such expediency. Sr. Blohm was most helpful with logistics and with assisting us in numerous details during the establishment of our vehicles and equipment in Venezuela.

Edgardo Mondolfi first suggested that we consider the premontane forests of Guatopo National Park as a study area. Permission to work in Guatopo was facilitated by Sr. Rafeal Garcia. Accommodations and the freedom to work in the park were generously provided by Luis Escalona, Jefe, Parque Nacional Guatopo. Dr. Pedro Trebbau was extremely helpful in assisting us with study-site selections and aided us in the shipment of prehensile-tailed porcupines and crab-eating foxes to the United States. Juan Gomez-Nuñez and his wife Joan showed us very generous hospitality when we were in Maracay. Juan Gomez-Nuñez's great interest in small mammal population dynamics greatly aided us in developing several projects. Johani Ojasti provided thoughtful advice and shared his experiences. Rodrigo Parra provided many thoughtful discussions concerning habitat utilization by capybara. Mr. William H. Phelps Jr. was most generous with his time in discussing the distribution of Venezuelan birds. Paul Schwartz was helpful with sage advice on living and working at Venezuelan field sites. To all of our Venezuelan colleagues and friends, many thanks for your help in the first phase of this project.

Maintaining correspondence with those of us who were away in the field for prolonged periods of time fell to Ms. Tabetha Gilmore. All of us are very grateful to Ms. Betty Howser who coordinated shipments of materials and equipment. The manuscript for this volume was ably prepared by Mrs. Wyotta Holden. Sigrid James prepared the illustrations.

The development of our research in Venezuela was actively supported by S. Dillon Ripley, Secretary of the Smithsonian Institution. The research reported herein was in part supported by continuing grants from the IESP Program administered by the Office of the Assistant Secretary for Science for the Smithsonian Institution, David Challinor. Dr. Challinor took a keen interest in the project and was helpful in many ways. We are especially indebted to Ross Simons and Harold Michaelson for their assistance and encouragement at difficult phases of the administration of this project. The primate research was supported in part by NIMH Grant No. MH 28840. The Friends of the National Zoo supported the publication of this volume through a generous grant in aid.

List of Contributors

*PETER V. AUGUST
Department of Biology, Boston University,
2 Cummington Street, Boston, Massachusetts 02215

CHARLES A. BRADY
Department of Zoology, University of Ohio,
Athens, Ohio

*JOHN F. EISENBERG
National Zoological Park, Washington, D.C. 20008

LINDA K. GORDON
Department of Vertebrate Zoology
Museum of Natural History, Smithsonian Institution,
Washington, D.C. 20560

*KENNETH GREEN
National Zoological Park, Washington, D.C. 20008

CHARLES O. HANDLEY, JR.
Department of Vertebrate Zoology
Museum of Natural History, Smithsonian Institution,
Washington, D.C. 20560

*DEVRA G. KLEIMAN
National Zoological Park, Washington, D.C. 20008

*DAVID MACK
National Zoological Park, Washington, D.C. 20008

*EUGENE MALINIAK
National Zoological Park, Washington, D.C. 20008

*DALE MARCELLINI
National Zoological Park, Washington, D.C. 20008

EDGARDO MONDOLFI
Escuela de Biologia, Facultad de Ciencias,
Universidad Central de Venezuela,
Caracas, Venezuela

*EUGENE S. MORTON
National Zoological Park, Washington, D.C. 20008

MARGARET A. O'CONNELL
Department of Biological Sciences,
Texas Tech University, P. O. Box 4149, Lubbock, Texas 79409

*KENT REDFORD
Museum of Comparative Zoology,
Harvard University, Cambridge, Massachusetts

JOHN ROBINSON
National Zoological Park, Washington, D.C. 20008

*RASANAYAGAM RUDRAN
National Zoological Park, Washington, D.C. 20008

BETSY TRENT THOMAS
Tomas Enterprises, Apartado, 80844
Caracas 108, Venezuela

*RICHARD W. THORINGTON, JR.
Department of Vertebrate Zoology,
Museum of Natural History, Smithsonian Institution,
Washington, D.C. 20560

R. G. TROTH
Department of Botany,
University of Michigan, Ann Arbor, Michigan

RALPH WETZEL
Biological Sciences Group,
University of Connecticut,
Storrs, Connecticut 06268

* Attended the Workshop on December 14, 1978.

SECTION 1:

Habitats and Distribution Patterns

View of the llanos. The savanna habitats of northern South America include several different subdivisions. Fire, impeded vertical drainage, flooding, and seasonal aridity contribute to the different savanna forms.

Introduction

The northern neotropics, as defined for this volume, include Panama, Colombia, Venezuela, Guyana, Surinam, and French Guiana (Figure 1). The region is extremely diverse with respect to elevation and climate. The western portions of Colombia and Venezuela are dominated by three separate cordilleras of the Andes. Drainage from the Andes contributes in part to the formation of the Orinoco River and the Orinoco, taking its origin from the south, proceeds in a northerly direction until approximately mid-Venezuela when it cuts eastward to exit at the Caribbean. The second montane area dominating the northern neotropics is the Guyana Highlands. This is an area of dissected and eroded granite formations that give rise to high, mesa-like plateaus. The soils of this region tend to be nutrient-poor. The Rio Negro takes its origin in part from the Guyana Highlands and proceeds south to enter the Amazon. Part of the Orinoco takes its origin from the Guyana Highlands and major rivers entering it, such as the Caroni, Caura, and Ventuari, begin in these same highlands (Figure 2).

Much of the research reported in this volume was conducted at two major study sites in Venezuela—Guatopo National Park and Fundo Pecuario Masaguaral. Guatopo National Park is in the north coast range of Venezuela and is approximately 40 km south-southeast of Caracas. The topography is submontane to montane with elevations averaging around 700 meters. The rainfall in northern Venezuela is characterized by a marked dry season commencing in December and often extending to May. The north coastal range, however, receives considerable precipitation even during the dry season as the result of the prevailing winds and adiabatic cooling of the air mass as it passes over the mountain range, thus releasing water at higher elevations. The net result is that precipitation is more evenly distributed throughout the year in the

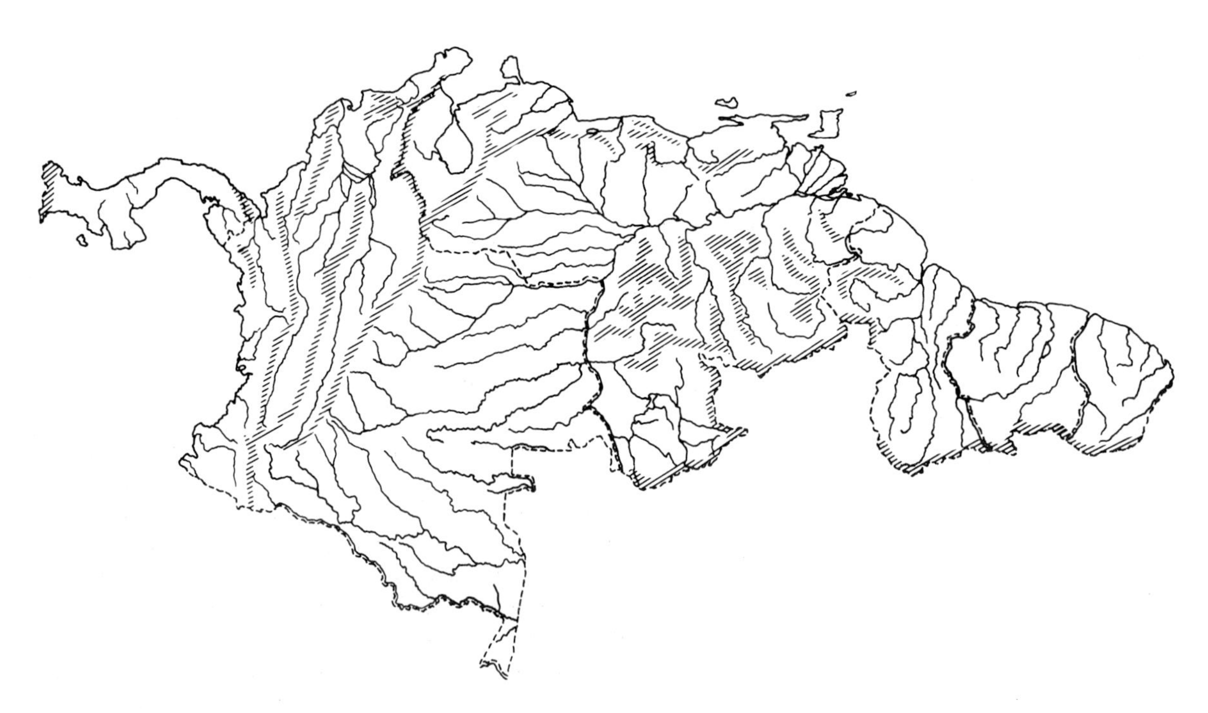

Figure 1. Region of the Northern neotropics used as a reference base for discussions in this volume. Major montane areas are indicated by cross-hatching. The Andean chain to the west has provided both access corridors and barriers to mammalian dispersal. In the northeast the Andes bracket Lake Maracaibo. The Orinoco River has acted both as a conduit and a barrier to dispersal. The Guyana Highlands to the south of the Orinoco have acted as a filter barrier.

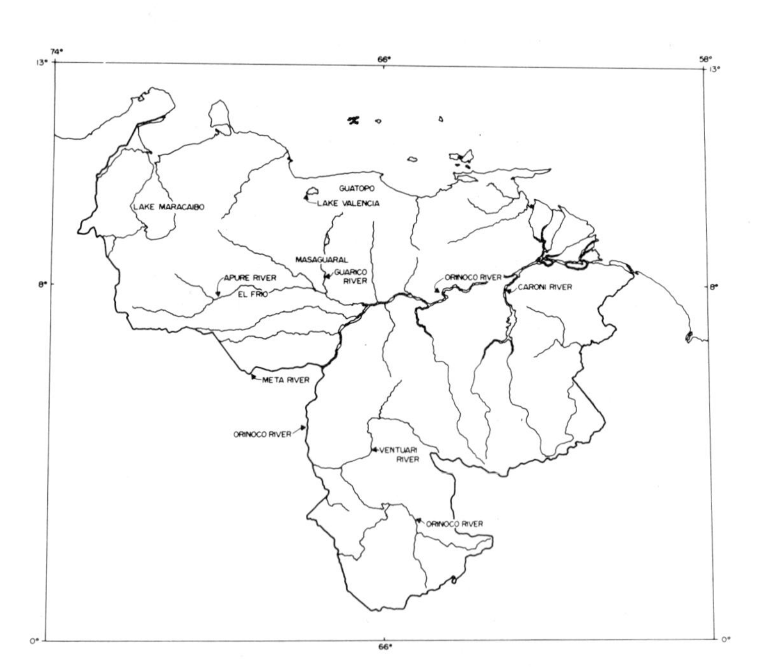

Figure 2. Political map of Venezuela. Major place names and rivers are indicated.

north coastal range than is the case in the seasonally arid llanos. However, a distinct dry season does occur annually in the north coastal range although the timing of its occurrence is somewhat unpredictable.

The region to the south of the north coastal range and to the east of the Andes, north of the Orinoco, is referred to as the llanos. This area is continuous into Colombia. The dominant feature is the low relief of the terrain. Land below the 100-meter contour is most typical of the region with vast tracts of open grasslands that are subject to seasonal inundation. The lack of tree growth is the result of a complex of factors including impeded vertical drainage during the rainy season and fire during the dry season. The problem of savanna formation and the classification of savanna forms for the neotropics has been undertaken by Sarmiento and Monasterio (1975). The vegetational mosaic typifying the high llanos is described by Troth in her consideration of the vegetation at Masaguaral.

In our studies of mammalian ecology and behavior in Venezuela, many of us were greatly aided by the existence of the Smithsonian-Venezuela mammal collection assembled by Charles O. Handley, Jr. during the 1960s. Handley's publication (1976) allowed us to analyze mammalian distributions in Venezuela. In the chapter by Eisenberg, and Redford, the contemporary patterns of mammalian distributions in Venezuela are analyzed with respect to seven defined biogeographical provinces. It is concluded that the llanos region contains a fauna which is derivative from the adjacent biogeographic provinces. One may consider the foothills of the Andes and the gallery forests along the major rivers as conduits for dispersal of forms endemic to other biogeographical provinces. The llanos has probably had a history of aridity alternating with seasonal inundation. Apparently, current conditions are more mesic than during the Pleistocene. As a result, invasions of mammalian forms have continued and the faunal complement in the llanos may be considered derivative. Our interpretations are consistent with those conclusions recently developed by Webb (1978).

The focus, then, in this section is to introduce the reader to the vegetational mosaic referred to as the llanos and to establish the hypothesis that the llanos is in several dynamic stages of succession which parallel the continuing deposition of sediments from the Andes and the Guyana Highlands and that current faunal distributions reflect successive periods of aridity in the distant past. The strong seasonal flux in rainfall imposes marked limitations on plant productivity throughout the annual cycle. Patterns of vertebrate reproduction are as profoundly influenced by the dry season as are the patterns of reproduction of vertebrates in the temperate zone influenced by the annual winter. Here in the geographically defined tropics, we find seasonality, succession, and dynamic distribution patterns which would be familiar to any scientist trained in the temperate zone. I trust by this preliminary presentation that we have laid to rest the notion of stability in the tropics as a generalized phenomenon.

J. F. E.

Literature Cited

Handley, C. O., Jr.
1976. Mammals of the Smithsonian Venezuelan Project. *Brigham Young Univ. Sci. Bull., Biol. Series*, 20(5):1–91.

Sarmiento, G. and M. Monasterio.
1975. A critical consideration of the environmental conditions associated with the occurrence of savanna ecosystems in tropical America. Pages 223–250 in *Tropical Ecological Systems*, edited by F. B. Golley and E. Medina. New York: Springer Verlag.

Webb, S. D.
1978. A history of savanna vertebrates in the New World. Part II. South America and the great interchange. *Ann. Rev. Ecol. Syst.*, 9:393–426.

R. G. TROTH
Department of Botany
University of Michigan
Ann Arbor, Michigan 48104

Vegetational Types on a Ranch in the Central Llanos of Venezuela

ABSTRACT

The study site is located in the central llanos of Venezuela, approximately 8° 34′ N 67° 35′ W. It includes gallery forest and savanna land. The four physiographic savanna units—Medano (sandhill), Banco (non-flooded, low ridge), Bajío (moderately flooded lowlands), and Estero (more deeply flooded lowlands)—are well represented on the ranch. They form a mosaic over the whole ranch except where the gallery forest occurs. Descriptions of the four savanna units and the gallery forest are given, with comments on productivity in the savanna.

RESÚMEN

El sitio del estudio está ubicado en los llanos centrales de Venezuela, 8° 34′N 67° 35′O en un area de bosque de galería y sabana. Los cuatro tipos fisiográficos de sabana están bien representados en el Hato: Médano (duna no inundada), Banco (no inundado, lomo bajo), Bajío (terreno bajo, medianamente inundado), Estero (terreno bajo, muy inundado). Con excepción del bosque de galería, estas unidades forman un mosaico sobre el Hato entero. Se da una descripcióne de todos los clases de tipos de sabana y del bosque de galería, esta presentada con comentarios sobre la producción de la sabana.

Study Site

The llanos of Venezuela cover an irregular area of approximately 200,000 km^2 in the central lowlands of the country. They are bordered to the south and east by the Orinoco River, to the north and northwest by the coastal mountains and the Andes, and extend southwestward into Colombia (Sarmiento, 1975; Sarmiento and Monasterio, 1975). The vegetation differs greatly across this area, both floristically and physiognomically, but savannas prevail; forests occur only in small patches or as gallery forests along rivers.

The study site for several of the papers in this volume is Fundo Pecuario Masaguaral, a working cattle ranch owned by Tomas Blohm. It is located in the western part of the State of Guarico, in the central llanos, about 45 km south of Calabozo, approximately 8° 34′ N, 67° 35′ W. It encompasses more than 3,000 hectares of savanna and gallery forest. The terrain is nearly flat, varying from 60 to 75 m above sea level and many parts of the ranch, in both savanna and forest, are flooded annually in the wet season. Vegetationally the ranch is in a transition zone between areas traditionally known as high llano (north of Calabozo) and low llano (south of San Fernando de Apure). In soil studies it has also been classified as an intermediate area (Sanchez and Fajardo, 1976).

Climate

This area of the central llanos has a strongly seasonal climate divided generally into six wet months (May through October), four dry months (December through March), and two transitional months (April and November). However, both the amount and duration of the rains are extremely variable (Table 1).

Temperature varies only slightly during the year; the greatest diurnal changes as well as the maximum and minimum temperatures occur during the dry season. Data from the Estación Biológica de los Llanos (SVCN) located near Calabozo show for a ten-year period an absolute maximum of 40.8° C (February 1978) and an absolute minimum of 17.3° C (December 1975). The average monthly maximum and minimum are found in Table 2. Temperatures are much the same on Masaguaral. In the first nine months of 1978, the maximum temperature on the ranch was 38.5° C (April) and the minimum, 17.5 C (January).

At the Estación Biológica, the prevailing winds are from the east-northeast or less frequently the east. During the wet season the winds are much reduced in velocity and frequency but the direction remains much the same. This same pattern occurs on the ranch.

Table 1. Summary of total annual precipitation for Corozo Pondo, a village 8 km south of Masaguaral[1]

Year	*Amount (mm)*	*Months with ≥50 mm*
1953	2041.0	8 (Apr–Nov)
1954	1535.0	8 (Apr–Nov)
1955	1829.0	8 (Apr–Nov)
1956	1453.0	6 (May–Oct)
1957	1152.0	7 (May–Nov)
1958	1588.0	6 (May–Oct)
1959	1084.0	7 (May–Nov)
1960	1495.0	9 (Apr–Dec)
1961	1167.0	6 (Jun–Nov)
1962	1600.0	6 (May–Oct)
1963	1821.7	8 (Apr–Nov)
1964	1346.0	8 (Apr–Nov)
1965	1548.8	7 (May–Nov)
1966	1509.7	8 (May –Dec)
1967	1515.0	6 (May–Oct)
1968	1346.6	7 (May–Nov)
1969	2283.8	10 (Feb–Nov)
1970	1555.3	7 (May ?–Nov)
1971	937.7	7 (Apr ?–Oct)
1972	888.0	7 (Apr–Oct)
1973	1424.2	8 (Apr–Nov)
1974	1271.6	6 (May–Oct)
1975	1349.9	7 (May–Nov)
1976	1742.6	8 (Apr–Nov)
1977	ca. 1069.5	7 (May–Nov)

[1] Data courtesy of M.A.R.N.R., Dirreccion General de Informacion e Investigacion del Ambiente, Centro de Informacion.

Vegetation Types

The mosaic of savanna types on the ranch exemplify the traditional physiographic units of Medano (sandhill), Banco (non-flooded, low ridge), Bajío (moderately flooded savanna), and Estero (more deeply flooded savanna). Ramia (1958, 1967) defined these traditional units more explicitly but within the perspective of the central llanos. The definitions used in this paper follow his but are slightly modified to reflect a more specific area of study. I have described these four savanna units plus the gallery forest to characterize the major habitats on the ranch.

In addition I have included phytosociological descriptions of specific sites as examples for most of the

Table 2. Average maximum and minimum temperatures (Estacion Biologica de los Llanos, SVCN)[1]

Month	*Maximum*	*Minimum*
January	36.2	19.8
February	36.8	20.9
March	37.7	22.1
April	37.7	22.4
May	36.0	21.9
June	34.0	21.7
July	32.9	20.8
August	33.1	21.1
September	33.8	21.4
October	34.4	21.4
November	34.6	20.5
December	34.8	19.0

[1] Data courtesy of M.A.R.N.R., Dirreccion General de Informacion e Investigacion del Ambiente, Centro de Informacion.

vegetation units. Tables 3, 4, 5, 7, and 8 list by layer the important plants for specific sites representing Medano, Banco, Bajío, and Estero. For each species of the herb layer, values are given for cover and presence. Cover value is the average percent cover in plots where a particular species occurred. Presence is defined here as the percent of plots in which a species occurred. For woody species density is given per 100 m^2. Each woody species listed has been given a qualitative abundance value. (The locations of the sites are given in Figure 1.)

MEDANO

The sandhills which never flood are irregularly distributed over the eastern half of the ranch. Some form uninterrupted grasslands while others are dominated by a shrub layer (Figure 2). The shrubs, primarily *Byrsonima crassifolia, Annona jahnii* and *Casearia mollis,* form clumps which stand 3 to 6 m tall and have characteristically dense canopies. Lianas and herbaceous vines are common. Trees are infrequent, but occasional small ones occur with the shrubs and even more occasionally, solitary giants are found. Most striking is the complete absence of the common palm. The herb layer in open sun is a mixture of grasses, sedges and tall (0.5 to 1.0 m) robust dicot herbs. There is a significant ground cover of small dicot herbs and dicot seedlings.

The specific Medano site analyzed photosociologically lies east-southeast of the ranch houses. Woody cover averages 47 percent in the plots. The shrubs are arranged in an average of six semi-discrete clumps. The canopy ranges from 3 to 5 m. Occasional slender trees occur, acting as emergents with a supra-canopy of 5 to 6 m. Table 3 lists the important species as determined by relative percent of cover, presence, or density. The soils, both from the open herbaceous patches and from under the shrub canopy, are sands, containing greater than 90 percent sand. They range in wet season pH values from 4.3 to 5.7, and apparently are some of the most basic soils on the ranch.

BANCO

These low ridges follow the line of old water courses that determined their origin (Figure 3). The soils are composed of heavy particles, the ones first deposited

Figure 1. Northern portion of Fundo Pequario Masaguaral outlining locations of the study sites described in the text.

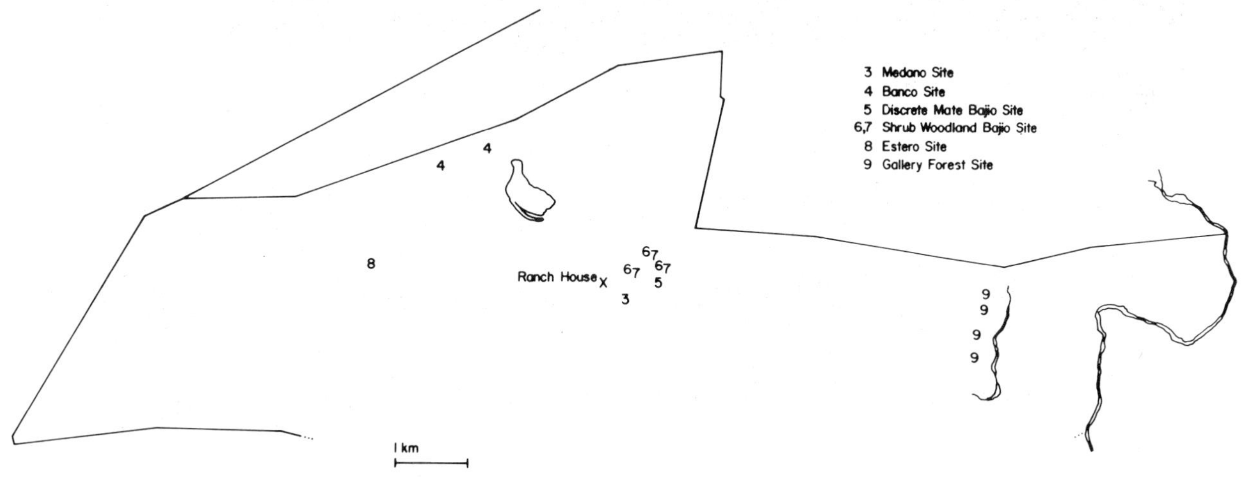

Figure 2. Medano, an area showing grasses and forbs interspersed with shrubs.

Figure 3. Banco. Palms dominate the ridge thrown up by a former water course.

Table 3. Important plants of a specific medano site

Layer/Family/Species		*Density (No./100 m²)*	*Cover (%)*	*Presence (%)*
Shrub				
Annonaceae	*Annona jahnii* Safford	11.00		
Flacourtiaceae	*Casearia mollis* H.B.K.	5.25		
Polygonaceae	*Coccoloba caracasana* Meisn.	0.67		
Herb				
Miscellaneous small dicots			16.0	100
Tall dicots (0.5–1.0 m)				
Euphorbiaceae	*Croton hirtus* L'Herit		28.0	90
Euphorbiaceae	*Croton miquelensis* Ferg.		25.0	70
Fabaceae	*Schrankia leptocarpa* DC.		4.8	60
Grasses and sedges				
Poaceae[1]	*Cenchrus pilosus* H.B.K.		10.4	50
Poaceae[1]	*Anthephora hermaphrodita* (L.) Kuntze		27.0	40
Cyperaceae	Unknown (RGT 1285); sedge ca. 1 m tall		33.0	30
Cyperaceae	*Cyperus brevifolius* (Rottb.) Endl. ex Hassk.		3.4	70

[1] Poaceae (= Graminiae).

by streams. Bancos are not common on the ranch. These areas are never truly flooded, but shallow puddles can be found there during the height of the rainy season. The herb layer reflects this mosaic of puddling, showing a distinct separation of communities. The communities in open sun on non-flooded soil resemble those on Medano sites: a mixture of grasses, sedges, and tall dicot herbs, with ground cover of small dicot herbs. On flooded soil the tall dicot herbs are absent. The woody vegetation is very open and is dominated by the common palm (*Copernicia tectorum*). Each palm usually has an associated strangler fig (*Ficus* spp.). Shrubs cluster openly at the base of the palms. Occasionally an independent tree or shrub will be seen. Herbaceous vines are common in the woody vegetation.

The specific Banco sites located several kilometers northwest of the houses have only 7–8 percent woody cover. Mature palms dominate the tree layer and young individuals dominate the shrub layer. Palms in the tree layer are 3–5 m tall; several have strangler figs well established around their trunks but their canopies are not extensive. Occasionally mimosoids are also found in the tree layer. Table 4 lists the phytosociologically important species. The soils are loams ranging from sandy loam to silty loam. This indicates 30–40 percent sand, 40–50 percent silt, and 10–20 percent clay. Wet season pH values range from 3.7 to 5.2, showing greater acidity than the Medano site.

BAJÍO

The Bajíos, patches of moderately flooded savanna, have the widest distribution on the ranch, occurring from the western edge of the gallery forest westward (Figures 4 and 5). Depending on the pattern of rainfall, the soils may be exposed to air and may begin to dry out several times during the wet season. At the height of the wet season, the water depth varies from 1 to 30 cm. The soils are always heavy, usually a clay. The herb layer is emergent, dominated by grasses, sedges and broad-leaved monocots. The woody vegetation is dominated by the palm and usually associated with it are the strangler fig and several shrubs. Despite this common element, however, small islands of woody vegetation clustered around a palm and known as core-tree matas, the individual Bajío sites show a

Table 4. Important plants of a specific banco site

Layer/Family/Species		*Density (no./100 m²)*	*Cover (%)*	*Presence (%)*
Tree				
Arecaceae[1]	*Copernicia tectorum* (H.B.K.) Mart.	0.67		
Moraceae	*Ficus* spp. (associated exclusively with palms)	0.24		
Shrub				
Arecaceae[1]	Young *C. Tectorum*	0.39		
Flacourtiaceae	*Casearia mollis* H.B.K.	0.39		
Rutaceae	*Zanthoxylum culantrillo* H.B.K.	0.33		
Polygonaceae	*Coccoloba caracasana* Meisn.	0.28		
Herb (dry)				
Miscellaneous small dicots			19.0	85
Tall dicots				
Malvaceae	*Sida acuta* Burm. f.		32.0	46
Fabaceae	*Cassia tora* (L.) Britt & Rose		43.0	54
Malvaceae	*Wissadula periplocifolia* (L.) Presl.		21.0	46
Grasses and sedges				
Poaceae[2]	*Panicum laxum* Swartz		19.0	54
"	*Axonopus compressus* (Swartz) Beauv.		12.0	23
"	*Eleusine indica* (L.) Gaert.		60.0	8
"	*Paspalum convexum* Humb. & Bonpl.		15.0	8
"	Unknown (RGT 1203)		20.0	8
Cyperaceae	*Scleria muhlenbergii* Steup.		12.5	31
Herb (wet)				
Miscellaneous small dicots			35.0	100
Grasses				
Poaceae[2]	*Panicum laxum* Swartz		26.0	75
"	*Cenchrus?*		12.5	25

[1] Arecaceae (= Palmae).
[2] Poaceae (= Graminiae).

greater variability than specific sites in any other unit. Floristically, they show the greatest diversity in woody species occurring in the savanna units. Physiognomically they vary from open stands of palms with few figs or shrubs to savannas with numerous discrete dense matas, to an almost closed woodland. Accordingly, I have divided the Bajíos into (1) palm Bajío, (2) discrete mata Bajío, and (3) shrub woodland Bajío.

PALM BAJÍO: The plots which are located on the far western part of the ranch have a depauperate woody community composed almost exclusively of palms. The density averages about 4.5 palms/100 m², half of which are mature. Occasionally associated with the mature palms are young strangler figs (*Ficus* spp.) and *Coccoloba caracasana*, a polygonaceous shrub. Of the palms 26 percent have figs attached; 26 percent have

Figure 4. Discrete mata bajio. Palms with associated stangler figs and shrubs interspersed with sedges characterize this habitat.

Figure 5. Shrub woodland bajio. Shrub cover is extremely extensive. Figs and legumes provide the most dominant canopy cover.

Table 5. Important plants of a specific site in the discrete mata bajio

Layer/Family/Species		*Density (no./100 m²)*	*Cover (%)*	*Presence (%)*
Tree				
Arecaceae[1]	*Copernicia tectorum* (H.B.K.) Mart.	0.67		
Moraceae	*Ficus* spp. (associated exclusively with palms)	0.42		
Shrub				
Arecaceae	Young *C. tectorum*	0.58		
Flacourtiaceae	*Casearia mollis* H.B.K.	0.83		
Rubiaceae	*Randia venezuelensis* Steyerm.	4.58		
Rutaceae	*Zanthoxylum culantrillo* H.B.K.	0.58		
Meliaceae	*Trichilia trifolia L.*	2.08		
	Ourtea cf. *guildingii* (Planch.) Urb.	0.58		
Herb				
Miscellaneous tall broad-leaved monocot				
Marantaceae	*Thalia geniculata* L.		14	28
Grasses and sedges				
Poaceae[2]	*Panicum laxum* Swartz		26	100
"	*Leersia hexandra* Swartz		21	93
"	*Luziola spruceana* Benth.		11	28
Cyperaceae	Unknown (RGT 1265); sedge ca. 1 m tall		28	21

[1] Arecaceae (= Palmae).
[2] Poaceae (= Graminiae).

C. caracasana nearby; 11 percent have both. The mature palms average 4–5 m but range from 3–9 m in height. The juveniles range from 0.25–3 m. Woody cover averages 16 percent, varying from a minimum of 2 percent to a maximum of 55 percent. Herbaceous vines are common in the wet season. There is standing water but it is shallow: 7–11 cm at the height of the wet season.

DISCRETE MATA BAJÍO: The plots, located due east of the ranch houses, average 52 percent woody cover. The tree layer, composed of palms, figs and a mimosoid, is ca. 12 m high. The shrub layer varies from 1 to 6 m. In 1200 m^2 there are all or part of seven matas. Herbaceous vines and lianas are common. The resulting canopy is very dense.

The emergent herbaceous vegetation is quite open, covering only ca. 75 percent of the plots; the remaining 25 percent is the area covered by water unshadowed by vegetation. The herb layer is dominated by grasses, sedges, and *Thalia geniculata* (Marantaceae). The list of important species can be found in Table 5.

At the height of the wet season, the water is 10–25 cm deep. Every palm has a slight mounding at its base. The shrubs are often rooted on that mound or on a hummock close to the base. The soils are usually clay with 45–65 percent clay. One sample was classified as clay loam with 33 percent clay. The wet season pH values range from 3.7 to 4.4.

SHRUB WOODLAND BAJÍO: In this form of the Bajío, the shrub cover is so extensive that the canopy (shrub layer) is sometimes 90–100 percent closed with only moderate light penetration. The palm is still the most abundant element of the tree layer but figs and legumes provide the most extensive canopy cover. The diversity in shrub and tree species is greater than in any other savanna unit. Tables 6 and 7 list the important tree and shrub species. The tree layer is 8–15 m high and the shrub layer 2–10 m. Lianas and herbaceous vines are common. The herb layer is less extensive but similar to that found in the Discrete Mata Bajío.

The water depth is quite variable ranging from 4 to

Table 6. Sample of large shrubs, shrub-like lianas and small trees of the shrub woodland bajío[1]

Family/Species		Relative abundance
Annonaceae	*Annona jahnii* Safford	common
	Annona sp.	abundant
Apocynaceae	*Rauvolfia ligustrina* R. & S.	rare
Cochlospermaceae	*Cochlospermum vitifolium* (Willd.) Spreng.	rare
Dilleniaceae	*Tetracera volubilis* L. ssp. *volubilis*	rare
Euphorbiaceae	*Sapium* sp.	rare
Fabaceae	*Machaerium moritzianum* Benth.	occasional
Flacourtiaceae	*Casearia mollis* H.B.K.	rare
	Hecastostemon completus (Jacq.) Sleum.	occasional
Meliaceae	*Trichilia trifolia* L.	abundant
Nyctaginaceae	*Guapira pacurero* (H.B.K.) Lundell	rare
	Guapira pubescens (H.B.K.) Lundell	rare
Ochanceae	*Ourtea* cf. *guildingii* (Planch.) Urb.	occasional
Rubiaceae	*Guettarda divaricata* (H. & B. ex R. & S.) Standley	rare
	Psychotria carthaginensis Jacq.	occasional
	Psychotria microdon (D.C.) Urban	occasional
	Randia venezuelensis Steyermark	ubiquitous
Rutaceae	*Zanthoxylum culantrillo* H.B.K.	very abundant
Verbenaceae	*Citharexylum* sp.	rare

[1] Data courtesy of R. Rudran.

Table 7. Sample of large trees in the shrub woodland bajío[1]

Family/Species		Relative abundance
Arecaceae[2]	*Copernicia tectorum* (H.B.K.) Mart.	ubiquitous (over 60% of sample)
Bignoniaceae	*Jacaranda obtusifolia* H. & B.	rare
Fabaceae (Caesalpinioideae)	*Cassia grandis* L.f.	occasional
Fabaceae (Mimosoideae)	*Albizia* cf. *colombiana* Britton ex Britton and Killip	rare
	Enterlobium cyclocarpum Griseb.	occasional
	Pithecellobium saman (Jacq.) Benth.	rare
	Pithecellobium tortum Benth.	occasional
Fabaceae (Papilionoideae)	*Platymiscium polystachyum* Benth.	abundant
	Platymiscium diadelphum Blake	abundant
	Pterocarpus acapulcensis Rose	occasional
Moraceae	*Ficus pertusa* L.f.	common
	Ficus trigonata L.	occasional
Rubiaceae	*Genipa americana* L.	abundant
Sterculiaceae	*Guazuma tomentosa* H.B.K.	occasional

[1] Data courtesy of R. Rudran.
[2] Arecaceae (= Palmae).

Figure 6. Estero. This area floods extensively during the wet season. The lack of woody vegetation is pronounced.

20 cm. Soils are clay and clay loam with a wet season pH value of ca. 4.

ESTERO

This more deeply flooded savanna is distinctive by its lack of woody vegetation (Figure 6). Only rarely is an isolated palm (*Copernicia tectorum*) or shrub (e.g., *Acacia farnesiana*) seen. *Ipomoea fistulosa*, a tall, robust emergent is found in patches throughout the Esteros. The patches of Esteros only occur on the western half of the ranch. The water depth at the height of the wet season generally varies from 20 to 60 cm but in occasional places it may reach a depth of 110 cm. The soils are heavy, not unlike the Bajío soils.

The herb layer is characterized by grasses, sedges, broad-leaved monocots, and floating dicots. Many of the species are colonial in habit. This habit combined with zonation along a gradient of increasing water depth results in a complex mosaic of herbaceous communities. The specific site is located in a wide Estero

Figure 7. Tropical deciduous forest typifying the area near the Rio Guarico.

about 5 km west of the ranch house. The water was 25–32 cm deep, putting the plots at the lower end of the gradient. No clear dominance was shown; the important species are listed in Table 8. The soils are clay (50–60 percent clay) and the wet season pH varies from 3.9 to 4.2.

Communities at greater water depths show a clearer dominance: at 35 cm, *Eleocharis mutata* and *Neptunia alerocea*; at 45 cm, *Eleocharis elegans* and *Hymenachne amplexicaulis*; at 46 cm, *Eleocharis mutata*; at 55 cm and 61 cm, *Hymenachne amplexicaulis* and *Thalia geniculata.*

GALLERY FOREST

A true forest exists along the Rio Guarico and Caño Caracol, parallel water channels that flow to the south (Figure 7). Within the borders of the ranch, it forms a rectangle ca. 4.5 km east-west and 2.5 km north-south. The river is a permanent water channel but the stream disappears at the height of the dry season. The forest is deciduous except for a few semi-evergreen patches along the river. The canopy layer is commonly 10 to 15 m high, occasionally as high as 20 m or as low as 10 m. The maximum canopy height recorded is 42 m. At the height of the wet season, the canopy is 90–100 percent closed with moderate light penetration. An emergent layer as such does not exist. The shrub layer is heterogeneous, in some places almost impenetrable; in others, quite open. Shrub height varies from 2 to 10 m. The herb layer is never dense but is more evident in the wet season, when some areas are carpeted with *Selaginella.* The soils range from clay to loam to loamy sand, with pH values during the wet season ranging from 3.9 to 4.9. The clay areas are characteristically flooded while the loamy sands are not. A mosaic of vegetational communities exists, the most floristically distinct being those which occur on the loamy sands. Table 9 lists most of the species of trees and large shrubs.

Productivity and Forage in the Banco-Bajío-Estero System

The four savanna units form a dissected pattern over the landscape, flooded areas juxtaposed to unflooded. A study on productivity and habitat utilization was done for the Banco-Bajío-Estero complex in the low llano (Gonzalez Jimenez, 1975). He found the Banco to be the least productive which was attributed partially to its shorter period of production. The Bajio was considered the most important unit because of its high productivity, intermediate period of production, greater area of palatability of grass species. The Estero showed the highest productivity per unit area, partly because of its extended growing season. Utilization was calculated for three grazers: horses, cattle and capybaras. Horses preferred the Banco sites year round; cattle used them to a lesser extent and capybaras hardly at all. Cattle were the main users in the Bajío; secondary use was equal for horses and capybaras. The capybaras were the prime users of the Estero but at the beginning of the dry season use by horses and cattle increased.

Various productivity studies have been done which indicate that the length of the wet season is certainly as important as the total annual precipitation (San José and Medina, 1975, 1976; Gonzalez Jimenez, 1975). In terms of the ranch, this trend should be most readily noticed on Medano and Banco sites, followed by Bajío and lastly Estero.

Table 8. Important plants of a specific estero site with water 25–32 cm deep

Layer/Family	*Species*	*Cover (%)*	*Presence (%)*
Herb			
Emergent dicot			
Acanthaceae	*Justicia laevilinguis* (Nees) Lindau	2	42
Floating dicot			
Fabaceae	*Neptunia oleracea* Lour.	5	50
Grasses and sedges			
Poaceae[1]	*Panicum laxum* Swartz	12	75
"	*Leersia hexandra* Swartz	6	58
Cyperaceae	*Eleocharis mutata* (L.) Roem. & Schult.	33	42
"	*Eleocharis elegans* (H.B.K.) Roem. & Schult.	30	8

[1] Poaceae (= Graminiae).

Acknowledgments

I wish to express my appreciation for assistance from Venezuelan colleagues and institutions. I am particularly grateful for the use of the facilities at the Instituto Botanico, Caracas. Drs. Ramia and Steyermark have been especially generous with their time in helping identify specimens. I am similarly indebted to Dr. Frómeta and the staff of the Estación Experimental de Calabozo, Fondo Nacional de Investigaciones Agropecuarias, for use of the laboratory and the assistance I have received in the soil analyses.

Table 9. Trees and large shrubs of the Gallery Forest[1]

Family/Species		*Relative abundance*	*Common name*
Anacardiaceae	*Anacardium excelsum* (Bert. & Balb.) Skeels	rare	Mijao
	Spondias mombin L.	common	Jobo
Annonaceae	*Annona jahnii* Safford	common	Manirito
Arecaceae	*Copernicia tectorum* (H.B.K.) Mart.	abundant	Palma
Bignoniaceae	*Godmania aesculifolia* (H.B.K.) Standl.	rare	Cornicabro
	Jacaranda obtusifolia H. & B.	common	Flor morada
	Tabebuia sp.	rare	Flor amarilla
Bombacaceae	*Bombacopsis* sp.	rare	Cedro dulce
	Ceiba pentandra (L.) Gaertn.	rare	Ceiba
Boraginaceae	*Cordia collococca* L.	common	Caujaro
Cactaceae	*Cereus hexagonus* (L.) Mill.	rare	Cardon
	Pereskia guamacho Weber	uncommon	Guamacho
Capparidaceae	*Capparis coccolobifolia* Mart.	common	Rabo palado
	Capparis odoratissima Jacq.	uncommon	Olivo
	Capparis sp.	uncommon	Cabo de vela
	Crataeva tapia L. (?)	rare	Toco
Cochlospermaceae	*Cochlospermum vitifolium* (Willd.) Spreng.	common	Carneval
Connaraceae	*Connarus venezuelanus* Baill.	uncommon	Conchagruesa
Dilleniaceae	*Curatella americana* L.	rare	Chaparro bobo
Ebenaceae	*Diospyros ierensis* Britton	common	Cacaito
Erythroxylaceae	*Erythroxylum* sp. (?)	rare	Jayito
Euphorbiaceae	*Sapium* sp.	uncommon	Lechero
Fabaceae (Caesalpinioideae)	*Caesalpinia coriaria* (Jacq.) Willd.	common	Divedive
	Copaifera officionalis H.B.K.	uncommon	Aceite
	Hymenaea courbaril L.	rare	Algorrobo
Fabaceae (Mimosoideae)	*Acacia articulata* Ducke	common	Una de gavilan
	Albizia cf. *colombiana*	uncommon	Caro, Carabali
	Albizia guachapele (H.B.K.) Dugand	uncommon	Masaguaro
	Enterolobium cyclocarpum Griseb.	uncommon	Caracara
	Pithecellobium carabobense Harms	abundant	Quiebrahacho
	Pithecellobium guaricense Pitt.	uncommon	Orore
	Pithecellobium daulense Spruce ex Benth.	common	Veramacho
	Pithecellobium lanceolatum (H. & B.) Benth.	rare	Taguapire
	Pithecellobium saman (Jacq.) Benth	common	Saman
Fabaceae (Papilionoideae)	*Lonchocarpus cruciarubierae* (?)	abundant	Menuito
	Lonchocarpus sericeus (Poir.) H.B.K.	common	Majomo negro
	Machaerium caicarense Pittier	common	Almendron
	Platymiscium polystachyum Benth	uncommon	Roble
	Pterocarpus acapulcensis Rose	common	Drago

Family/Species		*Relative abundance*	*Common name*
Flacourtiaceae	*Casearia mollis* H.B.K.	uncommon	Tapacondi
	Casearia sylvestris S. W. (?)	common	Casavito
	Hecastostemon completus (Jacq.) Sleum.	common	Barote
Lecythisaceae	*Lecythis ollaria* Loefl.	rare	Coco de mono
Malpighiaceae	*Malpighia* sp. (?)	uncommon	Cerezo
Meliaceae	*Trichilia trifolia* L.	common	Coloraito
Moraceae	*Brosimum* sp.	common	Charro
	Cecropia sp.	uncommon	Yagrumo
	Chlorophora tinctoria (L.) Gaudich.	uncommon	Mora
	Ficus pertusa L.f.	common	Matapalo
	Ficus trigonata L.	common	Higuerote
	Ficus sp.	common	Higuerote
Ochnaceae	*Ouratea* cf. *guildingii* (Planch.) Urb.	common	Casco de burro
	Ouratea sp. (?)	rare	Asta blanca
Polygonaceae	*Coccoloba caracasana* Meisn.	common	Uvero
	Ruprechtia sp. (?)	rare	Palo de agua
Rhamnaceae	*Zizyphys saeri* Pittier	uncommon	Limoncillo
Rubiaceae	*Chomelia spinosa* Jacq. (?)	common	Espinito
	Genipa americana L.	abundant	Caruto
	Guettardia divaricata (H. & B. ex R. & S.) Standl.	common	Punteral
	Psychotria anceps H.B.K.	common	Agallon
	Randia hebecarpa Benth.	abundant	Cachito
	Randia formosa (Jacq.) Schum.	uncommon	Jasmin
	Randia venezuelensis Steyermark	common	Diente de perro
Rutaceae	*Zanthoxylum caribeum* Lam.	uncommon	Mapurito
	Zanthoxylum culantrillo H.B.K.	abundant	Bosu
Sapindaceae	*Allophylus occidentalis* (SW.) Radlk.	common	Pata de danta
	Matayba sp.	rare	Zapatero
Sterculiaceae	*Guazuma tomentosa* H.B.K.	abundant	Guacimo
	Sterculia apetala (Jacq.) Karst.	uncommon	Camoruco
Verbenaceae	*Vitex compressa* Turcz.	common	Guarataro aceituno
	Vitex appuni Moldenke (?)	common	Guarataro pardillo
	Vitex orinocensis H.B.K. (?)	rare	Guarataro pardillo
Ignotaceae	—	uncommon	Canoito
	—	common	Casavito
	—	common	Zarcillo
	Protium sp. ?	rare	Guacharaco
	Eugenia sp. or *Myrcia* sp. ?	rare	Guayavito
	—	rare	Cacho

[1] Data courtesy of J. Robinson.

Literature Cited

Gonzalez Jimenez, E.
1975 *Tropical Grazing Land Ecosystems of Venezuela. III. Grassland Productivity, Organic Matter Consumption and Large Herbivores in Flooded Savannas.* World Report on Grasslands. Paris: UNESCO.

Ramia, M.
1958. Los Medanos del guarico occidental. *Bol. Soc. Ven. Cienc. Nat.*, 20(91):41–53.

Ramia, M.
1967. Tipos de sabanas en los llanos de Venezuela. *Bol. Soc. Ven. Cienc. Nat.*, 27(112): 264–288.

San José, J. J. and E. Medina.
1975. Effect of fire on organic matter production and water balance in a tropical savana. Pages 251–264 in *Tropical Ecological Systems*, F. B. Golley and E. Medina (eds.). New York: Springer-Verlag.

San José, J. J. and E. Medina.
1976. Organic matter production in the Tracmypogon Savanna at Calabozo, Venezuela. *Tropical Ecology*, 17(2): 113–124.

Sanchez, A. J. and C. M. Fajardo.
1976. *Los Suelos y las Aguas en la Region Llanera.* Maracay, Venezuela: Coplanarh.

Sarmiento, G.
1975. *Tropical Grazing Land Ecosystems of Venezuela. I. The Main Types of Grazing Land Ecosystems: Descriptions, Functioning and Evolution.* World Report on Grasslands. Paris: UNESCO.

Sarmiento, G. and M. Monasterio.
1975. A critical consideration of the environmental conditions associated with the occurrence of savanna ecosystems in tropical America. Pages 223–250 in *Tropical Ecological Systems*, edited by F. B. Golley and E. Medina. New York: Springer-Verlag.

Errata. Species nomenclature in Tables 4 and 9 should be changed as follows:

Table 4
Conchrus to *Paspalum* sp.

Table 9
Anacardium excelsum (Anacardiaceae) to *Himatanthus* sp. (Apocynaceae)
Albizia cf. *colombiana* to *Albizia* aff. *polycephala* (Benth.) Killip
Pithecellobium carabobense to *Pithecellobium tortum* Mart.
Pithecellobium lanceolatum to *Pithecellobium ligustrinum* (Jacq.) Klotzsch
Brosimum sp. to *Sorocea sprucei* (Baill.) Macbr.
Matayba sp. to *Cupania* sp.
Vitex appuni to *Vitex orinocensis* var. *multiflora* (Miq.) Huber vel.sp.aff.
Vitex orinocensis to *Vitex capitata* Vahl.

JOHN F. EISENBERG
National Zoological Park
Smithsonian Institution
Washington, D. C. 20008

KENT REDFORD
Museum of Comparative Zoology
Harvard University
Cambridge, Massachusetts 02138

A Biogeographic Analysis of the Mammalian Fauna of Venezuela

ABSTRACT

Distribution maps for the species of mammals found in Venezuela were individually prepared based on collecting records. Venezuela was divided into seven biogeographic regions based on topography, climate and vegetation. The arid Falcon region and the llanos exhibit the lowest species diversity. The number of species shared among the various biogeographic regions was then analyzed. The volant Chiroptera, with a greater potential mobility, are less useful for understanding the geographical isolation of the various subregions as defined for this paper. Non-volant mammalian distributions suggest times of faunal interchange among the regions followed by separations. The concept of refugia in the North Coast Range, Guyana Highlands, and in the Andean foothills is discussed. The data suggest that the Falcon peninsula has had a longer and more persistent history of aridity and that the llanos has a derived fauna from the adjacent biogeographic regions.

RESÚMEN

Los mapas de distribución para las especies de mamíferos que hay en Venezuela fueron preparados individualmente en base a recopilación de protocolos. Venezuela fué dividida en siete regiones biogeográficas basándose en la topografía, el clima, y la vegetación. La region árida de Falcón, y los llanos tienen la menor diversidad de especies. El número de especies repartido entre las regiones biogeográficas fue analizada. La Chiroptera volante, de mayor movilidad potencial es menos útil para la comprensión de estas regiones. La distribución de los mamíferos que no vuelan, indica épocas de intercambio de fauna entre las regiones, seguidas de separaciónes. El concepto de refugio en las montañas de la Costa del Norte, la region montañosa de Guyana y en las lomas de los Andes es discutido. Los datos indícan que la península de Falcón tiene una historia más larga y persistente de aridez y que los llanos tienen una fauna derivada de las regiones biogeográficas adyacentes.

The distribution of Venezuelan mammal species was advanced considerably with the publication of the collection data by Handley in 1976. Utilizing the study areas of Handley's collecting teams, we prepared maps for each individual species known to be in Venezuela. We incorporated additional location data from the bibliography in Handley's paper (1976) and data from our own observations.

For the purpose of faunal analysis, Venezuela was divided into seven provinces. In part, this division is based on natural geographic barriers as well as considerations of rainfall, vegetation, and altitude. The Holdridge vegetation map of Venezuela was used as a reference (Ewer and Madriz, 1968). A glance at the map of Venezuela will indicate that the Andean chain from the eastern Cordillera bifurcates just south of Lake Maracaibo with the extreme eastern range maintaining high elevations until approximately 10° north latitude (Figure 1). This spur of the eastern Cordillera, above an altitude of 1500 meters, was defined as the Andean Region. The depression to the north and west of the Andean Zone containing Lake Maracaibo is characterized by humid lowland forest and seasonally dry tropical forest. Because of its isolation from the rest of Venezuela by the Andean chain and its proximity to lowland rainforest in northern Colombia, we designated this the Maracaibo Region. The portion of land bordering the Gulf of Venezuela, which is in reality the entrance to Lake Maracaibo, is characterized by extreme aridity. This area includes the Falcon

Figure 1. Altitudinal map of Venezuela. Elevations exceeding 1,000 meters are indicated with bold cross-hatching. Elevations from 500 to 1,000 meters in dots.

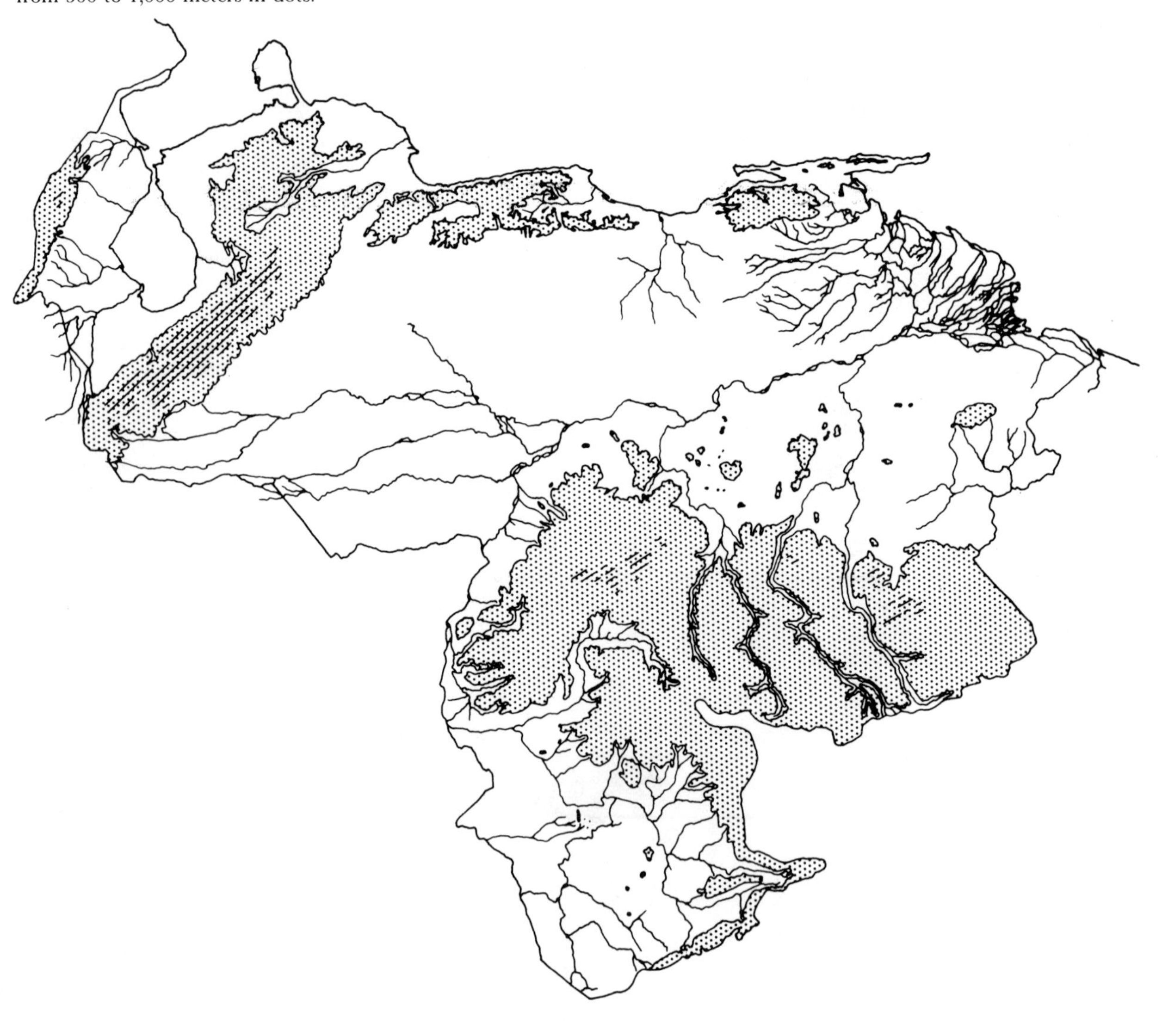

Peninsula. Because of its extremely low rainfall and distinctive vegetational patterns, this small area was designated the Falcon Region.

The region on the north coast of Venezuela, from approximately 62° west longitude to 69° west longitude, in the main lying north of 10° 30′ north latitude was designated as the North Coastal Range Region. This is a complicated habitat with some mountains rising to elevations in excess of 2000 meters. With the exception of some isolated pockets of aridity on the extreme north coast, these mountain ranges are characterized by humid rain forest, either lowland tropical or pre-montane. Some portions on the tops of the mountain ranges are true montane cloud forests.

The lowland areas north of the Orinoco River bounded on the west by the Andean Province and on the north by the North Coast Ranges we designated as the Northern Llanos Region. This area is of extremely low relief which, in a climax condition, would support seasonally arid forest but because of vast areas of impeded vertical drainage it is characterized by a mosaic of habitat types characteristically lacking large trees.

South of the Orinoco, we designated two general biogeographic regions. One, the Guyana Highlands, is dominated by intermediate to high elevation granitic formations from the Guyana Shield. This is a veritable habitat mosaic, including open savannas, lowland tropical forest, premontane forest, and montane forest. Its mosaic character cannot be emphasized enough because, in later discussions, it will be pointed out that this habitat type has acted as an effective filter barrier to mammalian faunal elements present in the northeast lowlands of South America. In the extreme southwest corner of Territorio Federal Amazonas, we noted tropical lowland forest which is continuous with forest in neighboring portions of Brazil and Colombia. This was termed the Amazonian Region because the rivers draining this area to a great extent ultimately empty into the Rio Negro and the Amazon system itself. A map of the defined biogeographic regions is included in Figure 2.

The distributional data for each species were then examined and its presence or absence in each of the defined zoogeographical regions was determined. In the following discussion, a species is considered endemic to a single region if it occurs only in that region of Venezuela. It is an index of endemism as defined by the political boundaries of Venezuela. The species may, however, occur in neighboring areas of Colombia, Brazil or Guyana. In the analysis, the following trends were established.

The Andean Region has the highest percentage of endemic species; 51 percent of its 39 species are found only in the Andean Region of Venezuela. The North Coast Range exhibits 16 species found only in that region; but this comprises only 10 percent of its total mammalian fauna. The Guyana Highland Region exhibits 14 species endemic to it, but this number includes only 9 percent of its total mammalian fauna. The Amazonian Region has 19 endemics, 12 percent of its entire faunal complement.

As one moves from north to south across Venezuela, the depauperate nature of the llanos mammal fauna can be readily noted (Figure 3). Considering its large area, the llanos have a small number of species. The

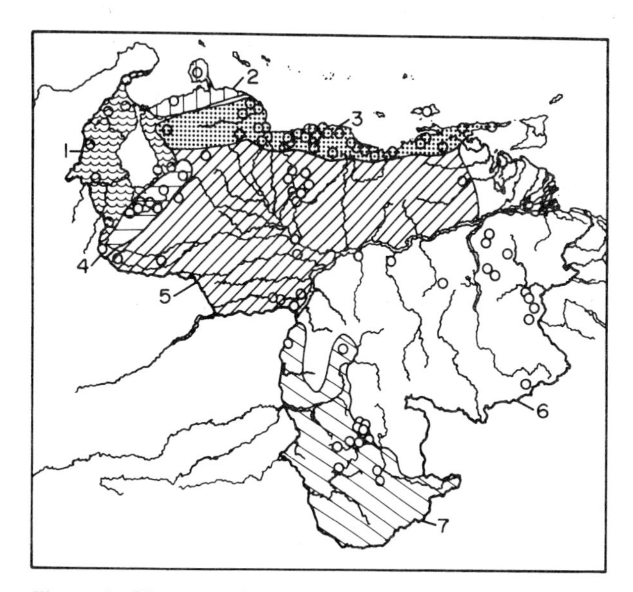

Figure 2. Biogeographic regions of Venezuela. Circles refer to collection stations as given in the publication by Handley (1976). 1 = Maracaibo Basin. 2 = Falcon arid zone. 3 = North coast range. 4 = Andean region. 5 = Llanos. 6 = Guyana Highlands. 7 = Amazon Lowlands. Boundaries of the regions are approximate, future research may define zone of intergradation. The region of the delta is excluded from consideration as is Isla Magarita and Trinidad.

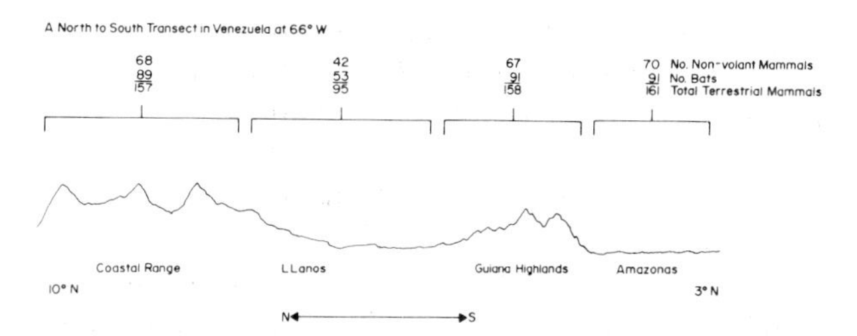

Figure 3. A north-south transect at approximately 66° 30′ W latitude indicating total species of mammals encountered in four biogeographic provinces. The scale is arbitrary. The Chiroptera are separated from the non-volant mammals in computing the totals. The more depauperate nature of the llanos fauna is immediately evident.

Table 1. Numbers of species present in biogeographic regions

Area	*Total no. species*[1]	*No. species per km*^{2}	*Estimated area (km*^{2}*)*
Maracaibo Lowlands	119	2.83×10^{-3}	42,000
Andean Highlands	39	3.25×10^{-3}	12,000
Falcon Arid Zone	31	2.52×10^{-3}	12,300
North Coast ranges and valleys	157	2.61×10^{-3}	60,000
Northern Llanos (north of the Orinoco)	95	3.40×10^{-4}	276,000
Amazon Lowlands	161	1.21×10^{-3}	132,000
Guyana Highlands	158	5.16×10^{-4}	306,000
			840,300

[1] Exclusive of Sirenia and Cetacea.

Table 2. Area tabulations for the mammals of Venezuela

Biogeographic region	*Total no. species*	*No. "endemic"*	*No. Chiroptera*
Maracaibo	119	5	63
Andean	39	20	10
North Coast Range	157	16	89
Llanos	95	0	53
Falcon Peninsula	31	4	17
Guyana Highlands	158	14	91
Amazon Lowlands	161	19	91

land area of each biogeographic region was estimated with the following results: The Andean Region and the Falcon Arid Zone have the smallest areas being only slightly larger than 12,000 km^2. The North Coast Range includes approximately 60,000 km^2 and the Maracaibo Basin, 42,000 km^2. The llanos includes approximately 276,000 km^2; the Guyana Highlands, 306,000 km^2; and the Amazon Lowlands, 132,000 km^2. These areas do not include the unstudied region of the delta and certain intermediate areas in the Maracaibo and Andean Regions.

When the number of species present in each of the biogeographic regions is expressed as number of species per km^2, we get some index of the relative diversity contained within the geographic regions. The number of species per unit area will decline with the increasing size of the base area surveyed. However, the number of species in comparably sized regions can vary tremendously. Relative to their size, the Falcon and llanos regions show the lowest species diversity (Table 1).

An inspection of Table 2 indicates that the percentage of species included within the order Chiroptera is remarkably uniform for all biogeographic regions except the Andean Highlands. Since bat diversity is adversely affected by increasing altitude or increasing latitude, this is not surprising. Only 26 percent of the mammalian fauna of the High Andes includes the species of the order Chiroptera. For all other areas, the value for the percent of Chiropteran species ranges between 53 percent and 58 percent. This value is slightly higher than that obtained for Panama from consulting Handley's species list published in 1966. Bats comprise 51 percent of the mammalian fauna of the Republic of Panama. However, the Venezuelan trends are in line with what we understand about neotropical mammalian faunal diversity; namely, that the diversity of bat species in the neotropics is far higher than the values determined for the Paleotropics (McNab, 1971).

Tables 3 and 4 give us an indication of how many species are shared for any biogeographic region with the remaining six. It should be immediately evident that bat species consistently are shared at a higher percentage than are the non-volant species. This no doubt results from the fact that bats have a higher mobility and greater dispersal range. Where this is not the case, for example in the Northern Llanos or in the Falcon Peninsula, this may be attributed to the fact that these areas have a lower diversity of vegetation hence a smaller number of feeding niches for bats.

The degree of faunal interchange among the biogeographic regions is probably best examined by looking at non-volant mammals rather than the Chiroptera. When this is done, the following trends are expressed. The Amazon Basin shares approximately 66 percent of its non-volant species with the North Coast Range and only 50 percent with the Guyana Highlands that are immediately adjacent to it. This suggests an older connection between the North Coast Range and Amazonian faunal elements. It further suggests that the Guyana Highlands themselves serve as an effective barrier to uniform dispersal of non-volant mammals, something that may be inferred from the mosaic nature of its habitat.

The North Coast, by contrast, shares 91 percent of its non-volant species with the Guyana Highlands indicating a persistent and long term exchange between the Guyana Highlands and the North Coast

Table 3. Percent of chiropteran species shared among the seven faunal provinces of Venezuela

	Maracaibo	*Andean*	*North Coastal Range*	*Llanos north of the Orinoco*	*Falcon Peninsula*	*Guyana Highlands*	*Amazon Lowlands*
Maracaibo	—	8%	78%	60%	1+%	76%	81%
Andean	40%	—	60%	10%	0%	50%	20%
Northern Coastal Ranges	55%	7%	—	42%	12%	67%	64%
Llanos North of the Orinoco	72%	2%	70%	—	15%	77%	85%
Falcon Peninsula	53%	0%	65%	47%	—	59%	53%
Guyana Highlands	53%	5%	66%	45%	11%	—	77%
Amazon Lowlands	64%	2%	63%	49%	10%	77%	—

Table 4. Percent of non-volant mammal species shared among the seven faunal provinces of Venezuela

	Maracaibo	*Andean*	*North Coastal Range*	*Llanos north of the Orinoco*	*Falcon Peninsula*	*Guyana Highlands*	*Amazon Lowlands*
Maracaibo	—	7%	73%	55%	14%	59%	71%
Andean	14%	—	38%	17%	0%	21%	10%
Northern Coastal Ranges	60%	13%	—	54%	13%	91%	46%
Llanos North of the Orinoco	71%	12%	88%	—	26%	79%	67%
Falcon Peninsula	57%	0%	64%	79%	—	36%	43%
Guyana Highlands	49%	9%	93%	49%	7%	—	58%
Amazon Lowlands	47%	4%	66%	40%	9%	56%	—

Region of Venezuela. It suggests further that the range of habitat types in the northern coastal portion of Venezuela is more suitable to a faunal interchange with the Guyana Highlands than is the faunal interchange permissible between the Guyana Highlands and the Amazon Lowlands. It may also suggest that faunal connections between the Guyana Highlands and the North Coast Ranges were established earlier in time than were the exchanges between the Amazon Lowlands and the Guyana Highlands themselves.

The Maracaibo Basin shares 71 percent of its mammalian fauna with the Amazon Basin, thus indicating the derivative nature of its fauna with respect to the Amazon basin and northern Colombia. The mammalian fauna of the North Llanos is entirely derivative; 88 percent of the non-volant mammals found in the Northern Llanos are shared in the North Coast Ranges; 71 percent are shared with the Maracaibo Basin, and 79 percent with the Guyana Highlands.

The analysis of these trends strongly suggests that the mammalian species in northern Venezuela are derived from several sources. One can imagine a reservoir of mammalian species in the higher portions of the eastern Amazon region (Peru and Bolivia) serving to populate northern South America and the Amazon Basin itself during the Pleistocene climate cycles. Successive alternation of aridity and pluvial conditions in the northern neotropics periodically converted the llanos into semi-desert conditions. Mountainous areas served as refugia and later as conduits for dispersal (Webb, 1978). During wet intervals, drainage conditions could favor the occupation of the Amazon Basin and the Orinoco Basin by terrestrially adapted forms. One can imagine then faunal interchange proceeding in cycles up the eastern side of the Andean foothills and successively repopulating the Maracaibo Basin, northern Colombia, and the Republic of Panama. Periodically, some spillover would temporarily invade the Northern Llanos.

The Andean chain could also serve as a conduit for high altitude forms, distributing them in the northwestern portion of Venezuela and allowing some of the more adaptable species to establish in the North Coast Range. The North Coast Range with its mosaic of habitats also served as a repository for specimens from the Maracaibo Basin as well as those moving along the foothills of the Andes from the east.

The Guyana Shield could also have acted as a reservoir permitting speciation somewhat in isolation from what was going on on the eastern side of the Andes and to the south. If we assume an early connection between the North Coast Range and the Guyana Highlands, a great proportion of species could establish themselves from the Guyana Highlands in the North Coast Range or vice versa. Given the peculiar nature of the topographic relief and the extraordinarily low fertility of soils over wide regions in the Guyana Highlands (Sioli, 1975; Janzen, 1974), the highlands themselves may have served as a filter barrier to faunal invasions from the Surinam Highlands. The maintenance of the Orinoco mouth adjacent to Trinidad might also have prevented an easy lowland dispersal from the Guyanas. These facts may account for the fact that tamarins (*Saguinus*) have never passed into northern Venezuela from the east nor have they passed over the Andes into northern Venezuela from the west. It may account for why the night monkey (*Aotus*) has successfully moved from the west into northwestern Venezuela and from the south into Amazonian Venezuela, but has never been able to pass the Guyana Highlands and colonize the Guyana lowlands from the west.

We are only beginning to understand faunal distributions in northern South America during the Pleistocene and can only vaguely interpret their movements, given the present day constellations. Certainly, Venezuela is a fertile crossroad to test certain hypotheses concerning mammalian faunal refugia in the Surinam and Guyana Highlands as well as in the foothills of the Andes.

Literature Cited

Ewell, J. J. and A. Madriz
1968. *Zonas de Vida de Venezuela.* Caracas: Ministario de Agricultura y Cria.

Handley, C. O., Jr.
1976. Mammals of the Smithsonian Venezuelan Project. *Brigham Young Univ. Sci. Bull., Biol. Ser.*, 20(5):1–91.

Janzen, D.
1974. Tropical black water rivers, animals and mast fruiting by the Dipterocarpaceae. *Biotropica*, 6(2):69–103.

McNab, B. K.
1971. The structure of tropical bat faunas. *Ecology*, 52:353–358.

Sioli, H.
1975. Tropical rivers as expressions of their terrestrial environments. Pages 275–288 in *Ecological Studies, Vol. 11, Tropical Ecological Systems*, edited by F. B. Golley and E. Medina. New York/Berlin: Springer Verlag.

SECTION 2:

The Edentata and Marsupialia

Dasypus sabanicola. This small armadillo is typical of the Orinoco llanos system. Similar sized species have evolved in the savanna ecosystems south of the Amazon.

Introduction

The living Edentata are the sole survivors of an ancient lineage which once exhibited a wider range of forms adapted to an array of ecological niches. The group is endemic to the neotropics and must surely have been one of the earliest isolated eutherian lineages in South America (Patterson and Pasqual, 1972). The survivors are grouped conveniently into three families: the armadillos or Dasypodidae, the sloths or Bradypodidae, and the anteaters or Myrmecophagidae. Ancient endemics such as the Edentata have interested anatomists and physiologists for some time, but progress in our understanding of their reproductive cycles and ecology has lagged somewhat. The three-toed sloth, *Bradypus infuscatus*, and the two-toed sloth, *Choloepus hoffmanni*, extend well into Panama and it is in Panama where the ecology of these forms was studied intensively by Montgomery and Sunquist (1975, 1978) and Sunquist and Montgomery (1973). *Bradypus infuscatus* occurs in northern Venezuela and was present in our study area at Guatopo National Park. *Choloepus hoffmanni* does not extend beyond the Maracaibo Basin in northern Venezuela. The giant anteater, *Myrmecophaga tridactyla*, was found at our primary study site at Hato Masaguaral. The lesser anteater, *Tamandua tetradactyla*, was found both at Masaguaral and at Guatopo National Park. The nine-banded armadillo, *Dasypus novemcinctus*, occurred at both Guatopo and Hato Masaguaral. The smaller *Dasypus sabanicola* occurs only in the llanos. Montgomery and Lubin (1977, 1978) have been intensively studying the ecology and behavior of the anteaters, both at Panama and in Venezuela.

The work of Montgomery and Sunquist (1975) clearly demonstrates that in mature rainforests, showing only a brief period of season aridity, the sloths may achieve very high densities with a significant impact on nutrient cycling in mature rainforests. In regions with more prolonged dry seasons, sloth density does not appear to be so high. The two-toed sloth apparently always exists at a density much lower than that of the three-toed. The research by Montgomery and Lubin (1977) suggests that the anteaters exist at densities commensurate with those of equivalent sized carnivores. Numerical density and biomass values are rather low. The giant anteater by virtue of its size and conspicuousness as well as its low densities is probably rather vulnerable to extermination through the activities of man. The lesser anteater, with its slightly greater numerical abundance, may be less vulnerable initially, but certainly will be profoundly affected by land-clearing patterns.

Our lack of knowledge concerning the Dasypodidae is well illustrated in the article by Wetzel and Mondolfi (see p. 43). Therein they review what is known concerning the biology of *Dasypus novemcinctus* and proceed to discuss the relationships of the species within the genus *Dasypus* itself. As it turns out, *Dasypus* as a genus can be organized into three subgenera. *Dasypus kappleri* apparently represents a somewhat conservative form both in its patterns of reproduction and in its distribution. *Dasypus novemcinctus, D. sabanicola,* and *D. septemcinctus* can be considered recent derivative forms. The smaller species represent an adaptation to savannas or the llanos and may exist sympatrically with the larger *D. novemcinctus. Dasypus pilosus* represents an extraordinary adaptation to high elevations. Here is an armadillo that has apparently re-evolved an almost complete coat of hair, undoubtedly an adaptation to high altitude life. This is an almost unknown species except for a few skins and would represent a challenge to any ecologist interested in the limits of adaptive radiation by this genus.

The versatile *Dasypus novemcinctus* appears to do well in a variety of habitats and can achieve moderate biomass values under protection (Eisenberg and Thorington, 1973). The rarer and more myrmecophagous armadillos, such as *Priodontes* and *Cabassous*, are to be found in certain parts of northern Venezuela, but we know little of their biology. It may be concluded from their apparently low densities that they are as vulnerable to extinction as are the larger anteaters. Much more research is necessary on the biology of the edentates before reasonable recommendations can be made concerning their conservation.

The living marsupials have two main centers of adaptive radiation: the neotropics and Australia. It is now widely accepted that the marsupials became distributed in the neotropics at a very early time and may probably have passed from South America to Australia via Antarctica before the complete separation of these continents (Keast, 1977). The radiation of the Marsupialia has had a long history in the neotropics and the fossil record has been amply reviewed by Patterson and Pasqual (1972). One species of neotropical marsupial, *Didelphis virginiana*, has become widely distributed in the southern and eastern United States. All other extant genera in the neotropics are confined to Central or South America.

The volume edited by Hunsaker (1977) summarizes our knowledge concerning the anatomy, ecology and behavior of New World marsupials. Clearly, only the genera *Didelphis* and *Marmosa* are reasonably well known and, even within these two genera, there are many species that have not been studied. In fact, the article by Handley and Gordon (see p. 65) demonstrates that much is still to be learned concerning speciation in the genus *Marmosa*. Two new forms are described for the first time.

The article by O'Connell takes a major step in elucidating the ecology of neotropical marsupials. In her chapter, she compares the patterns of reproduction exhibited by marsupials at our two major study sites in Venezuela, Guatopo National Park and Hato Masaguaral. The seasonality of reproduction is demonstrable and clearly the timing of reproduction is in some way geared to the abundance of fruit and insect prey which, in turn, are governed by rainfall patterns. Seasonality appears to be more strict in the llanos than is the case in Guatopo National Park.

The greater diversity of marsupials in Guatopo National Park suggests that more niches are available in this structurally complex second-growth forest where the study was executed than is the case in the llanos. In both areas, *Didelphis marsupialis* is the most abundant large marsupial. The smaller and more highly insectivorous genus, *Marmosa*, coexists with *Didelphis*, with *Marmosa fuscata* being the most abundant in Guatopo and *Marmosa robinsoni* the numerical dominant in the llanos.

The reproduction rates differ widely among the various species of marsupials. Litter sizes tend to be smaller for *Caluromys*. Litter sizes are quite large for *Marmosa robinsoni*. It would appear that, although a female *Marmosa* may successfully rear two litters in a given year, female survivorship beyond a year is highly improbable. Although *Didelphis marsupialis* may produce two litters in a given year, the survivorship of a female in the wild beyond her first season of reproduction is low. Both of the common species of marsupials, then, appear to be geared to reproduce in a single season; hence their high productivity during the year of their sexual maturity. *Caluromys*, with a smaller litter size, may possibly live longer and be able to reproduce in more than one year. This remains to be proved since the number of trapped females has been too low to make this a firm conclusion. It would appear that all species of didelphine marsupials studied to date exhibit a phase during the rearing of the young when the litters are left in a nest after detaching from the nipple, and nursed by the mother for some period of time prior to their dispersal. Thus, there is a phase of teat-attachment followed by a nest phase. This was confirmed in the field for *Marmosa robinsoni* in the case of a female that undertook to rear her litter in a nest box established as part of our bird surveys (see O'Connell, p. 73).

O'Connell discusses the location of trapping and the behavior upon release for the didelphid marsupials. In combining her results with similar data from the collections assembled by Handley (1976), she rightly concludes that different degrees of arboreality are displayed under conditions of high sympatry. This is an important first step in determining the degree to which habitat utilization patterns allow for niche separation where several species of didelphids occur in micro-sympatry. J. F. E.

Literature Cited

Eisenberg, J. F. and R. W. Thorington, Jr.
1973. A preliminary analysis of a neotropical mammal fauna. *Biotropica*, 5:150–161.

Handley, C. O., Jr.
1976. Mammals of the Smithsonian Venezuelan project. *Brigham Young Univ. Sci. Bull., Biol. Series*, 20(5):1–91.

Handley, C. O., Jr., and Gordon, L. K.
1979. New species of mammals from northern South America. Mouse possums, genus *Marmosa* Gray. Pages 65–72 in *Vertebrate Ecology in the Northern Neotropics*, edited by John F. Eisenberg. Washington, D.C.: Smithsonian Institution Press.

Hunsaker, D., II.
1977. *The Biology of Marsupials*. New York: Academic Press.

Keast, A.
1977. Historical biogeography of the marsupials. Pages 69–95 in *The Biology of Marsupials*, edited by B. Stonehouse and D. Gilmore. Baltimore: University Park Press.

Montgomery, G. G. and Y. D. Lubin.
1977. Prey influences on movements of neotropical anteaters. Pages 103–131 in *Proceedings of the 1975 Predator Symposium*, edited by R. Phillips and C. Jonkel. Missoula: Montana Forest and Conservation Experiment Station.

1978. Impact of anteaters (*Tamandua* and *Cyclopes*, Edentata, Myrmecophagidae) on arboreal ant populations. Abstr. # 111. Abstracts of Technical Papers, 58th Annual Meeting of the American Society of Mammalogists. Athens: Georgia.

Montgomery, G. G. and M. E. Sunquist.
1975. Impact of sloths on neotropical forest energy flow and nutrient cycling. Pages 69–98 in *Tropical Ecological Systems*, edited by F. B. Golley and E. Medina. New York: Springer-Verlag.

1978. Habitat selection and use by two-toed and three-toed sloths. Pages 324–360 in *The Ecology of Arboreal Folivores*, edited by G. G. Montgomery. Washington, D. C.: Smithsonian Institution Press.

O'Connell, M. A.
1979. Ecology of didelphid marsupials from northern Venezuela. Pages 73–87 in *Vertebrate Ecology in the Northern Neotropics*, edited by John F. Eisenberg. Washington, D. C.: Smithsonian Institution Press.

Patterson, B. and R. Pasqual.
1972. The fossil mammal fauna of South America. Pages 247–309 in *Evolution, Mammals and Southern Continents*, edited by A. Keast, F. C. Erk and B. Glass. Albany: State University of New York Press.

Sunquist, M. E. and G. G. Montgomery.
1973. Activity patterns and rates of movements of two-toed and three-toed sloths (*Choloepus hoffmanni* and *Bradypus infuscatus*). *J. Mammal.*, 54(4):946–954.

Wetzel, R. M. and E. Mondolfi.
1979. The subgenera and species of long-nosed armadillos, genus *Dasypus* L. Pages 43–63 in *Vertebrate Ecology in the Northern Neotropics*, edited by John F. Eisenberg. Washington, D.C.: Smithsonian Institution Press.

RALPH M. WETZEL
Biological Sciences Group
University of Connecticut
Storrs, Connecticut 06268

EDGARDO MONDOLFI
Escuela de Biología
Facultad de Ciencias
Universidad Central de Venezuela
Caracas, Venezuela

The Subgenera and Species of Long-nosed Armadillos, Genus *Dasypus* L.

ABSTRACT

The taxonomy, distribution, and recent literature on the natural history of the long-nosed armadillos, genus *Dasypus* L., are reviewed. New data on embryo counts and male genitalia of *D. novemcinctus*, *D. sabanicola*, and *D. kappleri* are presented. The following subgenera, with valid species, are proposed: **subgenus *Dasypus*** Linné—*D. hybridus* (Desmarest), *D. novemcinctus* L., *D. sabanicola* Mondolfi, and *D. septemcinctus* L.; **subgenus *Hyperoambon*** Peters—*D. kappleri* Kraus; and **subgenus *Cryptophractus*** Fitzinger—*D. pilosus* (Fitzinger). The three small, sibling forms occupying nonforest centers in South America, *D. septemcinctus, D. hybridus, D. sabanicola*, are found to be separable as species by both multivariant and bivariant comparisons. On the basis of type material examined, *D. mazzai* Yepes proved to be a composite. Lectotypes are selected for *D. fenestratus* Peters, *D. kappleri* Kraus and *D. novemcinctus mexianae* Hagmann.

RESÚMEN

Se revisa la taxonomía, distribución, y literatura reciente de la historia natural del *Dasypus*. Se presentan nuevos datos sobre los numeros de embriones y los organos genitales de los machos de *D. novemcinctus*, *D. sabanicola*, y *D. kappleri*. Se proponen los siguientes subgéneros, con especies válidos: subgénero *Dasypus* Linne—*D. hybridus* (Desmarest), *D. novemcinctus* L., *D. sabanicola* Mondolfi, y *D. septemcinctus* L.; subgénero *Hyperoambon* Peters—*D. kappleri* Kraus; y subgénero *Cryptophractus* Fitzinger—*D. pilosus* (Fitzinger). Las tres pequeñas formas hijas que ocupan las regiones no boscosas en América del Sur, *D. septemcinctus*, *D. hybridus*, *D. sabanicola*, son especies separables por examen de comparación tanto multivariante como bivariante. Ejemplares de *D. mazzai* Yepes, examinados, resultaron ser un compuesto. Se seleccionaron lectotypos para *D. fenestratus* Peters, *D. kappleri* Kraus, y *D. novemcinctus mexianae* Hagmann.

Introduction

Despite a surging interest in Neotropical mammals, distribution and taxonomy of many common genera have received little modern examination. This is the more striking where a standard component of Neotropical fauna is concerned, the 18th century prototype of all armadillos, the genus *Dasypus* Linné. A review of this genus, for which we offer the descriptive name of long-nosed armadillos, seems timely to us. Many of the distributional and taxonomic data that have helped form our opinions are based upon collections made in the past two decades and were not available to early authors. Some examples are collections made or directed by Charles O. Handley, Jr., in Panama; Ernesto Barriga, Jorge Hernández-Camacho, and Ronald Boyce MacKenzie in Colombia; Omar Linares, Gonzalo Medina, Edgardo Mondolfi, Andrés Musso, and Juhani Ojasti in Venezuela; A.M. Husson in Surinam; Cory T. de Carvalho and P.E. Vanzolini in Brazil; Sydney Anderson, Karl Koopman, and Richard G. Van Gelder in Bolivia and Uruguay; Alfredo Langguth, Alvaro Mones, and Alfredo Ximénez in Uruguay; Jorge A. Crespo in Argentina; and Philip Myers and R. M. Wetzel in Paraguay. These and many other collections, both old and new, in approximately forty institutions in Europe, North and South America provided our primary data. The junior author's observations of embryos of *Dasypus kappleri* and of male genitalia in that species, *D. novemcinctus* and *D. sabanicola* first suggested to him the subgeneric separation that we present here.

Methods

In the accounts of species only partial synonymies are presented. Emphasis is given names for which holotypes or syntypes have been examined, names used in reviews of the genus during the past five decades and other literature subsequent to that reviewed by Talmage and Buchanan (1954).

All measurements and, excepting a few cases which are noted, all conclusions on distribution are based upon wild-captured specimens that we have examined. The specimens used for measurements were further restricted to those with crania sufficiently matured to have a fusion of the occipital-basisphenoid suture. All cranial measurements were made by Wetzel; all linear measurements are in millimeters and weights are in kilograms. Weights and body measurements were taken from labels prepared by the field collector and are presumed to have been taken from the newly killed animal. In order to confine any undetected bias from damaged tails to only one measurement, we present length of head plus body rather than total length. Length of hindfoot includes the longest claw; length of ear is from notch. Counts of movable bands were made along the middorsal line. Counts of scutes on the fourth movable band include the left and right marginal scutes. Counts of teeth are given as fractions with the maxillary pairs as the numerator and mandibular pairs as denominator. As bilateral asymmetry was not uncommon, counts are the means of the left and right toothrows.

Of 15 cranial measurements that were taken, only the following are cited: *Condylonasal length* (CNL): distance from most posterior faces of occipital condyles to most anterior tips of nasal bones.

Palatal length (PL): length at midline of undamaged hard palate.

Rostral length (RL): a diagonal length from lacrimal foramen to tip of nasal bones.

Rostral length, adjusted (RL adj.): distance along midline of rostrum from a line through lacrimal foramina to tip of nasal bones. RL adj. is derived as the altitude of a triangle in which RL is the hypotenuse and one-half the distance between lacrimal foramina is the base.

Ratios or percentages were derived as follows: length of palate to length of skull = PL/CNL; length of rostrum to length of skull = RL adj./CNL.

Other symbols include: $\bar{Y}$ = mean; s = standard deviation of sample; O.R. = observed range of sample; N = number in sample.

Genus *Dasypus* Linné

Long-nosed Armadillos, Long-eared Armadillos, Mulitas

Type-Species.—*Dasypus septemcinctus* Linné, 1758: 51.

Range.—See map (Figure 1).

Diagnosis.—Carapace chiefly dark in adults; small scales and scutes of scapular and pelvic shields irregular circles or rosettes; 6–11 movable bands (= dorsals of Pocock, 1924), each having slender, acute triangular scales with apex of scale lying on midline of each bony scute directed anteriorly and flanked by scales with apices directed posteriorly, the latter scales overlapping the margins of adjoining bony scutes. Ears relatively long, 39–52 percent of length of skull (CNL). Tail long, over 55 percent of the length of head + body, and with very slender tip; the proximal two-thirds consisting of rings decreasing in diameter distally, each ring formed by two or more rows of scales and scutes. Forefoot with four claws except for vestigial fifth claw in *D. kappleri*; longest claws on second and third digits; hindfoot with five claws, longest on third digit.

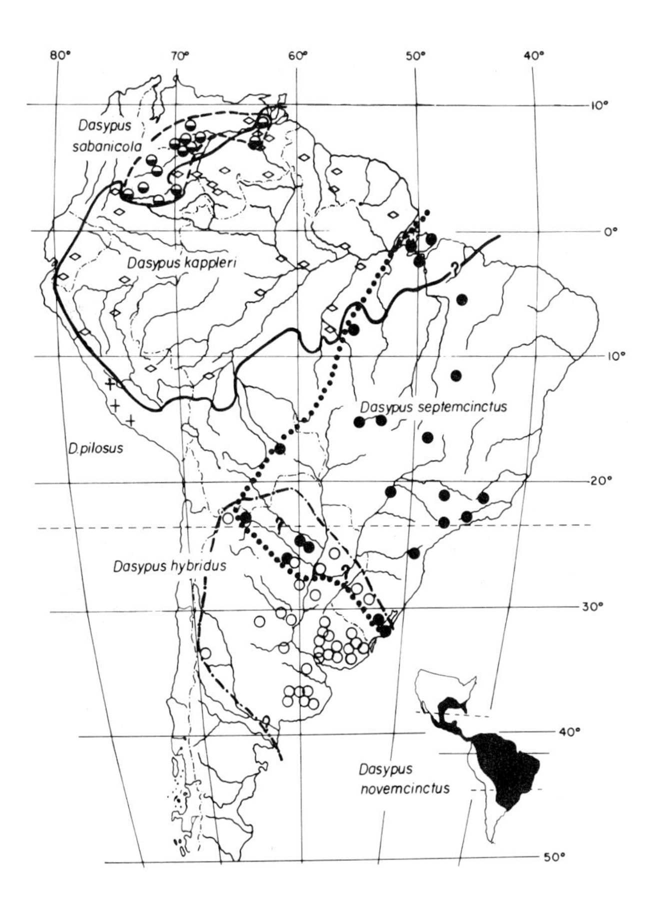

Figure 1. Estimated range of long-nosed armadillos, genus *Dasypus*. Insert, lower right, range of *D. novemcinctus*. Larger map, distribution and specimens examined, remaining five species: - - - - - and ◒ = *D. sabanicola* (N 35); ——— and ◇ = *D. kappleri* (N 54); + = *D. pilosus* (N 4), montane Peru; · · · · · · · · and ● = *D. septemcinctus* (N 32); —— · —— · and ○ = *D. hybridus* (N 82). Note band of sympatry of *D. septemcinctus* and *D. hybridus*.

Skull with tympanic ring; rostrum long and slender, constituting 55% or more of length of skull; palate long, 61–76 percent of length of skull and pterygoids forming posterior surface of palate except at midline. Teeth small and reduced to 7 or 8 maxillary and 7 or 8 mandibular pairs, with occasional gaps in toothrows of older specimens. Mandible slender and without distinct fossa in maxilla.

Four teats; polyembryony observed for subgenus *Dasypus* and likely for rest of genus.

Comparisons.—In *Dasypus* the three regions of the carapace are more distinct than in other genera except *Tolypeutes* and, except for the latter, no other armadillo has so few movable bands. The slender, triangular scales on the movable bands, with the posteriorly directed pairs overlapping adjoining bony scutes, are unique; other armadillos have rectangular scutes which are matched with overlying scales. The scales and scutes of the scapular and pelvic shields usually form rosettes or other types of circles rather than the rectangles, squares, or polygons which occur in other genera. The head is proportionally more narrow and the relatively longer ears are more closely set together than in other armadillos. The tail is proportionally much longer, with means of samples ranging from 56–81 percent of the head + body length, as compared to less than half the head + body length except for *Euphractus* with 51 percent (N 23). The tail has a slender tip and distinct rings on the proximal two-thirds, rather than being bluntly tapered and without rings as in other armadillos. The four toes of the forefoot contrast with five complete toes in other armadillos. The claws of both manus and pes are rounded dorsally in transverse section, rather than being laterally unequal and flattened and with some claws of the forefoot scimitar-shaped as in Priodontini and, although less so, in *Tolypeutes* and Euphractini.

A tympanic ring occurs in *Dasypus*, as well as in *Tolypeutes* and Priodontini, rather than tympanic bullae as in Euphractini and Chlamyphorini. The proportionally longer rostrum forms 55–67 percent of the length of the skull as compared to these lower percentages in other armadillos: *Priodontes* 53, *Zaedyus* 52, *Tolypeutes matacus* 48, *Euphractus* 48, *Chaetophractus villosus* 48, *Cabassous* 48, *Chaetophractus vellerosus* 45, *Chlamyphorus* 51 (N 1); *Burmeisteria* 44 (N 1). The palate extends far posterior to the toothrows and the pterygoids contribute to the posterior surface of the palate except at the midventral point of the posterior border; in other armadillos the palate ends nearer the posterior teeth and the pterygoids are confined laterally as pterygoid ridges.

The teeth are fewer in number and, except for *Priodontes*, smaller in proportion to the cranium than in other armadillos; means of pairs of adult teeth are 7 or 8/7 or 8 in *Dasypus* species as compared to 8 or 9/9 in *Tolypeutes* species, 9/10 or 8/9 in Euphractini, 9/8 in *Cabassous*, and 18/19 in *Priodontes*.

The mandible is proportionally more slender than in other armadillos including Priodontini, and is further distinguished from Priodontini by the height of coronoid process being greater than the condyloid (articular) process. The proximal segment of the anterior wing of the hyoid (ceratohyal or hypohyal of Basc, 1967) is directed toward the midline rather than pointing laterally from the basihyal as in *Cabassous* and *Euphractus* (ibid., Fig. 294, for *D. novemcinctus*).

Four teats have been noted for all of the genus whereas the inguinal pair is absent in other genera. Polyembryony, with multiple births of from 4 to 12 young, does not occur in other genera of armadillos where one or two young are the rule. Chromosomes in *Dasypus novemcinctus* and *D. hybridus* are 2N 64, NF 82–88, as compared to: *Priodontes maximus* 2N 50, NF 76; *Cabassous centralis* 2N 62, NF 78; *Euphractus sexcinctus* 2N 58, NF 104; *Chaetophractus villosus* 2N 60, NF 92 (Benirschke et al., 1969).

COMMENTS.—Although many scientific names of armadillos are based on the number of movable bands, it is proposed here that for vernacular names we discontinue using this variable characteristic and base names upon unique or more consistent features.

The genus is divided in this study into three subgenera with species as follows: **subgenus *Dasypus*** L.—*D. novemcinctus*, *D. septemcinctus*, *D. hybridus*, and *D. sabanicola*; **subgenus *Hyperoambon*** Peters—*D. kappleri*; and **subgenus *Cryptophractus*** Fitzinger—*D. pilosus*.

The few comments and studies of taxonomy and phylogeny made subsequent to the literature on *Dasypus* summarized by Talmage and Buchanan (1954) include: a discussion of I. von Olfer's names of armadillos (Hershkovitz, 1959:340); restriction of the type-locality of *D. longicaudis* Wied (Avila-Pires, 1965:12); characterization of *D. novemcinctus* (Carvalho, 1965:32–33); a description of a new species from the savannas of Venezuela and Colombia, *Dasypus sabanicola* Mondolfi (1968); and the relationships of four species of *Dasypus* based upon cranial proportions (Moeller, 1968). Additional distributional and natural history data on the genus for South America subsequent to Cabrera (1958) follow: **Argentina**—Crespo (1974:12–13), Olrog (1976:8). **Bolivia**—Crespo (1974:13). **Brazil**—Carvalho and Toccheton (1969:220), Pine (1973:61–62). **Guyana**—Hanif and Poonai (1968:26). **Paraguay**—Schade (1973:86), Wetzel and Lovett (1974:209). **Peru**—Soukup (1960:161), Grimwood (1969:27). **Surinam**—Housson (1973:10). **Uruguay**—Ximénez et al. (1972:13), Gonzalez (1973:9). **Venezuela**—Röhl (1959:128), Musso (1962:173–174), Ojasti and Mondolfi (1968:433), Handley (1976:44–45), Mondolfi (1976:122).

Subgenus *Dasypus*

Dasypus Linné, 1758:18, in part.
Subgenus *Cachicamus* McMurtrie, 1831:163.
Subgenus *Muletia* Rhoads, 1894:113, based upon *Muletia* Gray, 1874:246.

Type-Species.—*D. septemcinctus* L.

RANGE.—See *D. novemcinctus* and Fig. 1.

DIAGNOSIS.—Carapace with very sparse hair; movable bands 6–9; without enlarged, projecting scales on hindleg; forefoot with four toes; posterior palate with rounded lateral margins and posterior margin indented medially; rostral length chiefly below 62% of length of skull. Penis (as seen in *D. novemcinctus*, Figure 2, and *D. sabanicola*) somewhat thickened at base and with trifid appearance due to distal pair of lateral lobes and a pointed tip projecting from between the lobes. Polyembryony resulting in monovular quadruplets in *D. novemcinctus* (see Buchanan, 1957, and Storrs and Williams, 1968, for range of variation in embryos); quadruplets in *D. sabanicola* (Mondolfi, 1968:160); 4, 8, or rarely 12 young in *D. hybridus* (summarized by Moeller, 1968:414, as *D. septemcinctus*).

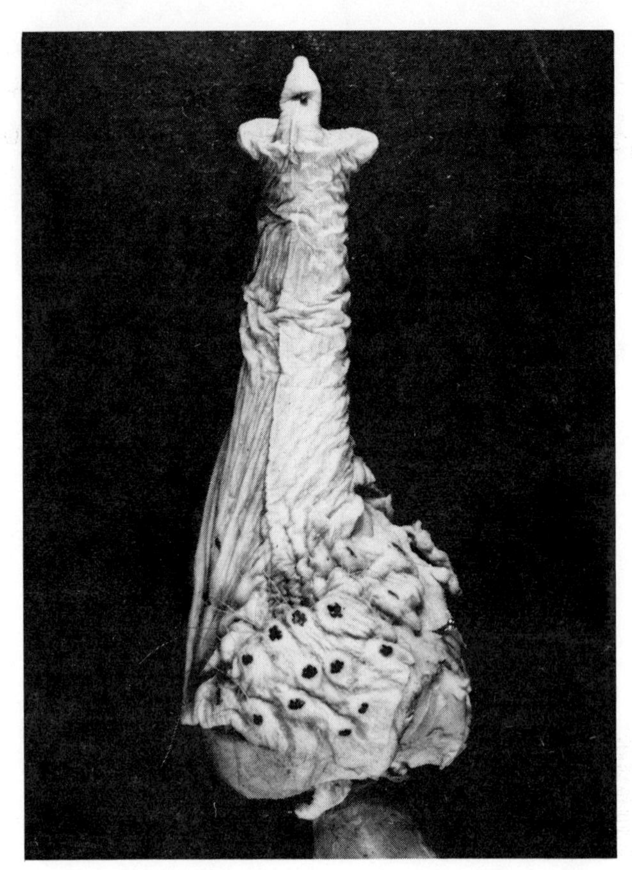

Figure 2. Penis, *Dasypus novemcinctus*, ventral view. Note distal lateral enlargements, small rounded termination of glans, and similar ventral plane of both shaft and glans. Photograph by E. Mondolfi.

COMPARISONS.—As in the subgenus *Hyperoambon*, the carapace has only sparse hair, not obscuring the scutes and movable bands as in the subgenus *Cryptophractus*. The number of movable bands, although not subgenerically distinct between *Dasypus* and *Hyperoambon*, is fewer than the 11 of *Cryptophractus*. The length of scales on the posterior surface of the hindleg never approaches the length found in *Hyperoambon* (greatest length of individual scales in largest specimens of *D. novemcinctus* 6.1 mm vs. 17 mm or more in *D. kappleri*). Size ranges from the smallest species of the genus, *D.*

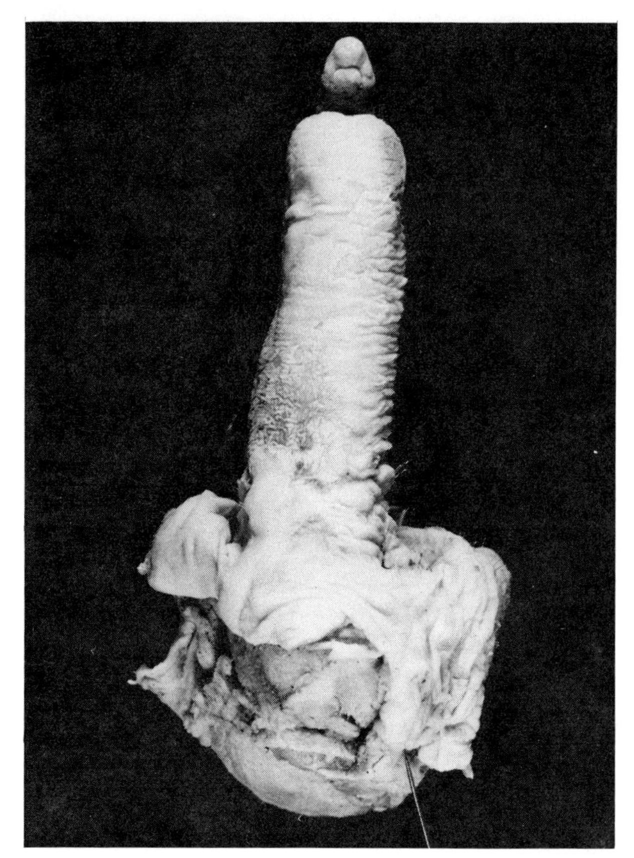

Figure 3. Penis, *Dasypus kappleri*, ventral view. Note glans arching ventrally, blunt termination of glans, and absence of lateral enlargements near distal end. Photograph by E. Mondolfi.

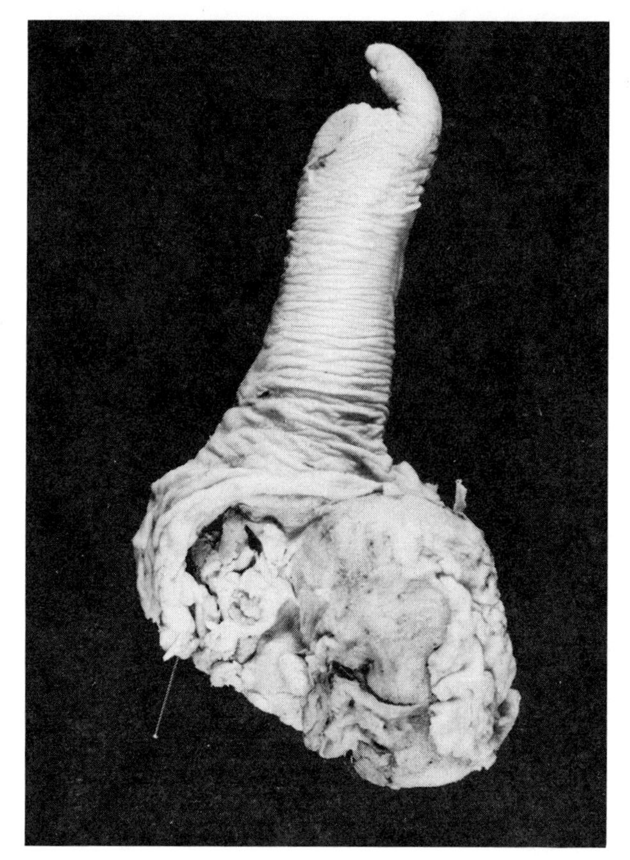

Figure 4. Penis, *Dasypus kappleri*, lateral view. Photograph by E. Mondolfi.

septemcinctus, to the second largest, *D. novemcinctus*, in which the largest specimens approach the size of small adults of *D.* (*Hyperoambon*) *kappleri*. The posterior palate has rounded lateral margins and the posterior margin indented medially vs. the keeled lateral margins and a straight posterior margin in *Hyperoambon*. The ratios of the rostrum and palate to length of skull are below those of the other subgenera (Tables 1–2–3). The penis is distally trifid and the smaller terminal portion has a still smaller protuberance at its tip; in *Hyperoambon* the penis is distally rounded and the smaller terminal portion is blunt, without a further protuberance. See Figures 3 and 4.

COMMENTS.—Moeller (1968) concluded from allometric comparisons of skulls that *D. septemcinctus* (including *D. hybridus*) was more similar to *D. pilosus* than to *D. novemcinctus*. Our placement of *D. septemcinctus*, *D. hybridus*, and *D. sabanicola* in the same subgenus with *D. novemcinctus* while placing *D. pilosus* in a separate subgenus, indicates major disagreement with Moeller's findings. Our reasons for separating *D. pilosus* are discussed under subgenus *Cryptophractus*. Both our bivariant and multivariant comparisons of *D. novemcinctus* with *D. septemcinctus* and allies indicate differences between the two groups, but far below the magnitude of subgeneric separation used in this study. This is also borne out, we believe, by Moeller's bivariant comparisons for, despite her conclusions, she showed that *D. novemcinctus* differs from *D. septemcinctus* chiefly by occupying different locations along the same size gradient. It is *D. pilosus* and *D. kappleri*, species we separate from the subgenus *Dasypus*, that show different curves in most of her graphs.

Table 1. Cranial and body measurements, adult *Dasypus novemcinctus*. $\bar{Y}$ and s on first lines, O.R. and N on second lines within parentheses. Note the variation between samples.

	Brazil, Ilha Mexiana (Pará) (D. n. mexianae Hagmann)	*Brazil, S.E. highlands (Minas Gerais, Goias)*	*Venezuela, N. of Río Orinoco & E. of Andes (Anzoátegui, Miranda, Monagas, Sucre)*	*Surinam, Guyana, & Venezuela S. of Río Orinoco (Bolivar)*
Condylonasal length (CNL)	80.8, — (78.7–82.4, 3)	90.1, 3.4 (85.6–95.5, 37)	89.1, 5.0 (83.2–97.3, 10)	101.9, 5.6 (91.1–110.9, 27)
Rostral length, adj. (RL adj.)	45.6, — (43.6–47.4, 3)	53.1, 2.4 (49.1–57.2, 37)	53.5, 3.5 (49.0–59.7, 15)	61.6, 4.2 (51.9– 67.1, 24)
Rostral ratio (RL adj./CNL)	0.59, — (0.58–0.60, 3)	0.59, 0.02 (0.55–0.60, 38)	0.60, 0.02 (0.57–0.62, 10)	0.61, 0.01 (0.59– 0.63, 22)
Zygomatic width	3.61, — (34.5–37.1, 3)	39.6, 1.4 (36.9–42.5, 37)	39.1, 1.8 (37.4–41.6, 11)	44.2, 2.7 (38.8– 48.0, 20)
Mastoidal width	23.5, — (23.3, 23.7, 2)	26.8, 1.2 (24.6–30.8, 38)	26.3, 1.3 (24.1–28.4, 11)	28.8, 1.3 (25.7– 30.3, 24)
Length of head + body	—	408.1, 21.2 (365–475, 30)	436.4, 37.6 (395–490, 7)	471.8, 47.0 (410–573, 10)
Length of tail	—	322.4, 23.0 (265–360, 17)	313.8, 46.4 (290–355, 4)	381.2, 40.1 (330–450, 10)
Length of hind foot	—	92.6, 6.5 (80–100, 29)	90.5, 5.9 (85–100, 8)	97.9, 9.1 (80–110, 8)
Length of ear	—	44.3, 3.8 (36–50, 31)	36.5, 2.0 (35–39, 6)	46.9, 6.2 (40–57, 9)
Weight, kg	—	3.40, 0.56 (2.65–4.70, 21)	3.77, 0.48 (3.20–4.10, 6)	5.43, 0.79 (4.00–6.25, 8)
Movable bands	—	8.3, 0.5 (8–9, 24)	9.0, — (9–— 10)	9.0, 0.4 (8–10, 15)
Scutes, fourth movable band	—	60.4, 2.6 (56–65, 17)	57.6, 2.4 (54–60, 10)	58.8, 2.7 (56–64, 10)

Table 2. Cranial and body measurements of adult *Dasypus* species[1] other than *D. novemcinctus*. Ȳ and s on first lines, O.R. and N on second lines within parentheses.

	Dasypus septemcinctus	*Dasypus hybridus*	*Dasypus sabanicola*	*Dasypus kappleri*	*Dasypus pilosus*
Condylonasal length (CNL)	63.4, 3.8 (57.9–72.4, 15)	70.2, 2.2 (66.3–75.5, 26)	68.0, 2.9 (60.0–72.1, 27)	123.4, 4.5 (112.1–135.0, 47)	99.0, — (88, 110, 2)
Rostral length, adj. (RL adj.)	35.6, 2.8 (31.4–41.3, 15)	39.1, 1.6 (37.0–42.7, 26)	37.9, 2.0 (33.0–42.7, 27)	79.0, 3.7 (69.8–90.5, 47)	66.4, — (57.8, 75.1, 2)
Rostral ratio (RL adj./CNL)	0.55, 0.01 (0.53–0.57, 15)	0.56, 0.01 (0.55–0.58, 26)	0.56, 0.01 (0.54–0.58, 27)	0.64, 0.01 (0.62–0.67, 47)	0.67, — (0.66, 0.68, 2)
Zygomatic width	27.1, 1.8 (25.0–30.5, 15)	30.9, 1.3 (29.5–32.8, 26)	30.7, 1.6 (28.2–33.4, 27)	50.6, 2.3 (46.0–55.3, 48)	35.3, — (35.3– —, 1)
Mastoidal width	20.2, 1.2 (18.4–22.7, 15)	22.5, 0.7 (21.3–23.8, 26)	21.3, 0.8 (19.6–22.7, 27)	33.9, 1.3 (31.6–36.4, 48)	24.4, — (22.4, 26.5, 2)
Length of head + body	265.0, 23.4 (240–305, 8)	297.3, 10.6 (281–312, 11)	290.3, 14.9 (253–314, 26)	543.2, 22.2 (510–575, 18)	440, — (440– —, 1)
Length of tail	147.7, 13.5 (125–170, 7)	169.0, 9.8 (150–185, 9)	197.4, 13.3 (175–205, 26)	406.1, 45.6 (325–483, 15)	310, — (310– —, 1)
Length of hind foot	60.0,— (45, 75, 2)	66.3, 3.9 (57–69, 7)	63.0, 2.7 (60–70, 25)	119.9, 8.4 (110–135, 15)	66, — (62, 70, 2)
Length of ear	30.9, 4.1 (30–38, 7)	27.4, 2.6 (26–30, 11)	26.6, 1.6 (22–29, 25)	50.5, 3.1 (48–55, 15)	50, — (50– —, 1)
Weight, kg	1.45, — (1.45– —, 1)	2.04, — (2.0– —, 2)	1.49, 0.3 (1.0–2.0, 7)	9.68, 0.7 (8.5–10.5, 11)	—
Movable bands	6.5, 0.5 (6–7, 13)	6.9, 0.6 (6–8, 20)	8.0, 0.3 (7–9, 27)	7.8, 0.4 (7–8, 33)	11, — (11– —, 3)
Scutes, fourth movable band	46.2, 2.3 (43–50, 13)	52.6, 2.2 (50–62, 20)	50.0, 2.3 (46–55, 27)	56.4, 2.9 (51–62, 28)	— —

[1] Source of specimens (also see map, Fig. 1):

D. septemcinctus—**Argentina**, Formosa 3. **Bolivia**, Santa Cruz 1. **Brazil**, Mato Grosso 4; Minas Gerais 1; Amazon basin 14; Rio Grande do Sul 2.

D. hybridus—**Argentina** 10. **Brazil**, Rio Grande do Sul 5. **Uruguay** 11.

D. sabanicola—**Columbia** 15. **Venezuela** 12.

D. kappleri—**Brazil**, eastern Amazon basin 6. **Columbia**, eastern 12. **Ecuador** and **Peru**, western Amazon basin 10. **Surinam** 7. **Venezuela** 15.

D. pilosus—**Peru**, Huánuco 1; Junín 1.

Table 3. Comparison of relative length of palate in terrestrial Edentata.[1]

	Ȳ	s	O.R.	N	
Myrmecophaga tridactyla	.90	—	.89, .90	2	Giant anteater
Tamandua tetradactyla	.84	.01	.82–.85	13	Middle-sized anteater
Dasypus pilosus[2]	.73	—	.70, .76	2	Hairy long-nosed armadillo
D. kappleri	.72	.01	.68–.73	44	Greater long-nosed armadillo
D. novemcinctus	.69	.01	.66–.71	25	Common long-nosed armadillo
D. hybridus	.66	.02	.63–.70	26	Southern long-nosed armadillo
D. sabanicola	.66	.02	.63–.69	27	Northern long-nosed armadillo
D. septemcinctus	.63	.02	.61–.66	14	Lesser long-nosed armadillo
Priodontes maximus	.63	.02	.60–.66	29	Giant armadillo
Cabassous tatouay	.62	.01	.58–.64	25	Naked-tailed armadillo
C. unicinctus	.58	.01	.55–.61	35	Naked-tailed armadillo
C. centralis	.58	.01	.56–.60	16	Naked-tailed armadillo
Euphractus sexcinctus	.59	.01	.56–.62	25	Yellow armadillo, poyū
Chaetophractus villosus	.58	.01	.54–.60	25	Hairy armadillo, peludo
C. vellerosus	.57	.01	.55–.59	21	Hairy armadillo, peludo
Zaedyus pichiy	.56	.01	.54–.57	20	Pichi
Chlamyphorus truncatus	.63	—	.63	1	Pichiciego, fairy armadillo
Burmeisteria retusa	.59	—	.59	1	Pichiciego, fairy armadillo

[1] Means of ratios: palatal length/length of skull.

[2] Note the proportionally longer palate in most *Dasypus* species as compared to other armadillos.

Dasypus novemcinctus L. Common long-nosed armadillo.

Dasypus novemcinctus Linné, 1758:51; Yepes, 1928:468, 506; Hamlett, 1939:329, 334; Frechkop and Yepes, 1949:16, 20; Talmage and Buchanan, 1954:84; Cabrera, 1958:224; Hall and Kelson, 1959: 244; Moeller, 1968.

Dasypus fenestratus Peters, 1864:180. Lectotype: ZMB 3175, Costa Rica, San José, adult male, museum exhibit mount, coll. by Hoffmann, labeled "syn-typus," is proposed as lectotype from syntypes ZMB 3175 and 3176, the latter an immature specimen.

Dasypus novemcinctus mexicanus Peters, ibid. Holotype: ZMB 2743, Mexico.

Tatusia mexicana Gray, 1873:14. Holotype: BMNH 43.9.11.6, Mexico.

Tatusia granadiana Gray, ibid. Holotype: BMNH 73.3.12.2, Colombia, Antioquia, Concordia.

Tatusia leptorhynchus Gray, ibid.:15. Holotype: BMNH 65.5.18.42, Guatemala.

Tatusia brevirostris Gray, ibid. Holotype: BMNH 46.5.13.16, Brazil, Rio de Janeiro.

Tatusia leptocephala Gray, ibid.:16. Holotype: BMNH 47.4.6.2, Brazil.

Tatusia novemcinctus mexianae Hagmann, 1908:29. Lectotype: MZS 167, Brazil, Pará, Ilha Mexiana, adult cranium, is selected from the six MZS syntypes.

Dasypus novemcinctus hoplites G.M. Allen. 1911:195. Holotype: MCZ 8116, Island of Grenada, above Gouyave.

Dasypus novemcinctus aequatorialis Lönnberg. 1913:34. Holotype: NHR 25, Ecuador, Pichincha, Perucho.

Dasypus mazzai Yepes, 1933:226, in part; Hamlett, 1939, in part. Holotype: MACN 31.273, Argentina, Salta, Tabacal, shell only. See Table 4.

Dasypus novemcinctus davisi Russell, 1953:21. Holotype: TCWC 4952, Mexico, Morelos, Huitzilac, 8500 ft.

Type-Locality.—Restricted by Cabrera (1958:225) to Brazil, Pernambuco.

Range.—From southern United States—Florida, west to Louisiana, Arkansas, Oklahoma and Texas—southward with continuous distribution through eastern Mexico and from northern Sinaloa on the west (Guasave, MVZ 126886; San Miguel, Jones et al., 1962:155), through Central America to South America: the Pacific side of the Andes—Colombia, Ecuador, and northwestern Peru (Deptos. Piura and Tumbes, Grimwood, 1969:27); east of the Andes—all of South America south to Uruguay and Argentina to the provinces of Santiago del Estero (MACN 36.962), Santa Fe (Calchaqui, MACN 30.16), and Entre Rios (Sauce de Luna, MACN 13221). It has also been recorded for the island of Grenada, Lesser Antilles and for the continental islands of Margarita, Trinidad and Tobago. See map, Fig. 1.

DIAGNOSIS. A large *Dasypus* (*Dasypus*) with many yellow-tan scales on lateral carapace; movable bands varying from 7 to 10, but chiefly 8 in the southern part of range and 9 from basin of the Amazon northward; scutes on fourth movable band ranging from 54 to 65, but chiefly 60; ear long, nearly half the length of skull (ear/CNL 0.46, s 0.04, N 75); tail long, 70% or more of length of head + body (our sample of 60 adults selected for unbroken tails, Nicaragua to Argentina, $\bar{Y}$ 83%), with distal third gradually tapering; rostrum long, 55–63 percent of length of skull; palate long, 66–71 percent of length of skull; pairs of teeth 8/8 (O. R. 7–8/6–9, N 37). Also see Tables 1 and 3. Young are monovular quadruplets.

COMPARISONS.—The carapace is not as dark as in the smaller species of the subgenus and contains more pronounced lateral areas of pale yellow or tan. Movable bands standardly number one more at points of sympatry with other species of *Dasypus*; a possible exception, *D. pilosus* with 11 bands, may or may not be sympatric. *D. sabanicola* and *D. kappleri*, with 8 movable bands, occur in parts of the range where *D. novemcinctus* has chiefly 9 bands. *D. septemcinctus* and *D. hybridus*, with 7 bands, occur in southern areas of sympatry where *D. novemcinctus* has 8 bands. In areas of sympatry, the number of scutes on the fourth movable band is greater than in other members of the subgenus. A similar, although less consistent, separation can be made between *D. novemcinctus* and *D. kappleri* where they are sympatric. Ears are proportionally longer than those of *D. hybridus* and *D. sabanicola* but not significantly different from *D. septemcinctus*. The tail is proportionally as long as in *D. kappleri* and *D. pilosus* and is much longer and more gradually tapered than in other species of the subgenus. Rostral and palatal lengths are greater, both actually and proportionally, than in other members of the subgenus but are proportionally smaller than in *D. kappleri* and *D. pilosus*. For other differences between *D. novemcinctus* and the latter two species, see accounts of the subgenera.

COMMENTS.—*Dasypus mazzai* Yepes (1933), a composite of *D. novemcinctus* and other species, is discussed under *Comments*, *D. septemcinctus*, and in Table 4.

The following summary of the natural history of *D. novemcinctus* is from literature chiefly subsequent to reviews by Kalmbach (1944) and Talmage and Buchanan (1954).

METABOLISM AND ACTIVITY.—The armadillo's 24-hour body temperature cycle, with a range of 2.5°C at an ambient temperature of 25°C, is correlated with periods of activity and rest (Johansen, 1961). Its lability is not, per se, an indication of primitive temperature control, as this range is exceeded by a number of other mammals (ibid.; Goffart, 1971:3). Burns and Waldrip (1971), using armadillos ranging freely in a room with 23°C air temperature, found males to have a rectal temperature of 33.4°C and heart beats of 126 per minute, while females had significantly lower readings of 31.3°C and 84 beats per minute. As their field data for this same winter period indicated more males than females were active, the authors suggested that there may be seasonal metabolic differences between the sexes. Adaptations to high temperatures include an initial, lower body temperature than most mammals (Talmage and Buchanan, 1954:104), vasodilation, panting with respiratory rates rising from 30–40 per minute to 180, and resorting to burrows during the heat of the day (Johansen, 1961). Adaptation to cold temperatures is made possible by vasoconstriction, shivering, increase in body temperature due to increased muscular work, and behavioral responses that include using burrows with bedding of hay, rolling into a "... ball-like posture" (ibid.) and, we would speculate, huddling where family units (such as mother and young) are occupying the same burrow. Experimental interruption of breathing in armadillos resulted in bradycardia, suggesting to Scholander et al. (1943) that armadillos are capable of digging under anaerobic conditions.

REPRODUCTION.—In North America there is a period of delayed implantation of three to four months (Talmage and Buchanan, 1954:118), followed by a gestation period of active development of four and one-half months (Hamlett, 1935). The estrous cycle has a duration of four days (D'Addamio et al., 1977).

FEEDING AND DIET.—The studies of food habits of *Dasypus novemcinctus*, reviewed by Talmage and Buchanan (1954:29–37) and Patterson (1975:221–223), indicate that its diet is predominantly animal material, of which over 75 percent of the total diet is insects. Kalmbach (1944:33) found beetles in all but two of 169 stomachs examined, with Coleoptera forming 28 percent of the total food volume. However, 40,000 ants of three species were taken in one meal by one armadillo and 13,000 termites, comprising one-third of the food volume for another armadillo, were found in that study. Termites were found in 126 of 169 stomachs examined. *D. novemcinctus* and other members of the genus are probably more dependent upon food requiring less mastication, such as softer and/or smaller items, than are most other armadillos. The more slender mandibles, reduced dentition with frequent gaps in adult toothrows, and poorly defined mandibular fossae support this speculation. If comparative propor-

Table 4. Specimens attributed to *Dasypus mazzai* Yepes (1933) and other specimens illustrating Yepes's treatment of *Dasypus*

MACN number	*Provenance—collector*	*Nature of specimen*	*Identification, this study*	*Comments*
31.273 holotype *D. mazzai*	Argentina, Salta, Tabacal—S. Mazza	Carapace only	*D. novemcinctus*	This, or the specimen below, was considered by Hamlett (1939:335) to be *D. novemcinctus*. 8 movable bands; 62 scutes on fourth band.
31.273 "Repetido"	Argentina, Salta, Tabacal—S. Mazza	Cephalic shield, carapace, tail	*D. novemcinctus* immature	Duplication of accession numbers (caused by confusion of "1" and "7"?); either this or above specimen may be 31.213. Both 31.213 and 31.273 were listed by Yepes (p. 229) for the holotype of *D. mazzai*. 8 movable bands; 60 scutes on fourth band.
33.23	Argentina, Salta, Tabacal—S. Mazza	Cranium; body without carapace in alcohol	*D. septemcinctus*	Not mentioned, at least by this number, by Yepes. Smaller than measurements of holotype and paratype given by Yepes. Occipital-basisphenoid suture closed.
13222 (=28.225)	Paraguay, Alto Paraguay, Pto. Guarani (Crespo, corresp.)	Museum exhibit mount	*D. hybridus*	Probably the paratype figured by Yepes (Fig. 1) and considered by Hamlett to be the paratype of *D. mazzai* (Crespo, corresp.). 7 movable bands; 62 scutes on fourth band.
30.17	Argentina, Chaco, Pampa del Indio—J. G. Dennler	Cranium and carapace	*D. hybridus*	*D. septemcinctus* according to Yepes (p. 231, 232). 7 movable bands; 52 scutes on fourth band.
30.18	Argentina, Chaco, Pampa del Indio—J. G. Dennler	Cranium and carapace	*D. septemcinctus*	*D. septemcinctus* according to Yepes (ibid.) 7 movable bands; 46 scutes on fourth band.

tions of the palate are used as an index within Edentata as we have done in Table 3, the question can be asked if the long-nosed armadillos are more committed toward myrmecophagy than are other armadillos. Comparative food-habit studies are needed to verify or reject this speculation or that of Kühlhorn (as discussed by Patterson, 1975:221–223) who considered *Priodontes maximus* much more committed to myrmecophagy than *Dasypus*. If the numerous teeth of *Priodontes* are the result of their "... escaping from selection pressure" (ibid.: 223), reduction and loss (as with the teeth in *Dasypus*) are an alternate and more common evolutionary response to disuse.

Habitat and Distribution.—The burrows of *D. novemcinctus*, up to 24 feet in length and five feet in depth (Layne, 1976:10), and the use of nesting material (Eisenberg, 1961) have probably been among the factors permitting that species to extend its range northward into temperate North America. This is not, however, a clear-cut relationship as *D. novemcinctus* is not the most southerly distributed armadillo, as we might expect if such adaptations against colder climates are sufficient explanation of its northerly dispersal. Certainly *Zaedyus pichiy*, *Chaetophractus villosus*, *C. vellerosus*, *Chlamyphorus truncatus*, and *Dasypus hybridus* range much more southerly than does *D. novemcinctus*.

It is possible that the euphractine armadillos might have ranged farther north than *D. novemcinctus* if they had been able to cross the tropical rainforest barriers of northern South America and Central America. *D. novemcinctus* does not reach the altiplano of the Andes as does *Chaetophractus nationi*, nor the more arid zones of the lower altitudes. Greegor (1975) placed *D. novemcinctus* with arid-adapted mammals but considered it less adapted to aridity than *Chaetophractus vellerosus*. His study compared medullary thickness in kidneys to body weight as well as weight loss of the two species under conditions of water deprivation. Slaughter (1961) found *D. novemcinctus* in its northerly range to be restricted to areas with rainfall greater than 457 mm a year and yearly mean temperatures greater than 16°C. *D. novemcinctus* is probably the most abundant armadillo of tropical rainforests; it was second only to the three-toed sloth, *Bradypus tridactylus*, in mammalian biomass by Walsh and Gannon's data as presented by Eisenberg and Thorington (1973:152).

Other reports subsequent to Talmage and Buchanan (1954): **ecology** (Davis, 1966:249–252; Platt et al., 1967; Moore, 1968; Meritt, 1973; Layne and Glover, 1977), **anatomy and physiology** (Starck, 1959; Keil and Venema, 1963; Shackleford, 1963; Wilbur, 1964; Azzali and DiDio, 1965; Maller and Kare, 1967; Dhindsa et al., 1971; Prejean and Travis, 1971; Jackson et al., 1972; Nagy and Edmonds, 1973; Royce et al., 1975), **embryology** (Buchanan et al., 1956; Buchanan, 1957; Enders et al., 1958, 1959; Enders, 1960a & b, 1962, 1963), **chromosomes** (Beath et al., 1962; Grinberg et al., 1966), and as a tool in **leprosy research** (Storrs et al., 1974).

Additional distributional and natural history notes for North America, subsequent to Hall and Kelson (1959) follow: **Belize:** Kirkpatrick and Cartright (1975:137). **El Salvador:** Felton (1958:220), Burt and Stirton (1961:42). **Guatemala:** Ibarra (1959:63–4). **Mexico:** Leopold (1959:339); Baker and Greer (1962: 57,76); Selander et al. (1962:335), Alvarez (1963:418), Hall and Dalquest (1963:264), Goodwin (1969:119), Armstrong and Jones (1971:752), Jones et al. (1974:9). **Panama:** Handley (1966:776), Mendez (1970:95). **United States:** Cleveland (1970), Baccus (1971), Humphrey (1974), Lowery (1974:148–154), Stevenson and Crawford (1974), Findley et al. (1975:73).

***Dasypus septemcinctus* L.** Lesser long-nosed armadillo, tatú mirim.

Dasypus septemcinctus Linné, 1758:51; Lönnberg, 1928:8, in part; Hamlett, 1939:320; Frechkop and Yepes, 1949:25, in part; Talmage and Buchanan, 1954:81; Cabrera, 1958:226; Moeller, 1968, in part; Mondolfi, 1968:149; Gonzales, 1973:9; Pine, 1973:62.
Tatusia megalolepis Cope, 1889:134. Holotype: ANSP 4643, Mato Grosso, Chapada.
Dasypus propalatus, Yepes, 1928:468; Frechkop and Yepes, 1949:22.
?*Dasypus mazzai* Yepes, 1933; Hamlett, 1939:335; Cabrera, 1958: 224; Olrog, 1976:8; Greegor, 1975. See Table 4.

Holotype.—No. 4. Mus. Adolpho Fred., Linné Coll. No. 24, Zool. Inst., Uppsala Univ., a poorly preserved, dried specimen with 7 movable bands and estimated 53 scales on fourth movable band (Å. Holm, corresp. and photograph). See also, Lönnberg (1928: 8) who believed the type lost.

Type-Locality.—Although both references cited by Linné in 1758 (*D. cingulis septenis* Balk, 1749:281 and *D. cingulis septem* Linné, 1748:6) are to the holotype, neither suggests its geographical origin. The first citation of Balk is to *Dasypus Hernand. mex.* 314. However, the 7 movable bands and estimated 53 scales on the fourth movable band of the holotype fit samples assembled by Hamlett (1939:332) and ourselves for the smaller species of *Dasypus*, none of which occurs as far north as Mexico, and not for *D. novemcinctus* which does occur in Mexico. Both the probable identity of the holotype and stability of nomenclature favor ignoring this citation to Mexico which was also used by Linné (1748:6) for Var. B., *D. novemcinctus*. The only citation of Linné (1748:6) and the second citation of Balk (1749:281) for the 7-banded armadillo are to tatucte Marcgrav (1648:231). Therefore, we agree with Hamlett (1939) and Cabrera (1958) that Pernambuco, Brazil, is the type-locality.

Range.—From the mouth of the Amazon River (Cametá, MZUSP 511; Ilha de Marajó, BMNH 9.7.18.3, MPEG 33, 595, MN 2370, including Caldeirão, BMNH 23.8.10.25; Ilha Mexiana, ZSM no number) south through eastern Brazil to Rio Grande do Sul (BMNH 85.6.26.4, 93.10.20.2) and west to Mato Grosso, southeastern Bolivia (Santa Cruz, San José de Chiquitos, ZSM 1925–584, 607; Tarija, Villa Montes, ZMB 31034), and northern Argentina (Salta, Tabacal, MACN 33.23; Formosa, La Victoria, ZSM 1926–379; Misión Tacaaglé, ZSM 1925–590, 587, 1926–377; Chaco, Pampa del Indio, MACN 30.18). See map, Fig. 1 and comments below. Moeller (1968:Abb. 3) showed this species occurring north of the Amazon River, apparently near the northern border of Brazil and French Guiana.

Diagnosis.—A small *Dasypus* with more dark than yellow-tan areas on lateral carapace; movable bands 6–7; scutes on fourth movable band 43–50; ear long, nearly half the length of skull (CNL); tail relatively long, over half the length of head + body; rostrum long, 53–57 percent of length of skull; palate long, 61–66 percent of length of skull; pairs of teeth in adults 7/7 (O.R. 6–8/6–8, N 15). Also see Tables 2 and 3.

Comparisons.—*D. septemcinctus, D. hybridus,* and *D. sabanicola* are the smallest members of the genus (see

Tables 1 and 2). A number of authors have synonymized the first two species and, indeed, all three are so similar that the question of whether they are sibling species or merely geographic differences at the subspecific level must be raised. Their habitat preference for grassland, at least for *D. sabanicola* and for *D. hybridus* over much of its range, is also similar. The preference of *D. septemcinctus* is not so apparent, due to both the dearth of habitat information about the specimens examined and fewer widespread, contiguous areas of grass within its range. However, the major part of its range is in the campos or shrub-palm-savanna of central Brazil as used by Sauer (1950:343 and Map 10). The distribution of *D. septemcinctus,* as well as *D. hybridus* and *D. sabanicola,* approximates certain nonforest centers of Müller (1973).

In addition to size and coloring of the carapace described under diagnosis; above, these three species differ from *D. novemcinctus* in the following: movable bands of carapace and scutes on fourth band fewer in number and statistically separable from sympatric or near-sympatric samples of *D. novemcinctus;* tail shorter than in *D. novemcinctus,* less than 70 percent of the length of head + body; diameters of rings of the tail decrease more abruptly from base distally; ratio of rostral to cranial length significantly less than in *D. novemcinctus;* lateral edges of posterior palate less rounded.

The three small species may be compared in Figure 5 and Table 2. *D. hybridus* and *D. sabanicola* at two geographical extremes are more similar to each other than either is to *D. septemcinctus,* the geographically central species. Character displacement is particularly noticeable between *D. septemcinctus* and *D. hybridus* which are sympatric along their mutual borders. Except for the length of ear, *D. septemcinctus* is smaller than the other two species (Table 2) and the number of scutes on the fourth movable band is fewer, with a mean of 46.2 ± 2.3 (O.R. 43–50). In the one instance in our sample where this count in *D. hybridus* was as low as 50, the specimen could be grouped with other *D. hybridus* on the basis of its short ears and larger body and cranial dimensions. Our data agree with Hamlett's (1939:329) that the ears of *D. hybridus* are shorter than those of *D. septemcinctus;* the ears of *D. sabanicola,* which was described subsequent to Hamlett's work, are also shorter than in *D. septemcinctus* and often shorter than in *D. hybridus.* The ratios of length of ear/length of skull (CNL), s, and N follow: *D. hybridus* 0.39 ± 0.05, 10; *D. sabanicola* 0.40 ± 0.02, 22; *D. septemcinctus* 0.49 ± 0.03, 5.

Hamlett (1939:329) stated that the tail of *D. septemcinctus* (80–100 percent of body length) is longer than in *D. hybridus* (67–70 percent of body length). Our data do not agree with this. In our sample, the ratios of length of tail/head + body, s, and N are as follows: *D. septemcinctus* 0.57 ± 0.07, 4; *D. hybridus* 0.57 ± 0.03, 9; *D. sabanicola* 0.68 ± 0.06, 22. Here the proportion of length of tail to head + body is the same in *D. hybridus* and *D. septemcinctus,* although *D. hybridus* has the longer tail. The tail of *D. sabanicola* is both actually and proportionally longer than in the other two species. However, because of the frequency of damaged tails in armadillos, this is not an entirely satisfactory taxonomic character.

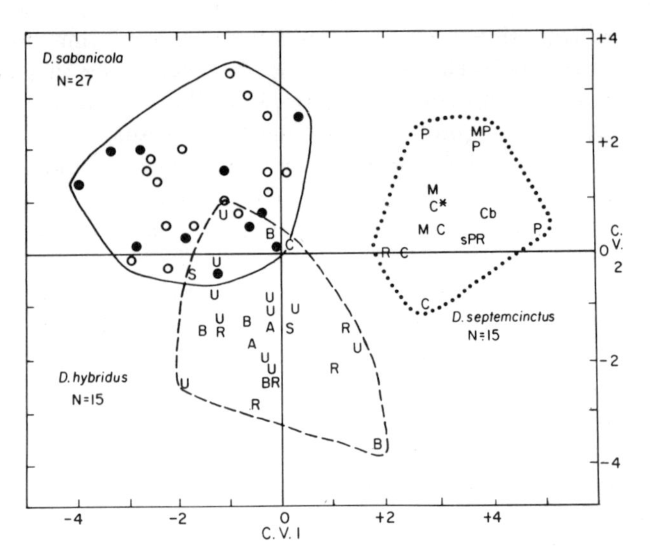

Figure 5. Canonical analysis, small species of *Dasypus,* 13 cranial measurements. Provenance of specimens:

D. sabanicola. ○ = Colombia. ● = Venezuela.

D. hybridus. ARGENTINA: **A** = no other locality data; **B** = Buenos Aires (Buenos Aires, Juárez, Laprida, Rauch); **C** = Chaco (Pampa del Indio); **S** = Santa Fe (no other data and Las Palmares). BRAZIL: **R** = Rio Grande do Sul (no

D. septemcinctus. ARGENTINA: **C** = Chaco Austral, Formosa (La Victoria, Misión Tacaaglé); Salta (Tabacal*). BOLIVIA: **Cb** = Chaco Boreal, Santa Cruz (San José de Chiquitos). BRAZIL: **M** = Mato Grosso (Chapada, São Domingos on Rio das Mortes); **P** = Pará (Cametá, Ilha de Marajó) and Maranhão (Barra do Corda); **R** = Rio Grande do Sul (São Laurenço do Sul); **sP** = São Paulo (Três Lagoas).

In the canonical analysis shown in Figure 5, the samples of *D. septemcinctus* are well divided from *D. hybridus* and *D. sabanicola.* Canonical variables overlap slightly for the latter two species although they are geographically widely separated. Because the two species that are geographically adjacent are well separated by this analysis, we support the integrity of the three species. The evidence of sympatry between *D. hybridus* and *D. septemcinctus,* although based upon only a few specimens, favors this treatment.

COMMENTS.—See Table 4. As the identity of *Dasypus mazzai* Yepes (1933) has been questioned by Hamlett (1939) yet accepted by Cabrera (1958), Greegor (1975) and Olrog (1976), all the extant original specimens were examined by Wetzel. Hamlett reported that the holotype, MACN 31.273, a carapace only from Salta, Tabacal, collected by Salvador Mazza, was actually *Dasypus novemcinctus.* We have verified this, as has Crespo (corresp.). A duplicate number MACN 31.273, also a carapace only and from Tabacal, located for our examination by J. A. Crespo, proved to be an immature *D. novemcinctus.* Hamlett stated that the paratype represented a new species meriting proper description, but did not cite the accession number of the specimen. An additional specimen shows all the features of a paratype pictured by Yepes in his Fig. 1 (Crespo, corresp.); this mounted specimen, MACN 13222 (old number 28.225), we find referable to *D. hybridus,* a species that Yepes did not accept in 1933. There are other complications. Yepes (p. 229) listed two different accession numbers, MACN 31.273 and 31.213, for his holotype. Yepes (p. 231) gave cranial measurements for a holotype and paratype; the skulls are not available today but because of the size of the zygomatic arch, we consider them probably to have been immature *D. novemcinctus.* MACN 33.23, not cited by Yepes, was collected by Mazza in Tabacal; the cranium of this specimen is *D. septemcinctus.* Aside from his selection of a holotype that was an immature *D. novemcinctus,* Yepes was recognizing the fact that Argentina and Paraguay have one large and two smaller species of *Dasypus.* Both his available original material and his description force us to conclude that *D. mazzai* is a composite, a junior synonym of *D. novemcinctus* and *D. hybridus* and that use of *D. mazzai* may also infringe upon *D. septemcinctus.*

Dasypus hybridus (Desmarest). Southern long-nosed armadillo, tatú mulita, Tatou sixième, ou tatou mulet Azara, 1801:186.

Loricatus hybridus Desmarest, 1804:28, including description *Le tatou mulet* d'Azara, Desmarest, 1803:432.

Dasypus hybridus, Yepes, 1928:468; Hamlett, 1939:329, in part; Talmage and Buchanan, 1954:82; Cabrera, 1958:223; Mondolfi, 1968:149.

Dasypus septemcinctus, of authors, not of Linné, 1758:51: Lönnberg, 1928:8, in part; Yepes, 1933, in part; Frechkop and Yepes, 1949: 25; in part; Moeller, 1968, for Uruguay and in part for Argentina; Ximénez et al., 1972:13 and Gonzalez, 1973:9 for Uruguay.

?*Dasypus mazzai* Yepes, 1933: Figure 1, paratype MACN 13222, Paraguay, Alto Paraguay, Puerto Guaraní. Hamlett, 1939:335; Cabrera, 1958:224; Greegor, 1975; Olrog, 1976:8. See Table 4.

TYPE-LOCALITY.—Paraguay as based upon Azara (1801:186) and restricted by Cabrera (1958:223) to Paraguay, Depto. Misiones, San Ignacio.

RANGE.—From eastern Paraguay (Guairá, Villarrica, ZMB 40472; Misiones, Curupayty, NHMB 1450; no locality, MZS 1086d) and Argentina (Corrientes province, MACN 34.733 and probably Misiones), southern Brazil (Rio Grande do Sul, SMF 5268, BMNH 86.10.4.6–9) and Uruguay on the northeast, west through the Chaco Austral of Northern Argentina (Chaco, Pampa del Indio, MACN 30.17; Formosa, La Victoria, ZSM 1926-379, Misión Tacaaglé, ZSM 1925-590) to Jujuy province on the northwest (MACN 34.669, 35.148) and south to Mendoza (AMNH 40068) and according to Cabrera (1958:223), Río Negro. See map, Fig. 1.

DIAGNOSIS.—A small *Dasypus* with carapace as in *D. septemcinctus;* movable bands usually 7; scutes on fourth movable band 50–62; ear relatively short, 39 percent of length of skull (CNL); tail relatively long, over half the length of head + body; rostrum long, 55–58 percent of length of skull; palate long, 63–70 percent of length of skull; pairs of teeth in adults 7/8 (O.R. 6–8/7–8, N 26). Also see Tables 2 and 3.

COMPARISONS.—See *D. septemcinctus,* Figure 5, Tables 1–3.

COMMENTS.—*Dasypus mazzai* Yepes (1933), a composite of *D. novemcinctus* and other species, is discussed under *Comments, D. septemcinctus,* and Table 4.

Dasypus sabanicola Mondolfi. Northern long-nosed armadillo.

Dasypus sabanicola Mondolfi, 1968:151; Handley, 1976:45.

HOLOTYPE.—MEBRG 965.

TYPE-LOCALITY.—Venezuela, Apure, Achaguas, Hato Macanillal.

RANGE.—Llanos of Venezuela from states of Monagas and northeastern Bolívar (Guri, MBUCV 1669, 1670) west through Guarico to Apure; llanos of Colombia in Depto. Meta, Intend. Arauca, and Com. Vichada. The provenance of ICN 1624, skin, as Colombia, Cundinamarca, Tocaima, would extend the range of this species to west of the Cordillera Oriental but should be supported by additional data. Additional records in Venezuela to those given by Mondolfi (1968:153): Monagas, Pasu Nuevo, 30 km S. Temblador; Guárico, road between Las Mercedes and Santa Rita; Portuguesa and Barinas, between Arismendi and Guanarito. Several specimens at the collection of the University del Valle, Dept. Microbiology, collected in Colombia, Meta, Finca Chaviva, and Vichada, Finca la Arepa, were also examined.

DIAGNOSIS.—A small *Dasypus* with carapace as in *D. septemcinctus;* movable bands usually 8; scutes on fourth movable band 46–55; ear relatively short, 40 percent of length of skull (CNL); tail relatively long, over half the length of head + body; rostrum long, 54–58 percent of length of skull; palate long, 63–69 percent of length of skull; pairs of teeth in adults 7.5/8 (O.R. 6–8/6–8, N 27). See Tables 2 and 3.

COMPARISONS.—See *D. septemcinctus,* Figure 5, Tables 1–3.

Subgenus *Hyperoambon*

Hyperoambon Peters, 1864:180.
Praopus Gray, 1873:16.

TYPE-SPECIES.—*Hyperoambon pentadactylus* Peters, loc. cit.

RANGE.—From Colombia east of Andes, Venezuela south of Río Orinoco, Guyana, Surinam and French Guiana south through the Amazon basin of Ecuador, Peru, Brazil and northern Bolivia. See map, Figure 1.

DIAGNOSIS.—Carapace with very sparse hair; movable bands 7–8; large projecting scales or spurs arranged in transverse rows on proximal, posterior surface of hindleg (largest scale, first proximal row, may exceed 17 mm in length); small, vestigial fifth digit on forefoot; lateral borders of palate posterior to toothrows forming definite keels which extend ventrally well below main surface of posterior palate; posterior margin of palate straight; length of rostrum 62–67 percent of length of skull. Penis with subcylindrical shaft terminating in smaller blunt-ended projection curving ventrally from its base at dorsal plane of shaft. (Figures 3 and 4). Two young.

COMPARISONS.—These are the only members of the genus or the family with enlarged, projecting scales or spurs on the hindleg and with lateral borders of palate posterior to toothrows forming keels. It is the largest armadillo within the genus and is second in the family only to *Priodontes maximus* Kerr. The rostral and palatal lengths are proportionally larger than in subgenus *Dasypus* and smaller than in subgenus *Cryptophractus.* Movable bands and scutes on fourth movable band number fewer than in sympatric *D. novemcinctus* but are similar to *D. sabanicola* and *D. septemcinctus.* Tables 1–3 compare these and other measurements. The penis is not trifid and its smaller terminal portion is blunt-ended in contrast to the trifid appearance and the more pointed terminal portion as in subgenus *Dasypus* (Figures 2–4).

COMMENTS.—We agree with Moeller's (1968) conclusions that *D. kappleri* is widely separated from the other species of *Dasypus.* This distance is indicated by our erection of a separate subgenus.

Mondolfi has examined three sets of embryos of *D. kappleri* preserved in formalin (MEBRG) from three females collected on January 18 and 19 in Venezuela, Río Grande, El Palmar and Amanza Guapo. Each set consisted of two fetuses of the same sex. As the fetal membranes and placentae were not preserved, it could not be confirmed that they were monovular twins. The Smithsonian Venezuelan Project (USNM) has one record of a pregnant *D. kappleri* with two fetuses collected January 6 in T.F. Amazonas, Belén, Río Cunucunuma.

Dasypus kappleri Kraus. Kappler's or Greater long-nosed armadillo.

Dasypus kappleri Kraus, 1862:20, Taf. III, Fig. 1, 2; Yepes, 1928:468; Talmage and Buchanan, 1954:83; Cabrera, 1958:223; Moeller, 1968; Mondolfi, 1968:149, 1976:122; Handley, 1976:44.
Hyperoambon pentadactylus Peters, 1864:179, 180. Holotype: ZMB 3174, Guyana, museum exhibit mount.
Tatu pastasae Thomas, 1901:370. Holotype: BMNH 80.5.6.71, Ecuador, Pastaza, Sarayacu.
Dasypus kappleri peruvianus Lönnberg, 1928:10. Holotype: NHR 35, Peru, San Martín, Roque.
Dasypus kappleri beniensis Lönnberg, 1942:49. Holotype: NHR 46, Bolivia, Pando, Victoria.

LECTOTYPE.—Kraus (1862) listed four crania in his type series, one in the zoology museum at Tubingen and three at Stuttgart (SMN), the latter three and perhaps the skull then at Tubingen collected by A. Kappler in Surinam along the Marowijne River. Cranium SMN 285, adult, with tag reading "Kappler, 1846, Typus?," may well have been Kraus' specimen No. II, an adult cranium at Stuttgart, and is therefore proposed as the lectotype.

TYPE-LOCALITY.—Surinam, Marowijne River.

RANGE, DIAGNOSIS and COMPARISONS.—See subgenus *Hyperoambon.*

Subgenus *Cryptophractus*

Cryptophractus Fitzinger, 1856:123.

TYPE-SPECIES.—*Cryptophractus pilosus* Fitzinger, ibid.

RANGE: Known definitely only from the mountains of Peru (Fig. 1); holotype from unknown locality in Peru; other specimens: Junín, Acobamba, 2440 m (BMNH 27.11.1.235), Maraynioc (BMNH 94.10.1.13); Huánuco, Paso Carpish (LSU, 7 specimens, D. A. Tallman, corresp.); "Zapatogocha (Acomayo), 3000 m" (MHN). The holotype of *Praopus hirsutus* (= *D. pilosus*), reported by Burmeister (1862: 147–8) to be at MHN and from Guayaquil (no nation stated), is probably one of two museum exhibit mounts, either MHN 26 or 27. A museum exhibit mount in Brussels (IRSNB 319) was part of a collection from Chile purchased in 1846 (X. Misonne, corresp.). Although only "Santiago, Chile" was mentioned,

Frechkop and Yepes (1949:29) suggested that, perhaps, Peru, Piura, Santiago was a more likely provenance.

Diagnosis.—Dense, long, tan hair obscuring scales and bands of carapace; 11 movable bands; forefoot with 4 toes; rostrum 66–68 percent and palate 70–76 percent of length of skull; rostrum extremely slender, anterior rostral width only 7 percent of length of skull; posterior palate with rounded lateral margins and medial indentation on posterior margin. See Figures 6 and 7.

Comparisons.—These rare and unique armadillos are more densely haired than any other, including the so-called hairy armadillos, *Chaetophractus villosus* and *C.*

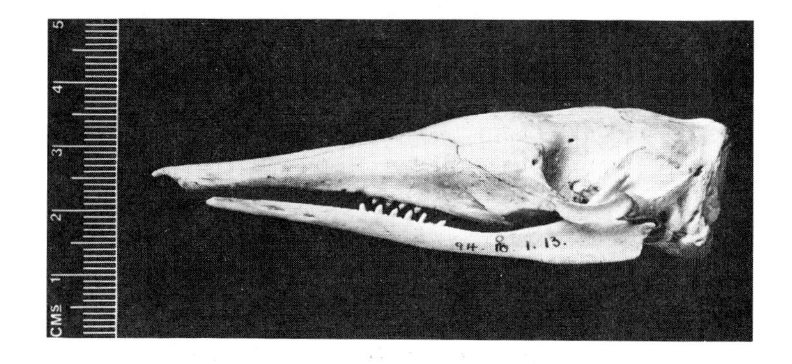

Figure 6. *Dasypus pilosus*, cranium in lateral view, BMNH 94.10.1.13, Peru, Junín, Acobamba. Note elongate, slender rostrum and mandible. Photograph courtesy of the British Museum (Natural History).

Figure 7. Skins of two long-nosed armadillos from Peru, dorsal view. Left: *Dasypus novemcinctus*, LSU 12306, Loreto, Balta. Right: *Dasypus pilosus*, LSU 14353, Huánuco, Paso Carpish. In *D. pilosus* note dense hair obscuring movable bands; hair on cephalic shield is short and sparse; large ears are dried and folded; tail is missing. Photograph by R. M. Wetzel.

nationi. The movable bands number more than in the other subgenera (10th free and 11th fused to pelvic shield—NMV 222; 11—BMNH 27.11.1.235, BMNH 94.10.1.13, and MHN). The rostrum and palate are proportionally the longest and most narrow in respect to length of the skull than in any other subgenus or, for that matter, any other armadillo. Tables 1–2–3 compare these and other measurements.

COMMENTS.—Moeller (1968) concluded that *D. pilosus* and *D. septemcinctus* are more closely allied to each other than either is to *D. novemcinctus.* We have argued against her wide separation of the latter two species (see subgenus *Dasypus*). A close alliance of *D. pilosus* with any other species of *Dasypus* ignores its unusual external characters which were not used in Moeller's comparisons of skulls—dense long hair and numerous movable bands. Even within the restriction of her study, the skull of *D. pilosus* is quite different and separable from other *Dasypus.* Where anterior portions of the skull are compared with posterior portions or with total lengths of skull, the proportionally long rostrum of *D. pilosus* places it off the curve for the rest of *Dasypus* in her study. (See, particularly, her comparisons of Condyloincisivlänge with Hirnhöhlenlänge, Abb. 24; Hirnhöhlenlänge with Medianlänge der Frontalia and with Medianlänge der Nasalia, Abb. 26; and Mediane Gaumgenlänge with Hirnhöhlenlänge, Abb. 27). The singularly slender cranium of *D. pilosus* would have caused additional displacement from subgenus *Dasypus* if Moeller had made more breadth × length comparisons. Moeller's comparisons of Hirnhöhlenlänge with Orale Gaumenbreite and with Aborale Gaumenbreite, Abb. 27, suggest only a part of this difference.

Dasypus pilosus (Fitzinger). Hairy long-nosed armadillo.

Cryptophractus pilosus Fitzinger, 1856:123.
Praopus hirsutus Burmeister, 1862:147. Holotype: MHN 26 or 27?
Dasypus pilosus, Yepes, 1928:468, as *D. pilosa;* Frechkop and Yepes, 1949:27; Talmage and Buchanan, 1954:84; Cabrera, 1958:226; Moeller, 1968.

HOLOTYPE.—NMV 222, museum exhibit mount.

TYPE-LOCALITY.—The holotype is labeled "Peru" and the only indication of the source of the specimen, as determined by Kurt Bauer, curator at NMV, is that it was purchased in 1833 from an animal dealer in London. Because of the more definite origin of other specimens we further restrict the type-locality used by Cabrera (1958:226) to montane Peru.

RANGE, DIAGNOSIS and COMPARISONS.—See subgenus *Cryptophractus.*

Acknowledgments

Support was given Wetzel for examination of specimens at approximately 40 museums by Charles O. Handley, Jr., through his Smithsonian Venezuelan Project (Contract DA-49-193-MD-2788 of the Medical Research and Development Command, Office of the Surgeon General, United States Army) and by grants to Wetzel from the National Geographic Society, the University of Connecticut Research Foundation, and the American Philosophical Society. We are grateful to Jorge Hernández-Camacho and Ronald Boyce MacKenzie for helpful comments about *Dasypus sabanicola* in the llanos of Colombia; to Å. Holm for examining, counting bands and scales and photographing the holotype of *Dasypus septemcinctus* L. at Zoologiska Institutionen, Uppsala Universitet; to C. O. Handley, Jr., for the record of one set of embryos of *D. kappleri;* to Jorge A. Crespo and Xavier Misonne for additional information on specimens in their care. We thank John F. Eisenberg, C. O. Handley, Jr., and Drew S. Wetzel for their critical reading of this manuscript and Gwilym S. Jones and Ronald H. Pine who read an earlier, more circumscribed manuscript.

We are especially indebted to the curators who generously permitted examination of collections during the course of this study. The following abbreviations are used to identify museums whose specimens are cited here; the curator was in charge of the collection at the time it was visited:

AMNH—American Museum of Natural History, New York. Sydney Anderson, Richard Van Gelder.

ANSP—Academy of Natural Sciences of Philadelphia. R. D. Estes.

BMNH—British Museum (Natural History), London. G. B. Corbet.

ICN—Instituto de Ciencias Naturales, Universidad de Colombia. Bogotá. Ernesto Barriga B.

IRSNB—Institute Royal des Sciences Naturelles de Belgique, Brussels. Xavier Misonne.

LSU—Louisiana State University, Museum of Zoology, Baton Rouge. G. H. Lowery, Jr. (deceased).

MACN—Museo Argentino de Ciencias Naturales "Bernardino Rivadavia," Buenos Aires. Jorge A. Crespo.

MBUCV—Museo de Biología, Instituto de Zoologia Tropical, Universidad Central de Venezuela, Caracas. Edgardo Mondolfi, Osvaldo A. Reig.

MCZ—Museum of Comparative Zoology, Harvard University, Cambridge. Barbara Lawrence, M. Edith Rutzmoser.

MEBRG—Estación Biológica Rancho Grande, Parque Nacional "Henry Pittier," Maracay. Gonzalo Medina P.

MHN—Museo de Historia Natural "Javier Prado," Lima. Hernando de Macedo.

MN—Museu Nacional, Universidade Federal do Rio de Janeiro. Fernando D. de Avila-Pires.

MPEG—Museu Paraense "Emílio Goeldi," Belém. Fernando C. Novaes.

MVZ—Museum of Vertebrate Zoology, University of California, Berkeley. James L. Patton.

MZS—Musee Zoologique de l'Universite et de la Ville, Strasbourg. F. Gouin.

MZUSP—Museu de Zoologia de Universidade de São Paulo. P. E. Vanzolini.

NHMB—Naturhistorisches Museum Basel. U. Rahm.

NHR—Naturhistoriska Riksmuseet, Stockholm. Ulf Bergström (deceased), Greta Vestergren.

NMBE—Naturhistorisches Museum, Bern. Peter Lüps.

NMV—Naturhistorisches Museum, Vienna. Kurt Bauer.

RMHN—Rijksmuseum van Natuurlijke Historie, Leiden. A. M. Husson.

SMF—Natur-Museum und Forschungsinstitut Senckenberg, Frankfurt am Main. Heinz Felten.

SMN—Staatliches Museum für Naturkunde, Stuttgart. F. Dieterlen.

TCWC—Texas Cooperative Wildlife Collection, Texas A & M University, College Station. Dilford C. Carter.

USNM—National Museum of Natural History, Smithsonian Institution, Washington. Charles O. Handley, Jr., Richard W. Thorington, Jr., Henry W. Setzer.

ZMB—Zoologisches Museum, Museum für Naturkunde der Humboldt-Universitat zu Berlin. Renate Angermann.

ZSM—Zoologische Sammlung des Bayerischen Staates, Munich. Theodor Haltenorth.

Literature Cited

Allen, G. M.
1911. Mammals of the West Indies. *Bull. Mus. Comp. Zool.*, 54:175–263.

Alvarez, T.
1963. The recent mammals of Tamaulipas, Mexico. *Univ. Kansas Publ. Mus. Nat. Hist.*, 14:363–473.

Armstrong, D. M. and J. K. Jones, Jr.
1971. Mammals from the Mexican state of Sinaloa. I. Marsupialia, Insectivora, Edentata, Lagomorpha. *J. Mammal.*, 52:747–757.

Avila-Pirez, F. D. de
1965. The type specimens of Brazilian mammals collected by Prince Maximilian zu Wied. *Amer. Mus. Nov.*, 2209:1–21.

Azara, F. de
1801. *Essais sur l'Histoire Naturelle des Quadrupèdes de la Province du Paraguay. Vol. 2.* Paris:

Azzali, G. and L. J. A. DiDio.
1965. The lymphatic system of *Dasypus novemcinctus* and *Dasypus sexcinctus*. *J. Morphol.*, 117:49–72.

Baccus, J. T.
1971. The mammals of Baylor County, Texas. *Texas J. Sci.*, 22:177–185.

Baker, R. H. and J. K. Greer.
1962. Mammals of the Mexican State of Durango. *Mich. State Univ., Publ. Mus., Biol. Ser.*, 2:25–154.

Balk. L.
1749. *Amoenitates Academicae. Vol. 1.* Stockholm:

Basc, J. P.
1967. Squelette hyobranchial. Pages 550–583 in *Traité de Zoologie*, edited by P.-P. Grasse. Paris: Masson et Cie.

Beath, M. M., K. Benirschke and L. E. Brownhill.
1962. The chromosomes of the nine-banded armadillo, *Dasypus novemcinctus*. *Chromosoma*, 13:27–38.

Benirschke, K., R. J. Low and V. H. Ferm.
1969. Cytogenetic studies of some armadillos. Pages 330–345 in *Comparative Mammalian Cytogenetics*, edited by K. Benirschke. New York: Springer Verlag.

Buchanan, G. D.
1957. Variation in litter size of nine-banded armadillos. *J. Mammal.*, 38:529.

Buchanan, G. D., A. C. Enders and R. V. Talmage.
1956. Implantation in armadillos ovariectomized during the period of delayed implantation. *J. Endocrin.*, 14: 121–128.

Burmeister, H.
1862. Beschreibung eines behaarten Gürtelthieres *Praopus hirsutus*, aus dem National-Museum zu Lima. *Abhandl. Nat. Ges.*, 6:147–148.

Burns, T. A. and E. B. Waldrip.
1971. Body temperature and electrocardiographic data for the nine-banded armadillo (*Dasypus novemcinctus*). *J. Mammal.*, 52:472–473.

Burt, W. H. and R. A. Stirton.
1961. The mammals of El Salvador. *Univ. Mich. Mus. Zool., Misc. Publ.*, 117:1–69.

Cabrera, A.
1958. Catálogo de los mamiferos de America del Sur. I. *Rev. Mus. Argentino Cienc. Nat. "Bernardino Rivadavia," Zool.*, (1957) 4:1–307.

Carvalho, C. T. de
1965. Comentários sôbre os mamíferos descritos e figurados por Alexandre Rodrigues Ferreira em 1790. *Arq. Zool. Est. São Paulo,* 12:7–70.

Carvalho, C. T. de and A. J. Toccheton.
1969. Mamíferos do nordeste do Pará, Brasil. *Rev. Biol. Trop., Univ. Costa Rica,* 15:215–226.

Cleveland, A. G.
1970. The current geographic distribution of the armadillo in the United States. *Texas J. Sci.,* 22:90–92.

Cope, E. A.
1889. Mammalia of southern Brazil. *Amer. Nat.,* 23:134–135.

Crespo, J. A.
1974. Comentarios sobre nuevas localidades para mamiferos de Argentina y de Bolivia. *Rev. Mus. Argentino Cienc. Nat. "Bernardino Rivadavia," Zool.,* 11:1–31.

D'Addamio, G. H., J. D. Roussel and E. E. Storrs.
1977. Response of the nine-banded armadillo (*Dasypus novemcinctus*) to gonadotropins and steroids. *Lab. Anim. Sci.,* 27:482–489.

Davis, W. B.
1966. *The Mammals of Texas.* Austin: Texas Parks and Wildlife Department.

Desmarest, A.
1803. Tatou. Pages 428–436 in *Nouveau Dictionnaire d'Histoire Naturelle, Vol. 21.* Paris: Chez Deterville.
1804. Tatou. Page 28, Section Mamiferes in *Nouveau Dictionnaire d'Histoire Naturelle, Vol. 24.* Paris: Chez Deterville.

Dhindsa, D. S., A. S. Hoverland and J. Metcalfe.
1971. Comparative studies of the respiratory functions of mammalian blood. VII. Armadillo (*Dasypus novemcinctus*). *Respiration Physiol.,* 13:198–208.

Eisenberg, J. F.
1961. Observations on the nest building behavior of armadillos. *Proc. Zool. Soc. London,* 137:322–324.

Eisenberg, J. F. and R. W. Thorington, Jr.
1973. A preliminary analysis of a neotropical mammal fauna. *Biotropica,* 5:150–161.

Enders, A. C.
1960a. Development and structure of the villous haemochorial placenta of the nine-banded armadillo (*Dasypus novemcinctus*). *J. Anat.,* 94:34–45.
1960b. Electron microscope observations on the villous haemochorial placenta of the nine-banded armadillo (*Dasypus novemcinctus*). *J. Anat.,* 94:205–215.
1962. The structure of the armadillo blastocyst. *J. Anat.,* 96:39–48.
1963. Fine structural studies of implantation in the armadillo. Pages 281–290 in *Delayed Implantation,* edited by A. C. Enders. Chicago: University of Chicago Press.

Enders, A. C. and G. D. Buchanan.
1959. Some effects of ovariectomy and injection of ovarian hormones in the armadillo. *J. Endocrin.,* 19:251–258.

Enders, A. C., G. D. Buchanan and R. V. Talmage.
1958. Histological and histochemical observations on the armadillo uterus during the delayed and post-implantation periods. *Anat. Rec.,* 130:639–667.

Felten, H.
1958. Weitere Säugetiere aus El Salvador. *Senckenb. Biol.,* 39:213–326.

Findley, J. S., A. H. Harris, D. E. Wilson and C. Jones.
1975. *Mammals of New Mexico.* Albuquerque: University of New Mexico Press.

Fitzinger, L. J.
1856. Tageblatt 32. *Versamml. Deutsch. Nat. Ärzte, Wien,* 6: 123.

Frechkop, S. and J. Yepes.
1949. Étude systématique et zoogéographique des Dasypodidés conservés à l'Institut. *Bull. Inst. Roy. Sci. Nat. Belgique,* 25:1–56.

Goffart, M.
1971. *Function and Form in the Sloth.* Oxford: Pergamon Press.

Gonzalez, J. C.
1973. Observaciones sobre algunos mamíferos de Bopicúa (Dpto. de Río Negro, Uruguay). *Commun. Mus. Mun. Hist. Nat. Río Negro,* 1:1–14.

Goodwin, G. G.
1969. Mammals from the State of Oaxaca, Mexico, in the American Museum of Natural History. *Bull. Amer. Mus. Nat. Hist.,* 141:1–269.

Gray, J. E.
1873. *Hand-list of the Edentate, Thick-skinned and Ruminant Mammals in the British Museum.* London.
1874. On the short-tailed armadillo (*Muletia septemcincta*). *Proc. Zool. Soc. London,* 1874:244–246.

Greegor, D. H., Jr.
1975. Renal capabilities of an Argentine desert armadillo. *J. Mammal.,* 56:626–632.

Grimwood, I. R.
1969. Notes on the distribution and status of some Peruvian mammals. *Spec. Publ. 21, Amer. Comm. Internat. Wildlife Protect. and New York Zool. Soc.*

Grinberg, M. A., M. M. Sullivan and K. Benirschke.
1966. Investigation with tritiated thymidine of the relationship between the sex chromosomes, sex chromatin, and the drumstick in the cells of the female nine-banded armadillo, *Dasypus novemcinctus. Cytogenetics,* 5:64–74.

Hagmann, G.
1908. Die Landsäugetiere der Insel Mexiana. *Arch. Rassen-Ges.-Biol., München,* 5:1–31.

Hall, E. R. and W. W. Dalquest.
1963. The mammals of Veracruz. *Univ. Kansas Publ., Mus. Nat. Hist.*, 14:165–362.

Hall, E. R. and K. R. Kelson.
1959. *The Mammals of North America, Vol. 1.* New York: Ronald Press.

Hamlett, G. W. D.
1935. Delayed implantation and discontinuous development in the mammals. *Quart. Rev. Biol.*, 10:432–451.
1939. Identity of *Dasypus septemcinctus* Linnaeus with notes on some related species. *J. Mammal.*, 20:328–336.

Handley, C. O., Jr.
1966. Checklist of the mammals of Panama. Pages 753–795 in *Ectoparasites of Panama*, edited by R. L. Wenzel and V. J. Tipton. Chicago: Field Museum of Natural History.
1976. Mammals of the Smithsonian Venezuelan Project. *Brigham Young Univ. Sci. Bull., Biol. Ser.*, 20(5):1–91.

Hanif, M. and N. O. Poonai.
1968. *Wild Life and Conservation in Guyana. Man and Nature Series No. 8.* Coconut Grove, Florida: Field Research Projects.

Hershkovitz, P.
1959. Nomenclature and taxonomy of the neotropical mammals described by Olfers 1818. *J. Mammal.*, 40: 337–353.

Humphrey, S. R.
1974. Zoogeography of the nine-banded armadillo (*Dasypus novemcinctus*) in the United States. *BioScience*, 24: 457–462.

Husson, A. M.
1973. Voorlopige lijst van zoogdieren van Suriname. *Zool. Bijdr., Rijksmus. Natuur. Hist. Leiden*, 14:1–15.

Ibarra, J. A.
1959. *Mamíferos de Guatemala.* Guatemala City: Ministerio de Educación Pública.

Jackson, C., C. M. Holcomb and M. M. Jackson.
1972. Strontium-90 in the exoskeletal ossicles of *Dasypus novemcinctus. J. Mammal.*, 53:921–922.

Johansen, K.
1961. Temperature regulation in the nine-banded armadillo (*Dasypus novemcinctus mexicanus*). *Physiol. Zool.*, 34: 126–144.

Jones, J. K., Jr., T. Alvarez and M. R. Lee.
1962. Noteworthy mammals from Sinaloa, Mexico. *Univ. Kansas Publ., Mus. Nat. Hist.*, 14:145–159.

Jones, J. K., Jr., H. H. Genoways, and J. D. Smith.
1974. Annotated checklist of mammals of the Yucatan Peninsula, Mexico. II. Marsupialia, Insectivora, Primates, Edentata, Lagomorpha. *Occas. Pap. Mus. Texas Tech. Univ.*, 26:1–12.

Kalmbach, E. R.
1944. *The Armadillo: Its Relation to Agriculture and Game.* Austin: Game, Fish and Oyster Commission.

Keil, A. and B. Venema.
1963. Struktur- und Mikrohärteuntersuchungen an Zähnen von Gürteltieren (Xenarthra). *Zool. Beitr., Neue Folge*, 9:173–195.

Kirkpatrick, R. D. and A. M. Cartwright.
1975. List of mammals known to occur in Belize. *Biotropica*, 7:136–140.

Krauss, F.
1862. Ueber ein neues Gürtelthier aus Surinam. *Arch. Nat., Berlin*, 28:19–34.

Layne, J. N.
1976. The armadillo, one of Florida's oddest animals. *Florida Nat.*, 49:8–12.

Layne, J. N. and D. Glover.
1977. Home range of the armadillo in Florida. *J. Mammal.*, 58:411–413.

Leopold, A. S.
1959. *Wildlife in Mexico.* Berkeley: University of California Press.

Linné, C. von.
1748. *Systema Naturae*, Ed. 6. Stockholm:
1758. *Systema Naturae*, Ed. 10. Stockholm:

Lönnberg, E.
1913. Mammals from Ecuador and related forms. *Ark. Zool.*, 8(16):1–36
1928. Notes on some South American edentates. *Ark. Zool.*, 20A(10):1–17.
1942. Notes on Xenarthra from Brazil and Bolivia. *Ark. Zool.*, 34A(9):1–58.

Lowery, G. H., Jr.
1974. *The Mammals of Louisiana and Its Adjacent Waters.* Baton Rouge: Louisiana State University Press.

Maller, O. and M. R. Kare.
1967. Observations on the sense of taste in the armadillo (*Dasypus novemcinctus*). *Anim. Behav.*, 15:8–10.

Marcgrav, G.
1648. *Historiae Rerum Naturalium Brasiliae.* Amsterdam.

McMurtrie, H.
1831. Pages 162–165 in *The Animal Kingdom Arranged in Conformity with Its Organization by the Baron Cuvier. Vol. 1. Vertebrata.* New York: Carvill.

Mendez, E.
1970. *Los Principales Mamiferos Silvestres de Panama.* Panama City: Privately printed.

Meritt, D. A., Jr.
1973. Edentate diets. I. Armadillos. *Lab. Anim. Sci.*, 23:540–542.

Moeller, W.
1968. Allometrische Analyse der Gürteltierschädel ein Beitrag zur Phylogenie der Dasypodidae Bonaparte, 1838. *Zool. Jb. Anat.,* 85:411–528.

Mondolfi, E.
1968. Descripcíon de un nuevo armadillo del género *Dasypus* de Venezuela (Mammalia-Edentata). *Mem. Soc. Cienc. Nat. La Salle,* 27:149–167.
1976. *Fauna Silvestre de los Bosques Húmedos de Venezuela.* Caracas: Sierra Club—Consejo de Bienestar Rural.

Moore, A. M.
1968. A radio location study of armadillo foraging with respect to environmental variables. Ph.D. Thesis, University of Texas.

Müller, P.
1973. *The Dispersal Centres of Terrestrial Vertebrates in the Neotropical Realm.* The Hague: W. Junk.

Musso, Q. A.
1962. Lista de los mamíferos conocidos de la Isla de Margarita. *Mem. Soc. Cienc. Nat. La Salle,* 22:163–180.

Nagy, F. and R. H. Edmonds.
1973. Morphology of the reproductive system of the armadillo. The spermatogonia. *J. Morphol.,* 140:307–319.

Ojasti, J. and E. Mondolfi.
1968. Esbozo de la fauna de mamiferos de Caracas. *Estudio de Caracas,* 1:411–561.

Olrog, C. C.
1976. Sobre mamiferos del noroeste Argentino. *Acta Zool. Lilloana,* 32:5–12.

Patterson, B.
1975. The fossil aardvarks (Mammalia: Tubulidentata). *Bull. Mus. Comp. Zool.,* 147(5):185–237.

Peters, W.
1864. Über neue Arten de Säugethiergattungen *Geomys, Haplodon* und *Dasypus. Monatsb. Königl. Akad. Wiss. Berlin,* (1865):177–181.

Pine, R. H.
1973. Mammals (exclusive of bats) of Belem, Para, Brazil. *Acta Amazonica,* 3:47–79.

Platt, J. J., T. Yaksh and C. L. Darby.
1967. Social facilitation of eating behavior in armadillos. *Psych. Reports,* 20:1136.

Pocock, R. I.
1924. The external characters of the South American edentates. *Proc. Zool. Soc. London,* 65:983–1031.

Prejean, J. D. and J. C. Travis.
1971. Clinical values in the nine-banded armadillo, *Dasypus novemcinctus mexicanus. Texas J. Sci.,* 22:245–246.

Rhoads, S. N.
1894. Description of a new armadillo, with remarks on the genus *Muletia* Gray. *Proc. Acad. Nat. Sci. Philadelphia,* (1894):111–114.

Röhl, E.
1959. *Fauna Descriptiva de Venezuela.* Madrid: Neuvas Graficas, S.A.

Royce, G. J., G. F. Martin and R. M. Dom.
1975. Functional localization and cortical architecture in the nine-banded armadillo (*Dasypus novemcinctus mexicanus*). *J. Comp. Neurol.,* 164:495–521.

Russell, R. J.
1953. Description of a new armadillo (*Dasypus novemcinctus*) from Mexico with remarks on geographic variation of the species. *Proc. Biol. Soc. Washington,* 66:21–26.

Sauer, C. O.
1950. Geography of South America. Pages 319–344 in *Handbook of South American Indians, Vol. 6,* edited by J. H. Steward. Washington, D. C., Government Printing Office.

Schade, F. H.
1973. The ecology and control of the leaf-cutting ants of Paraguay. Pages 77–95 in *Paraguay: Ecological Essays,* edited by J. R. Gorham. Miami: Academy of Arts and Sciences of the Americas.

Scholander, P. F., L. Irving and S. W. Grinnell.
1943. Respiration of the armadillo with possible implications as to its burrowing. *J. Cell. Comp. Physiol.,* 21: 53–64.

Selander, R. K., R. F. Johnston, B. J. Wilks and G. S. Raun.
1962. Vertebrates from the barrier islands of Tamaulipas, Mexico. *Univ. Kansas Publ., Mus. Nat. Hist.,* 12:309–345.

Shackelford, J. M.
1963. The salivary glands and salivary bladder of the nine-banded armadillo. *Anat. Rec.,* 145:513–519.

Slaughter, B. H.
1962. The significance of *Dasypus bellus* (Simpson) in Pleistocene local faunas. *Texas J. Sci.,* 13:311–315.

Soukup, J.
1960. Materiales para el catalogo de los mamiferos peruanos. *Biota,* 3:131–161.

Starck, D.
1959. Ontogenie und Entwicklungsphysiologie der Säugetiere. *Handbuch der Zoologie,* 8:1–276.

Stevenson, H. M. and R. J. Crawford.
1974. Spread of the armadillo into the Tallahassee-Thomasville area. *Florida Field Nat.,* 2:8–10.

Storrs, E. E., G. P. Walsh, H. P. Burchfield and C. H. Binford.
1974. Leprosy in the armadillo: New model for biomedical research. *Science,* 183:851–852.

Storrs, E. E. and R. J. Williams.
1968. A study of monozygous quadruplet armadillos in relation to mammalian inheritance. *Proc. Nat. Acad. Sci.*, 60:910–914.

Talmage, R. V. and G. D. Buchanan.
1954. The armadillo (*Dasypus novemcinctus*). A review of its natural history, ecology, anatomy and reproductive physiology. *Rice Inst. Pamphlet, Monogr. Biol.*, 41(2):1–135.

Thomas, O.
1901. New species of *Saccopteryx, Sciurus, Rhipidomys,* and *Tatu* from South America. *Ann. Mag. Nat. Hist., Ser. 7,* 7:366–371.

Wetzel, R. M. and J. W. Lovett.
1974. A collection of mammals from the Chaco of Paraguay. *Univ. Conn. Occas. Pap., Biol. Sci. Ser.,* 2:203–216.

Wilbur, C. G.
1964. Electrocardiogram of the armadillo. *J. Mammal.*, 45: 642.

Yepes, J.
1928. Los Edentata argentinos. Sistemática y distribucíon. *Rev. Univ. Buenos Aires, Ser.* 2a, 1:461–515.
1933. Una especie nueva de "mulita" (Dasipodinae) para el norte argentino. *Physis,* 11:225–232.

Ximénez, A., A. Langguth, and R. Praderi.
1972. Lista sistemática de los mamíferos del Uruguay. *An. Mus. Nac. Hist. Nat. Montevideo, Ser. 2,* 7(5):1–49.

CHARLES O. HANDLEY, JR.
LINDA K. GORDON
Department of Vertebrate Zoology
Museum of Natural History
Smithsonian Institution
Washington, D.C. 20560

New Species of Mammals from Northern South America: Mouse Possums, Genus *Marmosa* Gray

ABSTRACT

Two new species of the genus *Marmosa* are described. *Marmosa cracens* has been described from the moist montane region bordering on the Falcon arid zone. *Marmosa xerophila* was discovered in the arid portion of the Falcón peninsula and in the arid region at the head of Lake Maracaibo in the State of Zulia. *Marmosa cracens* shows affinities to *M. fuscata* while *M. xerophila* shows a close relationship to *M. robinsoni*.

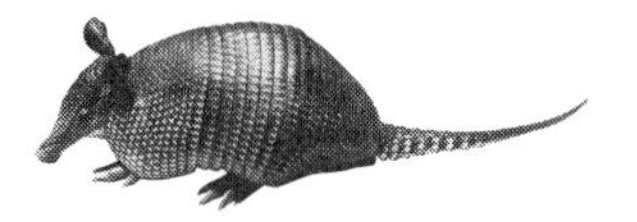

RESÚMEN

Se presentan dos especies nuevas del género *Marmosa*. *Marmosa cracens* ha sido descrita proveniente de la regíon montañosa humeda que limita con la zona árida en Falcón. *Marmosa xerophila* fué descubierta en la zona árida en la península de Falcón y en la zona árida cerca de la cabeza del Lago Maracaibo en el estado de Zulia. *Marmosa cracens* muestra similitudes con *M. fuscata*, mientras que *M. xerophila* muestra una estracha relacion con *M. robinsoni*.

Introduction

Mammals and their ectoparasites were collected in Venezuela between 1965 and 1968 by the Smithsonian Venezuelan Project (SVP), supported in part by a contract (DA-49-MD-2788) of the Medical Research and Development Command, Office of the Surgeon General, U. S. Army. Altogether 38,213 specimens of 270 species of mammals were obtained by the project.

To date the project collections of mammals and parasites have been the basis of about fifty papers. Data of the mammal collections were summarized by Handley (1976). New bats were described by Handley and Ferris (1972). Throughout these papers other new species of mammals discovered by the project have been referred to by alphabetical designations. Some of these are described and named here. A paper still in preparation will discuss departures from conventional mammalian nomenclature that have been used in publications of the project.

All measurements in this paper are in millimeters and weights are in grams. Hind feet were remeasured dry in museums. Coloration was determined under Examolites with natural light excluded. Capitalized color terms are from Ridgway (1912).

A New Narrow-Headed Mouse Possum

The common and widespread narrow-headed mouse possum of the mountains of northern Venezuela should, to judge by Tate (1933), be *Marmosa fuscata* Thomas. However, with much more abundant material now available, it is obvious that the northern Venezuelan possums are like *M. carri* J. A. Allen of Trinidad rather than the Andean *M. fuscata*. Allen and Chapman's (1897:28) measurements of the holotype of *M. carri*, Tate's (1933, Table 1, Sect. 6) measurements of other *M. carri*, and our remeasurements of the same material, fit nicely into the SVP northern Venezuelan series. Specimens from Trinidad, northeastern Venezuela, the Caracas region, and Rancho Grande are equally large and lacking posterior nasal expansion. Specimens from as far west as Carabobo are almost as large. There is no obvious geographic variation in this northern Venezuelan and Trinidadian population.

The holotype and topotypes of *M. fuscata* (both sexes) from the vicinity of Mérida in the Andes are consistently smaller than northern Venezuelan and Trinidadian *M. carri* (both sexes) in some measurements, and average smaller in most others, but have the posterior nasal breadth relatively greater (palatal breadth 9.3–9.8 *vs.* 10.0–11.8, m*1* - m*3* length 5.1–5.7 *vs.* 6.0–6.6, postorbital breadth 5.9 - 6.1 *vs.* 6.1–6.7, and posterior nasal width 3.1–3.4 *vs.* 2.6–3.4). An adult male (FMNH 22174) from La Azulita near Mérida, agrees with these measurements of *M. fuscata*, but has the nasals narrow posteriorly—2.9. A single specimen (US 372934) from Hda. Misisí, Trujillo, near the northern terminus of the Andes, has wide nasals (3.5) and is smaller than the average of Coast Range specimens in most other measurements. These characteristics relate it to *M. fuscata*, but otherwise it is indistinguishable from *M. carri*. We thus conclude that *M. carri* and *M. fuscata* represent subspecies of a single species. The older name is *M. fuscata*.

One of the surprises of SVP was the discovery of a miniature narrow-headed mouse possum on an isolated low mountain in the hot lowlands of northeastern Falcón State. All other members of this group live in higher mountain habitats. Geographically these Falcón possums are nearer to the large *Marmosa fuscata carri*, but in size they are much smaller than the more remote *M. f. fuscata*. They apparently represent an unnamed species.

Marmosa cracens new species

Holotype.—USNM 418503, young adult male (testis 10×5 mm) skin and skull, collected 11 November 1967, by Norman E. Peterson, F. P. Brown, Jr., and J. O. Matson, near La Pastora (11°12′N, 68°37′W), 150 m, 14 km ENE Mirimire, Falcón, Venezuela, on the ground in a clump of small shrubs, in a live trap baited with sardines, raisins, and rolled oats, original number SVP 14837.

Etymology.—Latin *cracens*, slender or neat, referring to the dorsal aspect of the face of this possum.

Distribution.—Known only from the steep, moist, north slope of an isolated, low mountain near La Pastora, Falcón, Venezuela. Captured on the ground; at the base of a tree and a rotting stump, and in a clump of small shrubs; in mature evergreen forest, with many epiphytes and vines and with a closed subcanopy at 10 m and an irregular upper canopy at 25–30 m; elevation range 125–170 m; Holdridge life zone (Ewel and Madriz, 1968), Premontane humid forest.

Description.—Dorsal mass-effect coloration grayish brown (near Chesnut Brown; pure color near Rood's Brown with suffusion of black); scattered short white over hairs on rump; flanks between Cinnamon Drab and Drab; ocular patch small, mainly anterior to eye, poorly defined; underparts white, with faint yellowish cast; slaty hair bases largely concealed by whitish tips; self-colored hairs confined to chin, lips, narrow median stripe on abdomen, and inguinal region; scrotum white, hands, feet, and insides of arms and legs whitish; ears fuscous; tail indistinctly bicolor, yellowish cream-color below, fuscous above. Hands and feet small, ears relatively large, tail medium, overall size small. Body fur short and smooth, extending about 5 mm on to tail; tail hairs short and inconspicuous; scales 27 per cm.

Skull long and narrow; rostrum relatively short; nasals barely exceeding premaxillae anteriorly, almost parallel sided, barely expanded in posterior third, posterior tips blunt and rounded; interorbital region narrow, parallel sided, without postorbital expansion, with edges rounded, not ledged or beaded (except for a short, weak, postorbital bead in holotype); temporal ridges weak and faint; braincase long, narrow, and shallow, not particularly inflated; postorbital process of zygoma rounded; palate relatively short and broad, with small fenestrae (narrow slits between P*3* and M*2*, and small pits opposite M*4*); I*2* not notably smaller than I*3*-I*5*; canines small, short, and compressed, lacking posterior basal cusps in males, but with weak cusps in female; P*1* not reduced; alisphenoid bulla small, slightly inflated, with strong anterior process; petrosal bulla with anteroexternal process; lower canine roughly rectangular and resembling premolars (angle of tip and hind edge less than 90°); coronoid process relatively wide; and angular process wide and short.

Measurements of *M. cracens* from Mirimire (holotype boldface) and *M. fuscata* from Montalbán, all age 2 (young adult), arranged in the following sequence: males—*M. cracens* **418503**, 418504 (*M. fuscata* 418515, 443787), females—*M. cracens* 442719 (*M. fuscata* 416936, 443783). Total length **233**, 237 (305, 300),—(280, 280); tail vertebrae **131**, 132 (172, 170),—(155, 152); hind foot (dry) **16**, 16 (19, 19),—(18, 17); ear from notch **24**, 26 (26, 25),—(24, 25); weight **24.0**, 26.5 (51.1, 42.0),—(35.7, 44.5).

Condylobasal length **28.4**, 28.9 (35.1, 34.1), 28.5 (32.6, 33.0); palatal length (anteriormost point of pmx to posteriormost point of palate, midline) **15.8**, 15.8 (19.4, 19.4), 16.1 (17.9, 19.0); palatal breadth **9.3**, 9.8 (10.7, 10.6), 9.1 (10.3, 10.6); greatest crown length m*1* - m*3* **5.3**, 5.6 (6.3, 6.1), 5.4 (6.2, 6.0); petrosal breadth **8.8**, 8.8 (10.0, 9.5), 8.5 (9.7, 9.4); zygomatic breadth **15.5**, 14.9 (18.2, 17.0), 14.8 (16.4, 17.3); nasal length (longest nasal) **13.1**, 12.4 (16.7, 16.3), 12.7 (15.2,—); greatest nasal breadth **3.0**, 2.7 (2.8, 3.1), 2.6 (3.0, 2.9); least postorbital breadth **5.9**, 5.7 (6.3, 6.2), 5.7 (—, 6.3); braincase breadth (between mastoid notches) **10.7**, 10.8 (12.1, 11.9), 11.0 (11.5, 11.8); braincase depth **8.0**, 8.2 (8.8, 8.5), 8.1 (9.0, 8.6).

COMPARISONS.—In almost every aspect of pelage and skull character, *Marmosa cracens* is a miniature replica of *M. fuscata*, distinguishable from it only by its small size and the short, broad shape of its palate. Both *M. cracens* and *M. fuscata* may be separated from any of the species compared below by the shape of the nasals, which are almost parallel-sided and barely expanded posteriorly, and by the rounded rather than angular nature of the postorbital process of the zygoma.

M. cracens may be easily distinguished from *M. marica* Thomas *M. dryas* Thomas by its larger size, gray-brown dorsal color (pale reddish-brown in *M. marica*, dark reddish brown in *M. dryas*), cream-colored underparts (deep cream-color in *M. marica*, pale tan in *M. dryas*), short, smooth hair, dark ears, and little feet. In comparison to *M. cracens*, the skulls of *M. marica* and *M. dryas* are small, with a globose, inflated braincase, large and inflated alisphenoid bullae, a narrow, very pinched rostrum, a narrow hourglass-shaped interorbital region, large palatal fenestrae, a weak anterior process on the alisphenoid bulla, narrow and attenuated coronoid and angular processes, with the lower canine caniniform instead of premolariform.

Compared to *M. parvidens* Tate, *M, cracens* is smaller and more gray in dorsal color (*M. parvidens* varies from pale reddish-brown at Urama to dark chocolate brown at Km 121) and has smooth hair, unlike that of *M. parvidens*, which is shaggy. *M. cracens* has a narrower interorbital and a rounded supraorbital, not weakly beaded as in *M. parvidens*. The palate of *M. cracens* is shorter and broader, I*2* subequal to I*3*–I*5*, not reduced as in *M. parvidens*, and the upper canine is larger than the tiny canine of *M. parvidens*.

M. cracens may be distinguished from *M. invicta* Goldman by the very different pelage color (slate gray or very dark brown dorsally, gray with white "frosting" ventrally in *M. invicta*), the smaller overall size, and the broader palate.

M. cracens differs from *M. impavida* Tschudi by its much smaller size, blunt and rounded posterior tips of the nasals, small palatal fenestrae, large P*1* size, and shorter and stouter coronoid and angular processes.

SPECIMENS EXAMINED:

Marmosa cracens.—Venezuela: FALCÓN, nr. La Pastora, 14 Km ENE Mirimire, 125–170 m, (3 US, including holotype).

Marmosa dryas.—Colombia: CUNDINAMARCA, Bogotá, Boqueron, San Francisco, 3000 m, (2 FMNH). Venezuela: MÉRIDA, La Montaña, 3 to 4 km SE Mérida, 2250–2600 m, (2 PNSN, 1 US); La Mucuy, 4.9 km E Tabay, 2400 m, (1 PNSN); Montañas Uchisera, 3000 m, (2 BM); Montes de la Serra, 3000 m, (1 BM); Selva Culata, 4000 m, (1 BM, holotype); 6 km ESE Tabay, 2630–2632 m, (2 US). TÁCHIRA, Buena Vista, nr. Páramo de Tamá, 41 km SW San Cristóbal, 2405–2410 m, (2 US). TRUJILLO, Hda. Misisí, 14 to 15 km E Trujillo, 2210–2360 m, (6 US).

Marmosa fuscata.—Trinidad: Caparo, (3 AMNH, including holotype of *M. carri*); Cumaca, (2 AMNH); St.

Augustine, (1 AMNH); Sangre Grande, Rio Grande Forest, 60 yds W Tree Station, (2 AMNH). Venezuela: ARAGUA, Est. Biol. Rancho Grande, 13 km NW Maracay, 1050 m, (1 AMNH, 7 US). CARABOBO, La Copa, 4 km NW Montalbán, 1513–1537 m, (15 US); La Cumbre de Valencia, (1 AMNH). DTO. FEDERAL, El Junquito, (1 UCV); Los Venados, 4 km NNW Caracas, 1443–1500 m, (5 US). Mérida, Cafetal de Milla, 1630 m (1 BM); La Azulita (1 FMNH); "Mérida", 1600 m, (2 BM); Rió Albarregas, 1630 m, (1 BM, holotype of *M. fuscata*). MIRANDA, Alto Ño León, 31 to 33 km WSW Caracas, 1750–2000 m, (5 US); Curupao, 5 km NNW Guarenas, 1160 m, (1 US); I.V.I.C., 15 km SW Caracas, 1460 m, (5 US); Pico Ávila, 5 km NNE Caracas, 1281–2232 m, (19 US). MONAGAS, San Agustín, 5 km NW Caripe, 1150–1339 m, (10 US); Santa Inés, 5 km ESE Caripe, 800 m, (1 UCV). SUCRE, Cerro Papelón, 900 m, (1 MHNLS). TRUJILLO, Hda Misisí, 14 to 15 km E Trujillo, 2210–2350 m, (2 US).

Marmosa impavida.—Colombia: CAUCA, Río Cauquitá, nr. Cali, 1000 m, (1 BM, holotype of *M. caucae*). MAGDALENA, Sierra Negra, nr. Villanueva, 1265 m, (5 US). Venezuela: TÁCHIRA, Buena Vista, nr. Páramo de Tamá, 41 km SW San Cristóbal, 2380–2415 m, (8 US).

Marmosa invicta.—Panama: BOCAS DEL TORO, Cylindro, 1220 + m, (1 ANSP). DAIRÉN, Cana, 610 m, (2 US, including holotype); Casita, 458 m, (1 US); Cerro Tacarcuna, 1464 m, (4 US); La Laguna, 975 m, (2 US). PANAMÁ, Cerro Azul, 640 m, (1 US).

Marmosa marica.—Colombia: CUNDINAMARCA, La Silva, nr. Bogotá, (1 BM). MAGDALENA, Las Marimondas, nr. Fonseca, Sierra de Perijá, 1000 m, (2 US). Venezuela: DTO. FEDERAL, Los Venados, 4 km NNW Caracas, 1500 m, (1 US); Pico Ávila, nr. Hotel Humboldt, 5 km NNE Caracas, 2124–2135 m, (2 US). FALCÓN, 19 km NW Urama, 25 m, (1 US). MÉRIDA, Cafetal de Chama, 1600 m, (1 ZMA); Cafetal de Mérida, 1600–1630 m, (1 US, 1 ZMA); Cafetal de Milla, 1600 m, (1 BM, 1 US); Cafetas de Mérida, 1630 m, (4 AMNH); Cafetas de Milla, 1630 m, (5 AMNH); Llano de Mérida, (1 BM); Mérida, (4 FMNH, 1 AMNH); Río Albarregas, 1630 m, (2 BM, including holotype). MONAGAS, Cerro Turimaquire, (1 FMNH); Hato Mata de Bejuco, 55 km SSE Maturín, 18 m, (3 US).

Marmosa parvidens.—Brazil: PARÁ, Belém, (5 US); Santarém, (1 BM). T. F. AMAPÁ, Serra do Navio, Rio Amapari, (6 US). Guayana: Hyde Park, 48 km up Demarara River, 6 m, (1 FMNH, holotype of *M. parvidens*): Venezuela: BOLIVAR, Auyantepuí, 460 m, (2 US); Auyantepuí, 1100 m, (1 US); Km 125, 85 km SSE El Dorado, 1032 m, (2 US). FALCÓN, 19 km NW Urama, 25 m, (1 US). T. F. AMAZONAS, Capibara, Brazo Casiquiare, 106 km SW Esmeralda, 130 m, (2 US).

REMARKS.—*Marmosa cracens* has been known as "*Marmosa* sp. A" in previous SVP publications.

A New Orange Mouse Possum

Marmosa robinsoni Bangs and its derivatives are the common mouse possums of the more xeric lowlands of northern Venezuela. In deciduous thorn forest these are sometimes the most abundant small mammals. In Venezuela these possums vary geographically in size, in width and shape of the interorbital region, in extent of white or cream-color on the underparts, and in coloration of the dorsum.

Dorsal color seems to be closely related to aridity of the environment and varies markedly where the climate is sharply zoned. For example, on Isla Margarita, parts of which are extremely arid, an isolated mountain, Cerro Matasiete, is capped with cloud forest. Eighty percent of the *Marmosa robinsoni* from the cerro (120–425 m elevation) are darker dorsally than 75 percent of the *Marmosa robinsoni* from the surrounding lowlands. The most pallid of the lowland animals is very much like those from the lowlands of the arid Península de Paraguaná, almost 700 km to the West. On Paraguaná there is also a cloud forest capped mountain, Cerro Santa Ana. *Marmosa robinsoni* is abundant at elevations of 500–615 m on the foggy mountaintop. Here, however, the mouse possum of the lowlands is not only more pallid in coloration, but also is smaller and has significant cranial differences from its montane counterpart. These small, pallid possums are abundant in the deserts and semi-deserts all around the Golfo de Venezuela. The *Marmosa robinsoni* of Cerro Santa Ana is an isolated relict, morphologically most like *M. r. mitis* Bangs of the Sierra de Perijá and Sierra Nevada de Santa Marta of W Venezuela and NE Colombia. It is less like the geographically much nearer *M. r. casta* Thomas which occurs on the mainland from the vicinity of Caracas to the shore of Lago de Maracaibo. The possum of the Golfo de Venezuela arid lands represents a new species which is described as follows:

Marmosa xerophila new species

HOLOTYPE.—USNM 443819, adult male, skin and skull, collected 28 June 1968, by Norman E. Peterson, F. P. Brown, Jr., and J. O. Matson, at La Isla, 15 m, near Cojoro, 37 km NNE Paraguaipoa, Dpto. Guajira, Colombia, in a tree in a live trap baited with sardines and rolled oats; original number SVP 23528.

Etymology.—Greek, *xeros*, dry, and *philios*, loving, referring to the habitat preference of this animal.

Distribution.—Desert shores of Golfo de Venezuela in the states of Falcón and Zulia, Venezuela, and the department of Guajira, Colombia. Usually captured in trees and bushes (81 percent), uncommonly on the ground (18 percent), and rarely in houses (1 percent). Almost always found in dry places (99 percent); rarely in moist sites (1 percent). Frequents thorn forest (98 percent), forest openings (2 percent), and evergreen forest (less than 1 percent). Elevation range 5–90 m. Life zones of Holdridge, Tropical thorny forest and Tropical very dry forest.

Description.—Dorsal mass-effect color pale sandy brown (from near Wood Brown in gray extreme to near Sayal Brown in brown extreme), paler laterally, and grayer on face above eyes and vibrissae; sides of neck, upper arms, and sometimes flanks and hind legs orange-buff (from Light Pinkish Cinnamon to Cinnamon); face mask reduced to narrow black eye rings and grayish projections into vibrissal areas; underparts and cheeks whitish, hairs monocolored except on sides of abdomen where whitish tips partly conceal gray bases; hands and feet small, white; ankles and wrists white; tail short, thick, gray-brown (varying from Light Mouse Gray to Light Drab) above, slightly paler below, covered with fine white hairs; scales 16 per cm.

Rostrum short, thick, and moderately inflated; nasals long, expanded posteriorly; both interorbital and postorbital areas much constricted; supraorbital ledges strongly beaded, forming angular postorbital processes in posterior third of interorbital-postorbital space, and continued posteriorward as weakly defined parietal ridges or sagittal crest; palate wide, with anterior pair of long rectangular fenestrae and posterior pair of roughly circular or square fenestrae, together forming exclamation mark-like openings; postpalatal constriction relatively broad; auditory bullae large, roughly globular in shape; teeth small, P3 and P4 subequal, and lower canine low but well differentiated from other teeth.

Measurements of holotype male, age 3 adult, followed in parentheses by US 443822, female, age 3 adult, also from the type locality: Total length 314 (285), tail vertebrae 180 (157), hind foot (dry) 20 (18), ear from notch 27 (28), weight 51.2 (41.3).

Condylobasal length 32.7 (31.3), palatal length (anteriormost point of pmx to posteriormost point of palate, midline) 18.2 (17.1), palatal breadth 10.6 (10.1), greatest crown length M*1*—M*3* 5.2 (5.0), petrosal breadth 9.4 (8.9), zygomatic breadth 18.7 (17.6), nasal length (longest nasal) 15.4 (14.5), greatest nasal breadth 3.7 (3.9), least interorbital breadth 5.0 (5.0), least postorbital breadth 5.0 (4.9), braincase breadth (between mastoid notches) 11.9 (11.1), braincase depth 8.5 (8.5). See also Table 1.

Comparisons.—*Marmosa xerophila* is closely related to *M. robinsoni* and probably is a fairly recent derivative of it. Populations around the Golfo de Venezuela could have been isolated by the mountains that rim it. *M. xerophila* probably first evolved on the mainland, perhaps on the Península de la Guajira where it is most differentiated today. Later it may have invaded the coast of Falcón and eventually the Península de Paraguaná, isolating there a relict population of *M. r. mitis*, on Cerro Santa Ana.

Guajira *M. xerophila* differ from *M. r. mitis* of the Sierra de Perijá region (Cerro Azul and Novito, Venezuela; Las Marimondas and Sierra Negra, Colombia) and from the holotype of *Marmosa mitis pallidiventris* Osgood in overall smaller size, including smaller teeth and smaller auditory bullae; interorbital region most expanded in the posterior third (*vs.* middle third) and projected into angular postorbital processes; underparts whitish rather than cream-colored, and the gray-based lateral abdominal hairs more concealed by whitish tips; and dorsum more pallid and more grayish. With the following exceptions these same characteristics distinguish *M. xerophila* of northern Falcón (Capatárida and Península de Paraguaná) from *M. r. mitis* of Cerro Santa Ana: auditory bullae large, not differing from those of *M. r. mitis*, and dorsal coloration averaging about the same in each species, darker and not so gray as in Guajira *M. xerophila*. See Table 1 for comparison of measurements of *M. xerophila* with those of other taxa.

From *M. r. casta* of southern Falcón (Mirimire, Urama, and Cerro Socopo) the northern Falcón *M. xerophila* differ in smaller size, except auditory bullae the same in both species, paler dorsal coloration, whitish rather than cream-colored underparts, and less completely veiled gray hair bases on abdominal flanks.

The only other *Marmosa* in the size range of *M. xerophila* which occurs near it is *M. murina* Linnaeus. From this species *M. xerophila* is easily distinguished by its overall pallid coloration; orangish side neck; whitish underparts; shorter, obviously hairy tail; reduced face patches; short, thick rostrum; narrow postorbital area; weak parietal ridges; wide palate, with two rather than one pair of fenestrae; broad postpalatal constriction; and globular auditory bullae.

Specimens Examined (all USNM except as noted):

Marmosa xerophila.—Venezuela: falcón, Capatárida and 16 to 18 km WSW and SSW Capatárida, 40–75 m, (135); Península de Paraguaná, 15 to 25 km SSW and SW Pueblo Neuvo, 13–90 m, (73). Colombia: guajira, and Venezuela: zulia, nr. Cojoro, 34 to 37 km NNE Paraguaipoa, 5–15 m, (38). Total 246.

Table 1. **Measurements of adult (age 3) males of *Marmosa xerophila, M. robinsoni,* and *M. murina*** [1]

Total length	*Tail vertebrae*	*Hind foot*	*Ear*	*Condylobasal length*	*Palatal length*	*Palatal breadth*	*M^1–M^3 length*
			Marmosa xerophila, La Isla				
294.1 ± 15.68	163.1 ± 11.38	19.1 ± .80	26.1 ± 1.02	33.2 ± 1.10	18.4 ± .60	10.5 ± .34	5.5 ± .08
260–314	144–181	17–20	24–28	32.5–35.4	17.8–19.5	10.1–11.2	5.4–5.7
(7)	(7)	(7)	(7)	(5)	(5)	(6)	(7)
			M. xerophila, Península de Paraguaná				
291.7 ± 12.64	156.8 ± 4.46	19.3 ± .42	25.5 ± .44	31.9 ± .86	17.6 ± .48	10.6 ± .38	5.5 ± .08
278–319	149–166	19–20	25–26	31.2–32.7	16.8–18.3	10.2–11.3	5.3–5.6
(6)	(6)	(6)	(6)	(3)	(5)	(5)	(6)
			M. robinsoni mitis, Cerro Azul and Novito				
331.0 ± 17.00	181.7 ± 14.34	22.0 ± 1.16	25.0 ± 1.16	35.9 ± .30	20.1 ± .34	11.8 ± .30	5.9 ± .14
322–348	174–196	21–23	24–26	35.7–36.0	19.8–20.3	11.5–12.0	5.8–6.0
(3)	(3)	(3)	(3)	(2)	(3)	(3)	(3)
			M. robinsoni mitis, Cerro Santa Ana				
316.6 ± 8.26	169.2 ± 5.28	20.5 ± .44	28.0 ± .46	34.7 ± .74	19.6 ± .38	11.4 ± .18	5.8 ± .12
292–344	155–184	19–22	27–29	33.2–36.5	18.7–20.6	11.1–12.0	5.5–6.3
(13)	(13)	(13)	(13)	(11)	(12)	(12)	(13)
			M. robinsoni casta, Urama, Mirimire, Cerro Socopo, and Río Socopito				
338.2 ± 20.92	184.8 ± 7.78	23.7 ± .98	27.0 ± 1.16	37.4 ± 1.54	20.8 ± .84	12.4 ± .30	6.3 ± .06
312–380	171–200	22–25	25–29	34.9–39.3	19.4–21.9	12.1–13.0	6.2–6.4
(6)	(6)	(6)	(6)	(5)	(5)	(6)	(6)
			M. murina waterhousei, El Rosario				
332.5	187.5	22.0	25.0	34.6	19.2	10.4	5.6
330–335	185–190	22–22	25–25	34.5–34.7	19.1–19.3	10.2–10.6	5.5–5.6
(2)	(2)	(2)	(2)	(2)	(2)	(2)	(2)

[1] For each measurement: line 1 includes the mean ±2 standard errors; line 2 = the extremes; line 3 = total number of specimens measured. Females, not tabulated, are proportionally smaller.

Marmosa robinsoni casta.—Venezuela: ARAGUA, 2 km NE Ocumare, 180 m, (16); 3 km S Ocumare, nr. sealevel, (3). BARINAS, Altamira, 697 m, (1). CARABOBO, nr. Montalbán, 562–1000 m, (9); San Esteban, (1 AMNH). FALCÓN, Cerro Socopo, 84 km NW Carora, 1258–1260 m, (2); nr. Mirimire and nr. La Pastora, 14 km ENE Mirimire, 90–250 m, (2); Río Socopito, 80 km NW Carora, 470 m, (1). LARA, Caserio Boro, 10 to 14 km NE and N El Tocuyo, 528–616 m, (3); La Concordia, 47 km NE El Tocuyo, 592 m, (1); Río Tocuyo, (10 AMNH). MIRANDA, Curupao, 5 km NNW Guarenas, 1160 m, (8); San Andrés, 16 km SSE Caracas, 1144 m, (6). TRUJILLO, La Ceiba, 52 km WNW Valera, 29 m, (1); 12 to 25 km N, NW, and WNW Valera, 90–930 m, (9). CARABOBO, FALCÓN, and YARACUY, 10 to 19 km NW Urama, 25 m, (22).

Marmosa robinsoni mitis.—Colombia: MAGDALENA, Bonda, (14 AMNH); La Concepción, (1 FMNH); Las Marimondas, Sierra de Perijá, nr. Fonseca, *ca.* 1000 m, (8); Mamatoco, (1 AMNH); Minca, (1 AMNH); Pal-

Zygomatic breadth	Nasal length	Nasal breadth	Inter-orbital breadth	Post-orbital breadth	Braincase breadth	Braincase depth
			Marmosa xerophila, La Isla			
18.5 ± .64	15.2 ± .72	3.9 ± .28	5.0 ± .20	5.1 ± .10	11.6 ± .36	8.6 ± .20
17.8–19.9	13.6–16.7	3.5–4.5	4.7–5.5	4.9–5.3	11.0–12.1	8.3–8.9
(6)	(7)	(6)	(7)	(7)	(6)	(6)
			M. xerophila, Península de Paraguaná			
18.9 ± .08	14.5 ± .66	4.5 ± .30	4.9 ± .28	4.7 ± .28	12.0 ± .30	8.6 ± .18
17.7–19.9	13.2–15.1	4.2–5.0	4.5–5.3	4.4–5.2	11.7–12.2	8.4–8.8
(4)	(5)	(5)	(5)	(5)	(3)	(4)
			M. robinsoni mitis, Cerro Azul and Novito			
20.7 ± .24	16.8 ± .58	5.0 ± .46	6.1 ± .18	6.1 ± .24	12.5 ± .10	9.6 ± .18
20.5–20.9	16.3–17.3	4.6–5.4	6.0–6.3	5.9–6.3	12.4–12.5	9.4–9.7
(3)	(3)	(3)	(3)	(3)	(2)	(3)
			M. robinsoni mitis, Cerro Santa Ana			
19.2 ± .40	16.2 ± .42	4.5 ± .14	5.4 ± .18	5.7 ± .18	12.2 ± .26	9.3 ± .18
18.3–20.1	15.2–17.9	4.0–4.8	5.0–6.0	5.0–6.4	11.4–12.9	8.9–9.9
(10)	(13)	(13)	(13)	(13)	(12)	(12)
			M. robinsoni casta, Urama, Mirimire, Cerro Socopo, and Río Socopito			
21.3 ± .78	17.9 ± .80	4.9 ± .30	6.3 ± .22	5.9 ± .18	12.6 ± .16	10.1 ± .34
20.2–22.5	16.9–18.7	4.4–5.3	5.9–6.6	5.6–6.2	12.3–12.9	9.5–10.8
(5)	(4)	(5)	(6)	(6)	(6)	(6)
			M. murina waterhousei, El Rosario			
19.7	15.7	5.3	6.2	6.3	12.4	9.7
19.6–19.7	15.5–15.8	5.1–5.4	6.0–6.3	6.2–6.3	12.3–12.4	9.6–9.7
(2)	(2)	(2)	(2)	(2)	(2)	(2)

omino, (1); Pueblo Viejo, (2 USNM, 2 FMNH); San Miguel, (1 FMNH); Sierra Negra, nr. Villanueva, 1265 m, (32); Taganga, (1 AMNH); Villanueva, upper Río Cesar, 274 m, (23). NORTE DE SANTANDER, 16 km N Cucuta, (1 FMNH, holotype of *Marmosa mitis pallidiventris* Osgood). Venezuela: FALCÓN, Cerro Santa Ana, Península de Paraguaná, 15 km SSW Pueblo Nuevo, 500–615 m, (62). MÉRIDA, Cafetos de Milla, 1680 m, (1); Mérida, (4 AMNH). ZULIA, nr. Cerro Azul, 35 to 40 km NW La Paz, 80 m, (7); Novito, 19 km WSW Machiques, 1132–1150 m, (2); Valera, (1 FMNH).

Marmosa robinsoni robinsoni.—Venezuela: NUEVA ESPARTA, Isla Margarita, 3 km NNE, NE, and S La Asunción, 37–425 m, (15). SUCRE, Cuchivano, 213 m, (1 AMNH); 16 to 21 km E Cumaná, 1–25 m, (10); San Antonio, 550 m, (2 AMNH).

REMARKS.—*Marmosa xerophila* has been known as "*Marmosa* sp. B" in previous SVP publications.

Acknowledgments

We are grateful to Sally DeMott for measuring the *Marmosa* skulls. Philip Hershkovitz was generous in allowing us to include in the comparisons previously unreported specimens in his northern Colombian collections. Curators of the following collections have kindly permitted us to include specimens under their care in these descriptions: American Museum of Natural History (AMNH), Academy of Natural Sciences of Philadelphia (ANSP), British Museum (Natural History) (BM), Field Museum of Natural History (FMNH), Museo de Historia Natural La Salle (MHNLS), Parque Nacional Sierra Nevada (PNSN), Universidad Central de Venezuela (UCV), U. S. National Museum of Natural History (US), Zoologisch Museum Amsterdam (ZMA). A portion of the SVP specimens will be returned to Venezuela.

Literature Cited

Allen, J. A. and F. M. Chapman.
1897. On a second collection of mammals from the island of Trinidad, with descriptions of new species., etc. *Bull. Amer. Mus. Nat. Hist.*, 9(2):13–29.

Ewel, J. J. and A. Madriz.
1968. *Zonas de Vida de Venezuela.* Caracas: Ministario de Agricultura y Cria.

Handley, C. O., Jr.
1976. Mammals of the Smithsonian Venezuelan Project. *Brigham Young Univ. Sci. Bull., Biol. Series*, 20(5):1–91.

Handley, C. O., Jr. and K. C. Ferris.
1972. Descriptions of new bats of the genus *Vampyrops*. *Proc. Biol. Soc. Washington*, 84:519–524.

Ridgway, R.
1912. *Color Standards and Color Nomenclature.* Washington, D. C.: R. Ridgway.

Tate, G. H. H.
1933. A systematic revision of the marsupial genus *Marmosa*. *Bull. Amer. Mus. Nat. Hist.*, 66(1):1–250.

MARGARET A. O'CONNELL
Department of Biological Sciences
Texas Tech University
Lubbock, Texas 79409

Ecology of Didelphid Marsupials from Northern Venezuela

ABSTRACT

The ecology of six species of didelphid marsupials has been investigated during a two year mark-recapture study in Northern Venezuela. During the study 255 individuals were captured a total of 668 times. Densities of the different species fluctuated during different times of the year; the density of some species was greater during the wet season, while that of others was greatest during the dry season. Movement and longevity data reflect the nomadic tendencies of these marsupials. The seasonality of rainfall appears to be an important factor in determining reproductive patterns of these didelphid marsupials. Most species studied appear to be seasonally polyestrus.

RESÚMEN

La ecología de seis especies de marsupiales didélfidos ha sido estudiada durante dos años con un método de captura con trampas y marcación en el norte de Venezuela. Durante el estudio, 255 animales fueron capturados 668 veces. La densidad de las diferentes especies varió durante las diferentes épocas del año. La densidad de algunas especies era mayor durante la época de lluvias, mientras que la de otras era mayor durante la época seca. Las tendencias nomades de estos marsupiales están reflejadas en los datos sobre movimiento y longevidad. Le época de lluvias parece ser un factor importante en la reproducción de estos marsupialas didélfidos. La mayoria de las especies estudiadas parecen ser poliestros.

Introduction

Despite an increasing interest in the study of marsupials (Tyndale-Biscoe, 1973; Hunsaker, 1977; Kirsch, 1977; Stonehouse and Gilmore, 1977), information on the ecology of these mammals, especially the New World forms, remains meager. Hunsaker (1977) has presented a comprehensive review of the ecology and behavior of the New World marsupials. While the North American opossum, *Didelphis marsupialis,* has been the subject of considerable investigation, both field and laboratory, the other New World marsupials have received far less attention. This is especially true of the South American marsupials. Few long term studies have investigated the ecology of South American marsupials.

Traditionally, marsupials have been considered "primitive" or "inferior" mammals, mainly due to their mode of reproduction. Several workers have proposed theories on the selective pressures acting on marsupial reproductive strategies (Kirsch, 1977; Parker, 1977; Low, 1978). These workers suggested that marsupial reproduction is not strictly the result of primitive phylogenetic position, but that it is an advanced condition conveying certain adantages that have evolved in response to selective pressures. However, many conclusions are drawn from short-term studies on relatively few taxa and, historically, special emphasis has often been on the Australian species (Low, 1978). Hunsaker (1977) suggested that more information is needed before many conclusions on the ecological strategies of the didelphid marsupials can be made.

The purpose of this paper is to present information on the ecology of six species of didelphid marsupials obtained during a two year study of small mammal populations in northern Venezuela. These species are: *Didelphis marsupialis, Marmosa cinerea, Marmosa fuscata, Marmosa robinsoni, Monodelphis brevicaudata,* and *Caluromys philander.* Demographic, reproductive, and movement data were collected for these opossums by means of mark-recapture procedures.

Study Areas

Populations of small mammals were studied at two localities in northern Venezuela. The first location, Parque Nacional Guatopo, lies in the mountainous region of northern Venezuela, approximately 40 km southeast of Caracas. The park encompasses about 100,000 square hectares and lies in the states of Miranda and Guárico. Elevations in the park range from 250 to 1500 m. The park was established in 1958 as a watershed area for Caracas; thus, much of the area is characterized by 16- to 20-year-old second growth with primary forest limited to the higher elevations and less accessible regions. Four vegetation types of the Holdridge Life Zone are found within the park boundaries: tropical humid forest, premontane humid forest, premontane dry forest, and tropical dry forest (Ewel and Madriz, 1968). Rainfall varies between the different areas in the park, but a dry season generally occurs from December through May.

The trapping grid was located at the site of an old finca in the premontane humid forest at an elevation of 710 m (see Eisenberg, et al., p. 187). Major trees in this area include: *Cecropia peltata, Erythrina poeppigiana, Ficus* sp., *Inga edulis, Castillia elastica.* Major ground plants in the area include: *Pilea pubescens, Phaesolus* sp., and *Tectaria martinicensis.* Rainfall from 1971 through 1977 averaged 1500 mm annually. Climatic data during the two years of my study are summarized in Figure 1. The average maximum temperature was 27.7°C and the average minimum temperature was 17.9°C. The two years differed considerably in amount of rainfall. The dry season of 1976 was relatively short, while that of 1977 was longer.

The second study area was located on the Venezuelan llanos, 45 km south of Calabozo in the state of Guárico. Trapping was conducted on the property of Sr. Tomas Blohm, Fundo Pecuario Masaguaral (see Eisenberg et al., this volume). Although Masaguaral is a working ranch, supporting cattle, goats, and pigs, it has also been established as a wildlife sanctuary. The ranch is about 90 m in elevation. Masaguaral falls in the tropical dry forest of the Holdridge Life Zone (Ewel and Madriz, 1968). The area is characterized by distinct wet and dry seasons with virtually all of the 1500 mm annual rainfall occurring from May through November. Impermeable soils over much of the area cause annual flooding. Where flooding is the greatest, open grasslands with scattered palms (*Copernicia tectorum*) occur. A combination of more permeable soils and control of fires has resulted in the growth of low stature forests. The trapping grid was located in such a forest (see Eisenberg, et al., p. 187). Major trees in the area are *Genopa americana, Ficus* sp., *Platymicium pinnatum, Copernicia tectorum* and *Cassia moschaia.* Dense thorny thickets composed of *Fagaia caribea, Annona* sp., and *Chomelia spinosa,* are found over parts of the grid. Open grass areas occur on other parts of the grid. Major grass and forb species include *Sida serrata, Cenchrus browni, Panicum laxum, Ipomea* sp., *Heliotropium indico, Cyperus* sp., *Leersia hexandra,* and *Axonophus purpusi.*

Climatic data for Masaguaral are summarized in Figure 2. Much more rain occurred during the wet season of 1976 than during that of 1977. Flooding was more extensive and prolonged in 1976. Monthly mea-

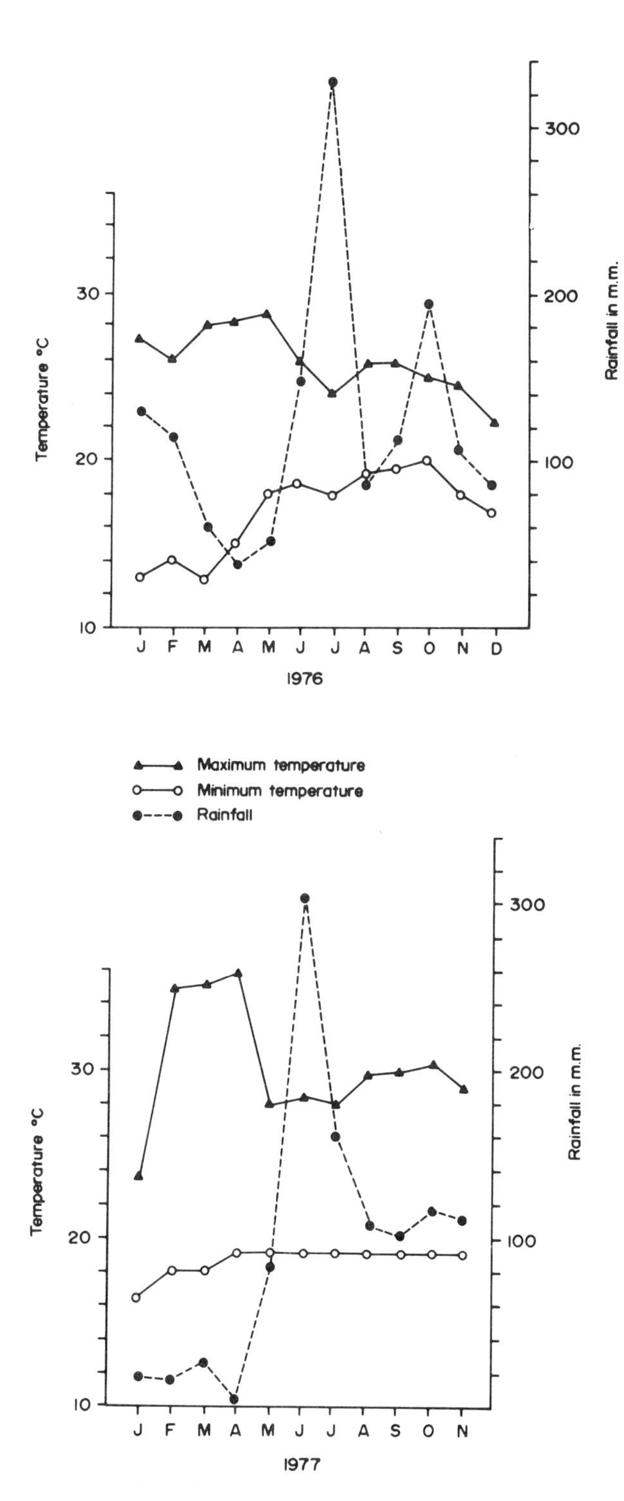

Figure 1. Temperature and rainfall variations for two years in Guatopo National Park (south end). Note the protracted dry period from January to April 1977 and the higher temperatures during the same period when compared with the preceding year.

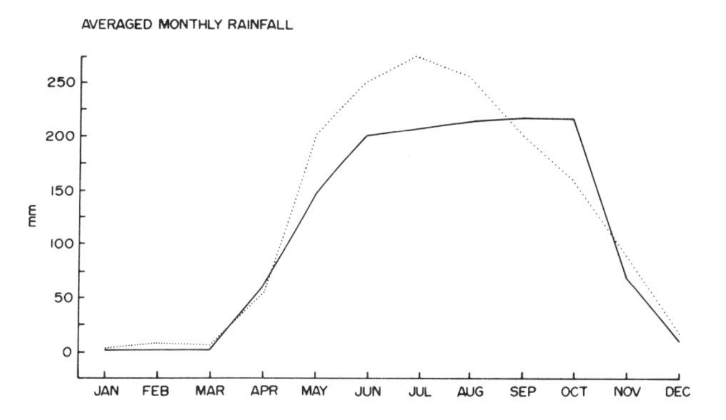

Figure 2. Average monthly rainfall for Corozo Pondo, 6 km south of Masaguaral. Dotted line = averages for 1952 to 1971; solid line = monthly averages for 1972 to 1977.

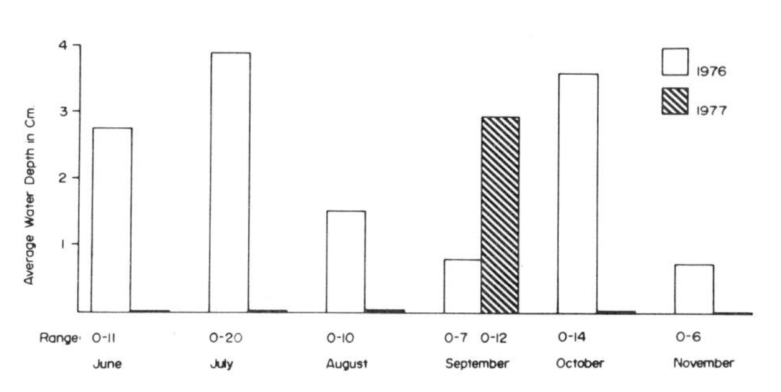

Figure 3. Comparison of water depths on study grid at Masaguaral during the wet seasons of 1976 and 1977.

surements of water depths at each grid station were recorded. The mean depths for the two years are shown in Figure 3.

Methods

A four hectare grid with 100 trapping stations at 20-m intervals was established at each locality. A Sherman live trap was placed at each station; a larger (18 × 6 × 6 in) National trap was placed at each station in Guatopo and at 25 stations at Masaguaral. In addition to the ground traps, smaller (16 × 5 × 5 in) National traps were placed in trees and vines on each grid; these traps were sufficiently sensitive to capture individuals weighing 25 g and large enough to handle animals up to 1.5 kg. Fifteen tree traps were used in Guatopo and 20 at Masaguaral. Traps were baited with banana and cracked corn.

Trapping was conducted from February 1976 through November 1977 in Guatopo and from October 1975 through November 1977 at Masaguaral. During the first year of study at each locality the sampling period was nine days each month, and reduced to seven the second year. Traps were checked and rebaited each morning. Traps were checked in the evening only during wet season in the llanos when either consumption of bait by ants or capture of the diurnal rodent *Sigmomys alstoni* made this necessary. Animals were marked by toe clipping, weighed, examined for sex, reproductive condition, ectoparasites, and injuries, and then released. Behavior upon release was recorded to help determine the arboreal tendencies of the different species. Data were analyzed using the Statistical Analysis System (Barr *et al.*, 1976).

Results

Species Description

Didelphis marsupialis—This species ranges from Central America, where it is sympatric with *D. virginiana* (Gardner, 1973), throughout most of South America below 1400 meters (Hunsaker, 1977). *D. marsupialis* is found throughout Venezuela except at high elevations and in desert regions (Handley, 1976). It was found at both study areas in this study.

This species is omnivorous and was observed feeding on a wide variety of fruits, insects, and land crabs. Although *D. marsupialis* is basically nocturnal, individuals were occasionally seen foraging during the day.

The arboreal tendencies of this species varied between sites. In Guatopo, *D. marsupialis* was captured on the ground 73 percent of the time. However, individuals were observed to climb during about 50 percent of the releases, whereas in the llanos, only 45 percent of the captures took place on the ground. An important determining factor in the arboreal habits of this opossum in the llanos was the seasonal flooding. During the extensive flooding of the 1976 wet season, individuals were captured in the trees 100 percent of the time. As the flooding receded in December 1976, individuals were again trapped on the ground. This shift was less pronounced during the wet season of 1977.

Marmosa—This is the largest genus of the family Didelphidae. At present it is comprised of approximately 50 species, but further study will probably revise this figure (Hunsaker, 1977). *Marmosa* is distributed in a wide variety of habitats through Mexico and most of Central and South America. Eleven species of *Marmosa* are known from Venezuela (Handley, 1976) and three, *M. robinsoni, M. fuscata,* and *M. cinerea,* were trapped on the permanent grids.

M. robinsoni occurs in Central America, Trinidad, Tobago, and the northwestern tropics of South America, where it is often associated with disturbed or second growth areas (Fleming, 1972). It is widely distributed in Venezuela at elevations generally less than 50 meters (Handley, 1976). During this study, this mouse opossum was found in the llanos and at lower elevations (250 meters) in Guatopo.

This nocturnal species is insectivorous and frugivorous. Individuals kept in captivity accepted a wide variety of fruits and insects. Moths with bright aposematic coloring were always rejected without being sampled. Females lack a true pouch and the young simply attach to the teats.

Individuals of *M. robinsoni* were captured in tree traps 59 percent of the time. Upon release, this mouse opossum climbed into vines and trees 95 percent of the time. Escape was generally to the nearest palm tree.

Marmosa cinerea is one of the largest of the mouse opossums. It is found from Belize south throughout most of tropical South America (Collins, 1973). In Venezuela, Handley (1976) reported finding *M. cinerea* in the coastal range, the state of Bolivar, and T. F. Amazonas. It was found from 25 to 1160 m, but generally below 325 m. During this study, it was trapped only in Guatopo. Like *M. robinsoni,* this nocturnal mouse opossum is insectivorous and frugivorous; it accepted a wide variety of fruits and insects while maintained in captivity. Like *M. robinsoni,* female *M. cinerea* lack a true pouch.

M. cinerea was one of the more arboreal didelphids found in Guatopo. It was captured in tree traps 50 percent of the time. Individuals escaped by way of trees and vines in 85 percent of the releases.

Marmosa fuscata occurs at higher elevations in Venezuela and northern and central Colombia. This species was found in the mountains of Venezuela at elevations from 1050 to 2350 m by Handley (1976), and it was trapped only at the higher elevations in Guatopo during this study. It appears similar to *M. robinsoni* in its food habits and activity patterns, but was very difficult to maintain in captivity.

Data from my study suggest that *M. fuscata* is one of the more terrestrial of the mouse opossums. It was captured on the ground 93 percent of the time and, upon release, almost always escaped on the ground. Even when placed in trees or on vines, individuals would climb down and then run along the ground. However Handley (1976) captured *M. fuscata* in trees and vines 71 percent of the time.

Monodelphis brevicaudata—Eleven species, distributed from eastern Panama to central Argentina, belong to the genus *Monodelphis* (Hunsaker, 1977). *M. brevicaudata* is found in Venezuela, the Guianas, Brazil, Paraguay and northern Argentina (Collins, 1973). This short-tailed opossum is found in discrete populations in northern Venezuela, the llanos, and southern Venezuela from elevations of 1 to 1160 m (Handley, 1976). During this study it was trapped both at low (250 m) and higher (710 m) elevations in Guatopo. *M. brevicaudata* is one of the more carnivorous of the smaller didelphids. It has been reported to feed on small rodents, insects, fruits, seeds and carrion (Hunsaker, 1977) and it is nocturnal in its habits. Female short-tailed opossums lack a true pouch.

During this study *M. brevicaudata* was captured on the ground 100 percent of the time. Handley (1976) reported trapping this species in trees and vines 4 percent of the time.

Caluromys philander—The three species of *Caluromys* range from Mexico through central South America (Hunsaker, 1977). *C. philander* is found in Venezuela, the Guianas, and Brazil (Collins, 1973). Handley (1976) reported this species in forested regions east of the Andes in Venezuela at elevations from 25 to 1600 m, but usually below 1200 m. During this study *C. philander* was found only in Guatopo. *C. philander* is probably one of the most arboreal of the didelphid marsupials in northern Venezuela. Handley (1976) reported capturing it in trees 94 percent of the time. This wooly opossum was found only during the initial three months of the study in Guatopo. For this reason it will only be discussed in limited sections of this paper.

Numbers Captured and Densities

During the two years, 255 individuals of the six species were captured a total of 668 times (Table 1). There was no apparent trend of more frequent recaptures for females, as Fleming (1972) has reported for Panamanian marsupials. Similar numbers of *D. marsupialis* were marked at Guatopo and Masaguaral. *M. fuscata* and *M. robinsoni* are similar in weight and body size; the number of individuals marked of these two species is almost identical, but the average number of captures per individual for *M. robinsoni* in the llanos is three times that of *M. fuscata* in Guatopo.

The sex ratios of all species except *M. cinerea* were equal ($P > .05$). Male *M. cinerea* significantly outnumbered females ($X = 7.14$; $P < .05$). Fleming (1972) reported a significantly higher number of male than female *M. robinsoni*. While more male *M. robinsoni* were marked than females at Masaguaral, the number is not significant.

The number of individuals captured each month is compared to the minimum number of individuals known to be alive in Table 2. Using direct enumeration (Fleming, 1971; Dalby, 1975) of the minimum number of adults known to be alive, density estimations were determined for each species for each month (Table 2).

An average of 3.7 individuals (range 1 to 10) of *D. marsupialis* was known to be present each month on the grid in Guatopo. The value is higher for *D. marsupialis* at Masaguaral where the average was 5.8 individuals per month and the range was 1 to 13. Numbers of individuals captured on the two grids were highest during the wet season and generally coincided with the appearance of juveniles on the grids. One exception to this was in February 1976 in Guatopo. The unseasonably wet weather (Figure 1) may have created stressful conditions for the animals and caused them to enter the traps more readily. Density estimations

Table 1. Numbers of individuals and captures of six species of didelphids on two grids in northern Venezuela

Species	*Site*	*Number captured*			*Number captured and recaptured*[1]		
		Male	*Female*	*TOTAL*	*Male*	*Female*	*TOTAL*
D. marsupialis	Guatopo	24	17	41	72 (3.0)	41 (2.4)	113 (2.8)
D. marsupialis	Masaguaral	29	20	49	73 (2.5)	53 (2.7)	126 (2.6)
M. cinerea	Guatopo	12	2	14	35 (2.9)	8 (4.0)	43 (3.1)
M. fuscata	Guatopo	32	32	64	48 (1.5)	49 (1.5)	97 (1.5)
M. robinsoni	Masaguaral	38	27	65	159 (4.2)	122 (4.5)	281 (4.3)
M. brevicaudata	Guatopo	7	11	18	18 (2.5)	19 (1.7)	37 (2.1)
C. philander	Guatopo	1	3	4	2 (2.0)	6 (2.0)	8 (2.0)

[1] Average per individual shown in parentheses.

Table 2. Population statistics and density for five species of didelphids from Northern Venezuela[1]

Month	*D. marsupialis*								*M. cinerea*				*M. fuscata*				*M. robinsoni*				*M. brevicaudata*			
	Guatopo				*Masaguaral*																			
1975 Oct.	—	—	—	—	1	1	1	0.25	—	—	—	—	—	—	—	—	3	3	3	0.75	—	—	—	—
Nov.	—	—	—	—	1	1	1	0.25	—	—	—	—	—	—	—	—	15	15	12	3.00	—	—	—	—
Dec.	—	—	—	—	1	1	1	0.25	—	—	—	—	—	—	—	—	17	18	17	4.25	—	—	—	—
1976 Jan.	—	—	—	—	5	5	5	1.25	—	—	—	—	—	—	—	—	22	22	17	4.25	—	—	—	—
Feb.	10	10	10	2.50	4	6	6	1.50	2	2	2	0.50	13	13	11	2.75	16	16	8	2.00	0	0	0	0.00
Mar.	5	6	6	1.50	3	6	6	1.50	3	3	3	0.75	10	13	13	3.25	14	14	9	2.25	2	2	2	0.50
Apr.	3	3	3	0.75	1	6	6	1.50	4	4	4	1.00	3	3	3	0.75	3	5	3	0.75	3	4	4	1.00
May	1	2	2	0.50	6	6	6	1.50	0	1	1	0.25	1	2	2	0.50	4	5	5	1.25	3	4	4	1.00
June	0	2	2	0.50	5	13	10	2.50	0	1	1	0.25	2	2	2	0.50	5	5	5	1.25	3	4	4	1.00
July	4	6	3	0.75	6	11	6	1.50	0	1	1	0.25	2	2	1	0.25	5	5	5	1.25	3	4	4	1.00
Aug.	3	6	3	0.75	3	9	8	2.00	0	1	1	0.25	6	6	3	0.75	5	5	4	1.00	1	1	1	0.25
Sept.	4	7	4	1.00	1	6	6	1.50	1	2	2	0.50	2	3	3	0.75	4	4	1	0.25	4	4	1	0.25
Oct.	4	5	2	0.50	1	6	6	1.50	1	2	2	0.50	5	5	5	1.25	5	5	5	1.25	0	0	0	0.00
Nov.	3	4	2	0.50	6	10	6	1.50	2	2	2	0.50	6	7	7	1.75	1	1	1	0.25	1	1	1	0.25
Dec.	2	3	2	0.50	10	12	10	2.50	2	2	2	0.50	6	6	6	1.50	2	2	2	0.50	0	1	1	0.25
1977 Jan.	1	2	2	0.50	5	11	11	2.75	3	3	3	0.75	5	5	5	1.25	9	10	8	2.00	2	2	2	0.50
Feb.	1	2	2	0.50	5	6	5	1.25	3	3	3	0.75	4	4	4	1.00	9	10	10	2.50	1	1	1	0.25
Mar.	1	3	3	0.75	3	5	5	1.25	2	2	2	0.50	2	3	3	0.75	5	9	9	2.25		1	1	0.25
Apr.	0	2	2	0.50	6	6	6	1.50	2	2	2	0.50	0	1	1	0.50	8	9	9	2.25	1	1	1	0.25
May	5	6	2	0.50	1	3	3	0.75	2	2	2	0.50	2	4	4	1.00	2	3	3	0.75	1	1	1	0.25
June	5	6	2	0.50	4	4	3	0.75	0	0	0	0.00	3	3	3	0.75	0	1	1	0.25	0	1	1	0.25
July	1	2	1	0.25	5	7	2	0.50	0	0	0	0.00	3	4	4	1.00	0	1	1	0.25	1	1	1	0.25
Aug.	1	2	1	0.25	4	5	2	0.50	1	1	1	0.25	1	1	1	0.25	3	3	1	0.25	2	2	1	0.25
Sept.	1	1	1	0.25	3	2	2	0.50	0	0	0	0.00	0	1	1	0.25	3	3	3	0.25	0	0	0	0.00
Oct.	1	1	1	0.25	3	3	2	0.50	0	0	0	0.00	1	1	1	0.25	5	5	5	1.25	0	0	0	0.00
Nov.	1	1	1	0.25	1	1	1	0.50	1	1	1	0.25	1	1	1	0.25	4	4	4	1.00	1	1	1	0.25

[1] First column = total number captured; second column = minimum number individuals known to be alive; third column = minimum number of adults known to be alive; fourth column = estimated density (No./ha).

during the different months of study ranged from 0.25 to 2.25 *D. marsupialis* per hectare in Guatopo and from 0.25 to 2.75 per hectare at Masaguaral. Assuming an average adult weight of one kilogram, biomass estimations are identical to those for density.

The average number of individuals of *M. robinsoni* captured each month was 5.6 (range 0 to 22) and the minimum number known to be alive averaged 6.9 individuals (range 1 to 22). Numbers of individuals were highest at the end of the wet season and during the dry season. As with *D. marsupialis,* the population levels were highest at the times juveniles appeared on the grid. Density estimates for *M. robinsoni* in the llanos ranged from 0.25 to 4.25 adults per hectare during the different months. The average adult weight is about 0.06 kilograms. Thus, biomass estimations varied from 0.015 kilograms to 0.255 kilograms per hectare.

Population fluctuations of the other three species are not as well defined. The average number of individuals of *M. fuscata* captured each month was 3.5 (range 0 to 13) and the minimum number known to be alive averaged 4 per month (range 1–13). The explanation for the high number of *M. fuscata* during February and March 1976 may be the same as for *D. marsupialis.* Density estimates for *M. fuscata* varied from 0.25 to 3.25 adults per hectare. With an average adult weight of 0.06 kilograms, these densities correspond to biomass estimations ranging from 0.015 to 0.195 kilograms per hectare. An average of 1.3 individuals (range 0 to 4) of *M. cinerea* were captued each month in Guatopo. The minimum number known to be alive ranged from zero to four, with an average of 1.6. More individuals were captured during the dry season. The density estimates for *M. cinerea* range from zero to one adult per hectare during the different months. The average adult weight of this larger mouse opossum is 0.15 kilograms. Thus, biomass estimates vary from zero to 0.15 kilograms per hectare. Population levels of *M. brevicaudata* were similar to those of *M. cinerea.* The average number of individuals captured each month was 1.3 (range 0 to 4) and the minimum number known to be alive averaged 1.4. The number of adult *M. brevicaudata* per hectare ranged from zero to one. The average adult weight of this species is about 0.07 kg and biomass estimates vary from zero to 0.07 kg per hectare.

Movements

The average and the longest distances between successive capture sites were determined for each species (Table 3). The average distances were compared using one way analysis of variance.

The average distance traveled by *D. marsupialis* ranged from 45.2 to 80.2 m and the longest distance ranged from 100 to 228 m. The small size of the grids has probably caused these movements to be greatly underestimated. At both sites, female *D. marsupialis* had longer average distance than males; however, only in Guatopo was the difference significant ($F = 6.99$; $P < .05$). There was no difference in the movements of individuals of *D. marsupialis* of the same sex between the two areas.

There was no significant difference in the movements of male and female *M. robinsoni* in the llanos, although males did move further than females. Nor was there a significant difference in the movements of *M. robinsoni* and *D. marsupialis,* although movements of *D. marsupialis* are probably underestimated.

Movements of *M. fuscata* in Guatopo are similar to those of *M. robinsoni* from the llanos. While male *M. robinsoni* moved further than females, the opposite was true for *M. fuscata;* females ranged further than males. However, there was no significant differences between movements of males and females for *M. fuscata.*

Data on the movements of *M. cinerea* and *M. brevicaudata* from Guatopo are rather limited, but indicate that differences in the movements of males and females of the same species are nonsignificant.

Longevity

The average number of days between first and last capture and the longest period between first and last captures are given in Table 4. Only individuals captured more than once were used in the calculations.

Table 3. Distances (in meters) between successive capture sites for five species of didelphids in northern Venezuela

Species	*Site*	*Sex*	*Number of movements*	*Average distance*	*Longest distance*
D. marsupialis	Guatopo	Male	38	46.6	128.1
		Female	18	80.7	228.0
	Masaguaral	Male	22	45.2	161.2
		Female	9	60.1	100.0
M. cinerea	Guatopo	Male	11	52.0	80.0
		Female	2	40.0	40.0
M. fuscata	Guatopo	Male	7	30.9	63.2
		Female	8	58.5	172.0
M. robinsoni	Masaguaral	Male	69	42.8	170.9
		Female	47	29.8	164.9
M. brevicaudata	Guatopo	Male	4	60.9	89.4
		Female	3	32.2	56.6

Table 4. Average number of days and longest period between first and last captures of five species of didelphids in northern Venezuela

Species	*Site*	*Sex*	*N*	*Average days between*	*Longest period between*
D. marsupialis	Guatopo	Male	12	38.5	86
		Female	8	110.1	530
	Masaguaral	Male	7	63.0	202
		Female	11	202.0	491
M. cinerea	Guatopo	Male	5	111.4	315
		Female	2	65.5	125
M. fuscata	Guatopo	Male	10	44.7	176
		Female	10	63.2	161
M. robinsoni	Masaguaral	Male	23	96.9	244
		Female	17	127.1	284
M. brevicaudata	Guatopo	Male	5	55.4	105
		Female	2	142.5	198

The means were compared using one way analysis of variance.

Female *D. marsupialis* remained on the grid an average of twice as long as males at both sites, but significantly so only in the llanos ($F = 8.57$; $P < .05$). The longest period between first and last captures was also much greater for females than for males.

Although the difference was not significant, female *M. robinsoni* remained on the grid in the llanos longer than males. The longest length of time on the grid was similar for males and females of this species. The longest period between first and last captures for both males and females may reflect the life span of this mouse opossum in the llanos. Both individuals were captured as young juveniles (5 g) and trapped continuously until their disappearance from the grid.

The length of time that *M. fuscata* remained on the grid in Guatopo was approximately half that of *M. robinsoni* from the llanos. As with *M. robinsoni,* the average length of time on the grid was greater for female as compared to male *M. fuscata,* but the difference was not significant.

The few data for *M. cinerea* and *M. brevicaudata* indicate that while male *M. cinerea* remain on the grid longer than females, the opposite is true for *M. brevicaudata;* females are more permanent than males.

Reproduction

Individuals of *D. marsupialis* were observed mating in late February. Females with pouch young were captured beginning in March of each year at both sites. The pouch young averaged 20 mm in crown-rump length and were estimated to be 16 to 20 days old (Collins, 1973). Females with pouch young continued to be captured at both sites throughout the wet season until October. There appear to be two peaks in reproductive activity, one in March and another in June (Figure 4). These peaks suggest that females breed at least two times each season. One female at Masaguaral had nine 23 mm pouch young in March 1976 and four 20 mm young in June 1976. The timing of the litters coincides with the 100-day period from birth to weaning reported by Collins (1973). Further evidence suggests that females produce at least two litters per season. The crown-rump lengths of pouch young as averaged for each month are presented in Table 5. Both sites and years were combined; additional information was obtained from road-kill records. The young found in March averaged 26.6 mm and were generally hairless except for cheek vibrassae, which corresponds to an approximate age of 20 days (Collins, 1973). The average length increases in April (41 mm) and in May (71 mm). In June the average length was

Figure 4. Reproduction condition of adult females of *Didelphis marsupialis*, *Marmosa cinerea*, *M. fuscata*, *M. robinsoni*, and *Monodelphis brevicaudata* during 1976 and 1977.

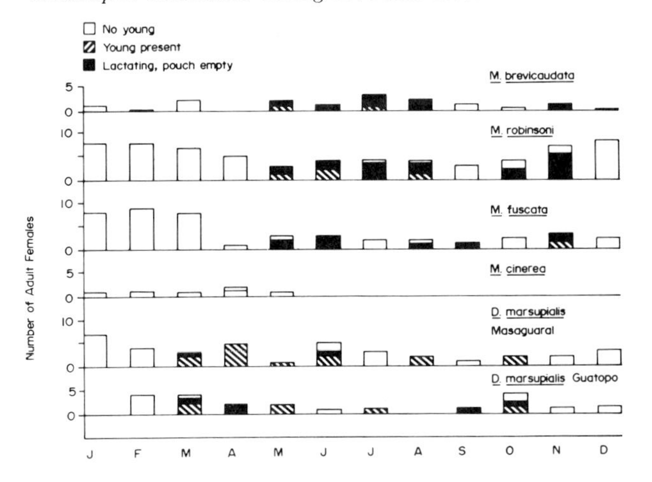

Table 5. Average crown-rump length of pouch young *D. marsupialis* during different months of reproductive season

Month	*Number*	*Average crown-rump length*[1]
March	5	26.5
April	7	41.0
May	4	71.6
June	5	20.0
July	3	50.0
August	4	70.0
October	4	46.3

[1] In mm.

20 mm. The appearance of small pouch young in October (Table 5) suggests that a third litter may be produced, but no female was known to have had three different litters in one season.

The reproductive life of *D. marsupialis* in nature is not firmly established (Hunsaker, 1977). Data from this study indicate that female *D. marsupialis* in northern Venezuela are reproductively active for at least two seasons. One female at Masaguaral was found with pouch young in April 1976 and again in April 1977. In Guatopo a female had pouch young in April 1976 and was lactating, but without pouch young, in May 1977.

Litter size averaged 7.5 (range 3 to 10) in Guatopo and 5.9 (range 2 to 10) at Masaguaral. The data suggest that the first litter of the season may be larger than subsequent ones. The one female captured with two litters in one season had nine young in the first litter and four in the next. The young of both litters were similar in age. The average litter size of females captured from March to May was 8.3, that of females captured from June through August was 7, and the average size of litters found in October was 4. Older females, as judged by tooth wear and general body condition, tended to have smaller litters than younger females.

Female *M. robinsoni* were lactating from May through October in the llanos. Females with attached young were found only four times in May, June (two females), and August 1976. The average crown-rump lengths of these litters were 15 mm, 10 mm, and 8 mm, respectively. All four litters were probably less than 20 days old (Collins, 1973). The data do not suggest definite peaks in reproductive activity as was found with *D. marsupialis.* Although no females were found with two consecutive litters, it is possible that *M. robinsoni* is polyestrus in the llanos. The length of the breeding season is sufficiently long to allow two litters, as the period from birth to weaning in captivity is 65 days (Eisenberg and Maliniak, 1967).

M. fuscata appears to have a breeding season similar to that of *M. robinsoni.* Lactating females were caught from May through November in Guatopo. However, only one female had young attached. This female was captured in November 1976 with five 9 mm young attached; one dead young was in the trap. Four days later the same female was retrapped, but without young.

Data on the reproduction of *M. cinerea* are meager, as only two females were captured during the study in Guatopo. One female was lactating in April 1977.

Data suggest that *M. brevicaudata* also breeds seasonally. Lactating females were found in May, June, July, August and November. Two females were captured with young attached. One caught in July 1976 had eight 12 mm young attached, the other trapped in May 1977 had seven 10 mm young. The latter was captured several times in May with the young intact. The same female was captured in July without young attached but in lactating condition and, in August, was nonreproductive.

Although *C. philander* was present for only the first three months of study in Guatopo, some reproductive data were obtained. In March 1976, three female *C. philander,* each carrying six pouch young, were captured. Crown-rump lengths of these litters averaged 15, 17 and 20 mm. Two of these females had their

Figure 5. *Didelphis marsupialis.* Widely distributed and versatile, it has adapted to both llanos and rain forest.

litters in April. One litter had grown an average of 18 mm (17 to 35 mm) and the other an average of 17 mm (15 to 32 mm).

Discussion

In a study of this nature, it is important to recognize the problems inherent in extrapolating information on densities, movements, and home ranges from trapping data (see Smith *et al.,* 1975). Nevertheless, intensive long term trapping of small mammal populations can yield much information on their dynamics. And many studies are based on trapping small mammals, thus allowing comparison of different regions and different species.

Population Densities

Densities of small mammal populations are affected by many factors: climatic conditions, food availability, and disruptive events, such as fires and floodings. The response of the animals to the traps alters apparent densities. Using average densities to describe resident populations can be misleading because the density of a species varies during different months and between years.

D. marsupialis was most abundant during the wet and the first part of the dry season, a pattern also found for this species in Panama by Fleming (1972). Density estimates for *D. marsupialis* in Venezuela are higher than Fleming reported for this opossum in Panama (Figure 5). He reported an average density ranging from 0.09 to 1.32 per hectare, depending on sex and location. However, the density of the Virginia opossum, *D. virginiana,* varies between different localities over its range (Fitch and Sandidge, 1953; Wiseman and Hendrickson, 1950; Lay, 1942; Verts, 1963; Homes and Sanderson, 1965; Stout and Sonenshine, 1974; Hamilton, Jr., 1958). It is likely that densities of *D. marsupialis* will be found to vary between localities as more studies are conducted.

Fleming's (1972) estimates of average density for *M. robinsoni* (0.31 to 2.25 per hectare) are comparable to estimates obtained during this study. Numbers of *M. robinsoni* were greatest at the end of the wet season and during the dry season in both northern Venezuela and in Panama (Fleming, 1972).

Few estimates of the other didelphids in this study are available for comparison (Hunsaker, 1977). Although *M. robinsoni* and *M. fuscata* are similar in size, density estimates for *M. fuscata* are lower. A factor probably contributing to the lower density of *M. fuscata* is that this species is generally found at elevations higher than the grid in Guatopo. Handley (1976) did not find *M. fuscata* below 1000 meters. It is possible that the grid in Guatopo, at 710 meters, represents the lower altitudinal limit of this species. *M. robinsoni* was captured at lower elevations (250 m) in Guatopo, but never on the permanent grid. Handley (1976) reported *M. robinsoni* from one to 1260 m, but generally below 500 m. Further study of the ecological requirements of these two species may reveal that *M. fuscata* replaces *M. robinsoni* at higher elevations.

The densities of *M. cinerea* and *M. brevicaudata* are similar and both are lower than the other two species of didelphids. *M. brevicaudata* is more carnivorous than the other didelphids in this study and requires a larger home range relative to its size (McNab, 1963), which explains its lower densities.

Movements

Opossums are often reported as nomadic animals that do not restrict their activities to specific areas (Hunsaker, 1977; Fleming, 1972). Thus, determining movements and home ranges for these marsupials becomes difficult, unless actually followed by means of radio tracking (Shirer and Fitch, 1970). The number of recaptures per individual (usually <20) was considered too low to warrant a valid analysis of home range (Koeppal, Slade, and Hoffman, 1975). To compare movements, the average distance and the longest distance between successive captures can be used.

The movements of female *D. marsupialis* are comparable between this study (Table 3) and Fleming's (1972) work in Panama. He reported average movements of 71.4 and 68.5 m for females from two localities. Females in this study averaged 80.7 and 60.1 m at the two localities. Movements of male *D. marsupialis* are probably greatly underestimated as mentioned. The average distance males traveled in this study is almost half the average distance Fleming (1972) reported for male *D. marsupialis* in Panama. The reason for this may be that males at both Guatopo and Masaguaral tended to move on the grids for short periods of time and then disappear, whereas females remained on the grid longer (Table 4).

Hunsaker (1977) presented home range estimates for various species of didelphids. In order to compare Fleming's (1972) data with other workers' findings, Hunsaker used the average distance between captures as the diameter of the home range. His estimated home range of 0.43 hectares for *D. marsupialis* in Panama is considerably smaller than home range estimates for *D. virginiana* in the United States, which range from 4.7 to 15.5 hectares (Lay, 1942) to 254 hectares (Fitch and Shirer, 1970). Applying the same procedure to movement data from this study, the home range of *D. marsupialis* is 0.27 hectares. However, the trapping grids used both in this study and Fleming's (1972) study were too small to accurately determine movements. Thus, the home range estimates based on movements on these grids are biased.

Fleming (1972) reported that *M. robinsoni* had longer average movements (46.6 to 60.4 m) than has been reported in this study (Table 3). However, the longest distances between captures are similar. In neither study were movements between sexes different.

Hunsaker (1977) used Fleming's (1972) movement data for *M. robinsoni* to calculate a home range of 0.22 hectares for this opossum. Applying the same procedure to data from this study, the home range of *M. robinsoni* in the llanos is 0.103 hectares.

The average movements of *M. fuscata, M. cinerea* and *M. brevicaudata* in Guatopo are all similar. Although *M. fuscata* and *M. brevicaudata* were the smallest marsupials on the grid in Guatopo, their movements were similar to those of the larger species. The relatively long movements of these smaller opossums may be explained by their more terrestrial habit. The larger species are more arboreal and were utilizing vertical space. In addition, because *M. brevicaudata* is more carnivorous, it requires a larger range of activity.

Longevity

The data on movements and especially the data on the average length of time individuals remain on the grid reflect the nomadic tendencies of these marsupials. Due to the size of the grid, the absence of dense trap lines surrounding the grid, and the length of the study, it is difficult to relate disappearance of individuals with mortality. It is likely that the disappearance rates reflect movements on and off the grid rather than survivorship.

There are few data available on the life span of tropical didelphids (Hunsaker, 1977). The average life span of the Virginia opossum is estimated at two years in the wild (Hunsaker, 1977). Captive individuals average three to five years (Collins, 1973). During this study several female *D. marsupialis* remained on the grids over a year. In Guatopo, one female was known to be alive for 530 days; she was captured initially in

May 1976 as an adult with pouch young. Assuming birth in the previous year, she would have been 2.4 years old at the time of her disappearance. In the llanos a female remaining on the grid for 491 days was also first captured as an adult. Her estimated age at the time of disappearance was 2.3 years.

Fleming (1972) reported that female *M. robinsoni* remained on the grid twice as long as males and suggested that females are more sedentary than males. Hunsaker (1977) also suggests that female *M. robinsoni* are more sedentary than males. Although the differences were not significant, female *M. robinsoni* remained on the grid longer than males during this study. Females that remained on the grid for long periods were generally trapped consistently from month to month on specific portions of the grid. In contrast, males were generally captured more erratically. Although females appear more sedentary than males, it has been reported that females do not utilize a nest continuously, rather that they tend to occupy a nest for a few days and then leave (Enders, 1935). My observations do not support this. During this study, a female with young occupied a nest box placed in a tree from June 1976 through August 1976. The young were weaned and foraging on their own when the next box was abandoned in August. While males and subadult females do appear to be more nomadic, adult females are more sedentary and females with young may utilize definite nest sites.

Hunsaker (1977) estimated the average life span of *M. robinsoni* in nature to be less than one year. In this study, data on the longest period individuals remained on the grid were obtained from individuals trapped as juveniles and recaptured continuously until their disappearance. The male was known to survive at the minimum 8.1 months and the female 9.5 months. These data are consistent with Hunsaker's (1977) estimates.

Few data are available on the life spans of the three other species, *M. fuscata, M. cinerea,* and *M. brevicaudata.* Data on the longest length of time on the grid were obtained in each case from individuals trapped initially as adults. Assuming that all these species are seasonal breeders, the month of birth for each of these individuals has been estimated and, from this, its age at time of disappearance was estimated. *M. fuscata* may have a life span similar to *M. robinsoni.* The male remaining longest on the grid was estimated to be 9.5 months old at time of disappearance and the female 10.8 months. Male *M. cinerea* may live at least 1.5 years in Guatopo. The estimated age of the female *M. cinerea* was 9.2 months at time of disappearance. The male *M. brevicaudata* remaining on the grid longest was estimated to be 10.5 months at time of disappearance and the female 13 months.

Reproduction

Litter size of *D. marsupialis* increases with increasing latitude (Fleming, 1973; Tyndale-Biscoe and Mackenzie, 1976). A comparison of litter size and latitude is presented in Table 6. Above 37°N, litter size averages above seven (Hamilton, 1958; Reynolds, 1952, Sanderson, 1961; Llewellyn and Dale, 1964). In the lower latitudes the average litter size is lower (Fleming, 1973; Tyndale-Biscoe and Mackenzie, 1976; Burns and Burns, 1957). My data on the litter size of *D. marsupialis* are consistent with this pattern. As latitude increases south of the equator, the litter size of *D. marsupialis* again increases (Davis, 1945; Hill, 1918).

The average litter size of *M. robinsoni* in Panama was reported to be ten, with a range from six to 13 (Fleming, 1973). In the llanos of Venezuela, the average litter size was 14 (range 13–15). The large production of young may offset two factors, high infant mortality and short reproductive life. Fleming (1973) reported the total loss of litters for three female *M. robinsoni* during his study. Enders (1966) reported that a captive female *Marmosa* gave birth to 14 young but only weaned five. During this study, a female *Marmosa* was found in a nest box with thirteen 12 m young in June 1976. In August, when the young were foraging on their own, six remained. Data suggest that the reproductive life of *M. robinsoni* is limited to one season. The combination of short reproductive life and high infant mortality has probably selected for a large initial litter size.

The few didelphid marsupials studied appear to be seasonally polyestrus. *D. virginiana* produces at least two litters each breeding season over different parts of its range (Hamilton, 1958; Wiseman and Hendrickson, 1950; Reynolds, 1945, 1952; Lay, 1942; Hartman 1928). *D. marsupialis* has also been reported as seasonally polyestrus (Davis, 1945; Biggers, 1966; Fleming, 1973; Tyndale-Biscoe and Mackenzie, 1976). At both Guatopo and Masaguaral, *D. marsupialis* is known to produce two, and perhaps three, litters per season. The other South American *Didelphis, D. albiventris,* also appears to be seasonally polyestrus (Tyndale-Biscoe and Mackenzie, 1976).

Few data are available for most species of mouse opossums. Enders (1935) reported that female *M. robinsoni* may produce only one litter per season in Panama. However, Fleming (1973) suggested that a second litter is possible. Although no conclusive evidence was collected during this study, my data suggest that *M. robinsoni* is seasonally polyestrus. The same may be true for *M. fuscata.* The length of the apparent breeding season (May through November) and the estimated time from birth to weaning support the possibility of two litters per season. Insufficient data

Table 6. The relation between litter size and latitude for the three species of *Didelphis*.[1]

Species	*Latitude*	*Locality*	*Litters/ year*	*Litter size*	*N*	*Reference*
D. virginiana	44°–42°N	New York State	2	8.4	346	Hamilton (1958)
D. virginiana	41.5°	Nebraska	?	8.6	23	Reynolds (1952)
D. virginiana	41°	Iowa	2	9.0	7	Wiseman and Hendrickson (1950)
D. virginiana	40°	Illinois	2	7.9	58	Sanderson (1951)
D. virginiana	39°	Missouri	2	8.9	42	Reynolds (1945)
D. virginiana	39°	Kansas	2	7.4	28	Fitch and Sandidge (1953)
D. virginiana	38.5°	Maryland	2	7.7	57	Llewellyn and Dale (1964)
D. virginiana	37°	California	2	7.2	44	Reynolds 1952)
D. virginiana	31°	Eastern Texas	2	6.8	65	Lay (1942)
D. virginiana	32°	Central Texas	2–3	6.2	33	Hartman (1928)
D. virginiana	30°	Florida	?	6.3	50	Burns and Burns (1957)
D. marsupialis	9°	Panama	2	6.0	29	Fleming (1973)
D. marsupialis	8°–10°	Northern Venezuela	2–3	6.7	32	Present study
D. marsupialis	6°–4°	Eastern Colombia	2	6.5	37	Tyndale-Biscoe and Mackenzie (1976)
D. marsupialis	3.5°	Western Colombia	2–3	4.5	41	Tyndale-Bisco and Mackenzie (1976)
D. albiventris	6°–2°	Colombia	?	4.2	10	Tyndale-Biscoe and Mackenzie (1976)
D. marsupialis	23°S	Teresopolis	2	8.5	4	Davis (1945)
D. marsupialis	23°S	Rio de Janeiro	2	7.1	16	Hill (1918)

[1] Adapted from Tyndale-Biscoe and MacKenzie (1976).

are available for female *M. cinerea* to draw conclusions on the reproductive patterns.

Data presented in this paper suggest that *M. brevicaudata* also breeds seasonally. Walker (1968), however, reported that the short-tailed opossum breeds throughout the year. It is possible that *M. brevicaudata* may produce two litters per season. The apparent length of the breeding season (May through November) would allow sufficient time for two litters to be weaned.

Fleming (1973) reported that another didelphid, *Philander opossum,* is also seasonally polyestrus in Panama. Limited data suggest that the same may be true for *Metachirus nudicaudatus* (Fleming, 1973).

The seasonality of rainfall in most of the New World tropics appears to be an important factor in determining the reproductive patterns of the didelphid marsupials. Breeding in the larger species, *Didelphis, Caluromys,* and *Philander,* commences in January-February and continues through September-October.The young begin foraging during the wet season when food levels are high (Smythe, 1970). The smaller didelphids, *Marmosa* and *Monodelphis,* seem to initiate breeding later in the dry season; but again, young begin foraging during the wet season.

Low (1978) examined the hypothesis that an important selective pressure influencing the evolution of marsupial reproductive strategies is environmental uncertainty. The unpredictability of rainfall patterns over much of Australia was cited as an example of environmental uncertainty. The low energy investment of marsupial females during the initial stages of reproduction (Parker, 1977; Kirsch, 1977; Low, 1978) allows termination of reproductive effort with minimal loss to the female if conditions become unfavorable (Low, 1978). Low (1978) stated that while most placentals are seasonal breeders, most marsupials are nonseasonal in their reproductive patterns. The continuous estrous cycle and facultative anestrus of most marsupials are used as evidence that marsupial reproductive strategies have evolved in response to environmental uncertainty.

Nonseasonality in breeding may be the case for some Australian forms (Tyndale-Biscoe, 1973; Newsome, 1965). The euro (*Macropus robustus*) and the red kangaroo (*Megaleia rufa*) inhabit the central area of Australia where rainfall is erratic. These macropods are opportunistic breeders. The females are in continuous estrus except under adverse conditions, when they enter facultative anestrus (Tyndale-Biscoe, 1973).

However, other Australian marsupials do exhibit seasonality in their reproductive patterns. The grey kangaroos (*Macropus giganteus* and *Macropus fuliginosus*) inhabit the eastern and southern regions of Australia where rainfall is more regular. These two macropods are seasonal breeders, producing young only in the spring and early summer (Tyndale-Biscoe, 1973). Various species of phalangerids (*Trichosurus vulpecula, Pseudocheirus peregrinus, Schoinobates volans*) are also seasonal breeders. An interesting example of a marsupial that is highly predictable in its reproductive patterns is the dasyurid, *Antechinus stuartii.* Female *A. stuartii* are monestrous. The estrous period, which coincides with the peak of reproductivity in males, occurs in July and August (Woolley, 1966). After this period the males become sterile and generally die within months after breeding (Woolley, 1966; Wood, 1970). The reproductive pattern of *A. stuartii* cannot be related to environmental uncertainty. In addition, the majority of the New World marsupials studied (Tyndale-Biscoe and Mackenzie, 1976; Fleming, 1973; Hunsaker, 1977; this paper) also exhibit seasonality in their breeding.

Just as placentals exhibit different reproductive patterns, marsupials have evolved multiple reproductive strategies in response to varying environmental conditions. An hypothesis concerning the selective pressures influencing the evolution of marsupial reproductive strategies must take into consideration the reproductive patterns of all marsupials.

Acknowledgments

During my stay in Venezuela many people helped in all aspects of my study. I would especially like to thank Edgardo Mondolfi, Venezuelan director of the Smithsonian Project, Juan Gomez-Nuñez, head of the Division of Fauna, Tomas Blohm, owner of Fundo Pecuario Masaguaral, Jose Ramon Orta and Jose Rafael Garcia, of the main office of the National Parks, Luis Escalona, director of Parque Nacional Guatopo, and the workers in the park. John F. Eisenberg, of the National Zoological Park, was a constant source of support, encouragement, and ideas. J.G. Hallett, P.G. Dolan, and R.J. Baker critically reviewed the manuscript.

My study was supported by a predoctoral fellowship from the Smithsonian Institution and by a Smithsonian grant to J.F. Eisenberg. Computer time was provided by The Museum, Texas Tech University.

Literature Cited

Barr, A. J., J. H. Goodnight, J. P. Sall, and J. T. Helwig.
1976. *A Users Guide to SAS 76.* Raleigh, North Carolina: Sparks Press.

Burns, R. K. and L. H. Burns.
1957. Observations on the breeding of the American opossum in Florida. *Revue Suisse de Zoologie,* 64:595–605.

Collins, L. R.
1973. *Monotremes and Marsupials, A Reference for Zoological Institutions.* Washington, D. C.: Smithsonian Institution Press.

Dalby, P. L.
1975. Biology of pampa rodents, Balcarce Area, Argentina. *Publ. Mus. Michigan State Univ., Biol. Ser.,* 5(3):1–271.

Davis, D. E.
1945. The annual cycle of plants, mosquitoes, birds and mammals in two Brazilian forests. *Ecol. Monogr.,* 15: 243–295.

Eisenberg, J. F., M. A. O'Connell and P. V. August.
1979. Density and distribution of mammals in two Venezuelan habitats. Pages 187–207 in *Vertebrate Ecology in the Northern Neotropics,* edited by John F. Eisenberg. Washington D. C.: Smithsonian Institution Press.

Enders, R. K.
1935. Mammalian life histories from Barro Colorado Island, Panama. *Bull. Mus. Comp. Zool., Harvard,* 78: 383–502.
1966. Attachment, nursing, and survival of young in some didelphids. *Symp. Zool. Soc. London,* 15:195–203.

Ewel, J. J. and A. Madriz.
1968. *Zonas de Vida de Venezuela.* Caracas: Ministerio Agriculture y Cria.

Fitch, H. S. and L. L. Sandidge.
1953. Ecology of the opposum on a natural area in northeastern Kansas. *Misc. Publ. Mus. Nat. Hist., Univ. Kansas,* 7:305–338.

Fitch, H. S. and H. W. Shirer.
1970. A radiotelometric study of the spatial relationships in the opposum. *Amer. Midl. Nat.,* 84:170–186.

Fleming, T. H.
1971. Population ecology of three species of neotropical rodents. *Misc. Publ. Mus. Zool., Univ. Michigan,* 143:1–77.
1972. Aspects of the population dynamics of three species of opossums in the Panama Canal Zone. *J. Mammal.,* 53:619–623.
1973. The reproductive cycles of three species of opossums and other mammals in the Panama Canal Zone. *J. Mammal.,* 54:439–455.

Hamilton, W. J., Jr.
1958. Life history and economic relation of the opossum (*Didelphis marsupialis virginiana*) in New York State. *Mem. Cornell Univ., Agric. Exper. Sta.,* 354:1–48.

Handley, C. O., Jr.
1976. Mammals of the Smithsonian Venezuelan project. *Brigham Young Univ. Sci. Bull., Biol. Ser.,* XX(5):1–91.

Hartman, C. G.
1928. The breeding season of the opossum (*Didelphis virginiana*) and the rate jof uterine and postnatal development. *J. Morphol.*, 46:143–215.

Hill, J. P.
1918. Some observations on the early development of *Didelphys aurita*. *Quart. J. Microscopical Sci.*, 63:91–139.

Holmes, A. C. and G. C. Sanderson
1965. Populations and movements of opossums in east-central Illinois. *J. Wildl. Manag.*, 29:287–295.

Hunsaker, D., II.
1977. Ecology of New World marsupials. Pages 95–156 in *The Biology of Marsupials*, edited by D. Hunsaker, III. New York: Academic Press.

Kirsch, J. A. W.
1977. The six-percent solution: Second thoughts on the adaptedness of the Marsupialia. *Amer. Sci.*, 65:276–288.

Koeppal, J. W., N. A. Slade and R. S. Hoffman
1975. A bivariate home range model with possible application to ethological data analysis. *J. Mammal.*, 56: 81–90.

Lay, D. W.
1942. Ecology of the opossum in eastern Texas. *J. Mammal.*, 23:147–159.

Llewellyn, L. M. and F. H. Dale.
1964. Notes on the ecology of the opossum in Maryland. *J. Mammal.*, 45:113–122.

Low, B. S.
1978. Environmental uncertainty and the parental strategies of marsupials and placentals. *Amer. Nat.*, 112: 197–213.

McNab, B. K.
1963. Bioenergetics and the determination of home range size. *Amer. Nat.*, 97:133–140.

Parker, P.
1977. An evolutionary comparison of placental and marsupial patterns of reproduction. Pages 273–285 in *Biology of Marsupials*, edited by B. Stonehouse and D. Gilmore, New York: Macmillan.

Reynolds, H. C.
1945. Some aspects of the life history and ecology of the opossum in central Missouri. *J. Mammal.*, 26:361–379.
1952. Studies on reproduction in the opossum (*Didelphis virginiana virginiana*). *Univ. Calif. Publ. Zool.*, 52:223–284.

Sanderson, G. C.
1961. Estimating opossums by markering young. *J. Wildl. Manag.*, 25:20–27.

Shirer, H. W. and H. S. Fitch.
1970. Comparison from radio movements and denning habits of raccoon, striped skunk, and opossum in northeastern Kansas. *J. Mammal.*, 51:491–503.

Smith, M.H., R. H. Gardner, J. B. Gentry, D. W. Kaufman and H. H. O'Farrell.
1975. Density estimations of small mammal populations. Pages 25–53 in *Small Mammals: Their Productivity and Population Dynamics*, edited by F. B. Golley, K. Petrusewicz and L. Ryszkowski. Cambridge: Cambridge University Press.

Smythe, N.
1970. Relationships between fruiting seasons and seed dispersal methods in a neotropical forest. *Amer. Nat.*, 104:25–35.

Stonehouse, B. and Gilmore (editors).
1977. *Biology of Marsupials.* New York: Macmillan.

Stout, J. and D. E. Sonenshine.
1974. Ecology of an opossum population in Virginia, 1963–1969. *Acta Theriologica*, 19:235–245.

Tyndale-Biscoe, H.
1973. *Life of Marsupials.* New York: American Elsevier Publication Company, Inc.

Tyndale-Biscoe, H. and R. B. Mackenzie.
1976. Reproduction in *Didelphis marsupialis* and *D. albiventris* in Colombia. *J. Mammal.*, 57:249–265.

Verts, B. J.
1963. Movements and populations of opossums in a cultivated area. *J. Wildl. Manag.*, 27:127–129

Walker, E. P.
1975. *Mammals of the World, Third Edition.* Baltimore: Johns Hopkins Press.

Wiseman, G. L. and G. O. Hendrickson.
1950. Notes on the life history and ecology of the opossum in southeast Iowa. *J. Mammal.*, 31:331–337.

Wood, D. H.
1970. An ecological study of *Antechinus stuartii* (Marsupialia) in a South-east Queensland rain forest. *Australian J. Zool.*, 18:185–207.

Woolley, P.
1966. Reproduction in *Antechinus* spp. and other dasyurid marsupials. *Symp. Zool. Soc. London*, 15:281–294.

SECTION 3:

The Primates

Alouatta seniculus, the red howler monkey. This species is adapted to a range of habitat types and can successfully survive in forest patches within the llanos of the northern neotropics.

Introduction

The study of primate ecology in the neotropics has been making steady progress (Smith, 1977; Klein and Klein, 1977; Kinzey, 1977). A long tradition of primate studies has been carried out on Barro Colorado Island in Panama beginning with Carpenter (1934) and culminating in the ecological study by Hladik and Hladik in 1969. Research on neotropical primate behavior and ecology has been summarized for the Callitrichidae in the volume edited by Kleiman (1978) and the research through 1976 has been reviewed in the chapter by Eisenberg (in press).

Three primate species (*Ateles belzebuth, Alouatta seniculus* and *Cebus nigrivittatus*) are widely distributed in the north coast range of Venezuela. The latter two are the pioneer species in the llanos habitats. This section opens with a discussion of the distribution of *A. belzebuth* in northern Venezuela which suggests that the central mountain range of northern Venezuela has served as a refugium for certain mammalian species, a view already put forth in the chapter by Eisenberg and Redford. The main focus of this section then turns to a long-term study of the *Alouatta seniculus* population at Hato Masaguaral. Thorington, Rudran and Mack discuss field techniques for obtaining accurate measurements on living primates. As a result of their efforts, they present weights and measurements for a sample drawn from our *A. seniculus* population and compare these results with data previously collected for *Alouatta palliata* and *A. caraya.* They conclude that definite differences are demonstrable among these three species with respect to degree of sexual dimorphism in size, degree of genital dimorphism, and speculate on the evolution of coat-color dimorphism in *A. caraya.* Clearly, different selective pressures have been acting on these three different species. *A. palliata* on Barro Colorado tend to be larger than the other two species, but the degree of sexual dimorphism between males and females with respect to size is less. This is a trend that would not easily have been predicted since, in general, larger mammals, if they are sexually dimorphic in size, tend to show a greater divergence in size between males and females. They raise the interesting hypothesis that the convergence in external genitalia between juvenile male and female *A. seniculus* may have resulted from the fact that females stand a better chance of surviving if they resemble males. This could follow from the idea that infanticidal tendencies shown by adult male *A. seniculus* may be differentially expressed with respect to the juvenile sexes. In *A. caraya,* surely some different selective pressure is involved in the maintenance of the color dimorphism. Young males of *caraya* tend to retain a coat color similar to that of young females and adult females. Does this reflect a differential susceptibility to socially imposed mortality when *A. caraya* is compared with *A. seniculus?*

R. Rudran analyzes the demography of the *Alouatta seniculus* population at Masaguaral. He describes the phenomenon of infanticide which results from the takeover of a troop by an alien male displacing the resident male. He describes the variations in the takeover process, their frequency and the percentage of infanticide accompanying such takeovers. He concludes that frequency of takeover and attendent frequency of infanticide in part is density dependent and he suggests that the imposed infant mortality at high densities may be a significant factor in contributing to population regulation. This derives directly from an hypothesis developed by Rudran from studies of the folivorous langur monkeys in Sri Lanka (Rudran, 1973). David Mack concludes with a discussion of the ontogeny of behavior in *Alouatta seniculus* and proposes a classification scheme based on known age animals thus clearly defining ontogenetic stages. This is the first time that such a scheme has been developed with real time coordinates for a free ranging species of New World primate.

John Robinson discusses his results deriving from a detailed study of a *Cebus nigrivittatus* population. *C. nigrivittatus,* in contrast to *Alouatta seniculus,* is an omnivore. Its feeding strategy is versatile and at certain times of the year it may spend over 50 percent of its time foraging on the ground. In an incisive analysis, Robinson discusses the phenomenon of urine-washing displayed by *Cebus* (a behavioral trait also displayed by other New World primates). He suggests that, in addition to functions in social communication, one of the primary functions may be to assist in thermoregulation. His data are strongly suggestive of a novel interpretation for this behavioral trait. J.F.E.

Literature Cited

Carpenter, C. R.
1934. A field study of the behavior and social relations of howling monkeys. *Comp. Psychol. Monogr.,* 19(2):1–168.

Eisenberg, J. F.
In press Habitat, economy and society: Some correlations and hypotheses for the neotropical primates. In *Ecological Influences on Social Organization: Evolution and Adaptation,* edited by I. Bernstein and E. Smith. New York: Garland Press.

Hladik, A. and C. M. Hladik.
1969. Rapports trophiques entre vegetation et primates dans la foret de Barro Colorado (Panama). *La Terre et la Vie,* 1:25–117.

Kinzey, W. G.
1977. Diet and feeding behavior in *Callicebus torquatus.* pages 127–152 in *Primate Ecology,* edited by T. Clutton-Brock, New York: Academic Press.

Kleiman, D. G. (editor).
1978. *The Biology and Conservation of the Callitrichidae.* Washington, D. C.: Smithsonian Institution Press.

Klein, L. L. and D. B. Klein.
1977. Feeding behaviour of the Colombian spider monkey. Pages 153–180 in *Primate Ecology,* edited by T. Clutton-Brock. New York: Academic Press.

Rudran, R.
1973. Adult male replacement in one-male troops of purple-faced langurs (*Presbytis senex senex*) and its effect on population structure. *Folia Primat.,* 19:166–192.

Smith, C. C.
1977. Feeding behaviour and social organization in howling monkeys. Pages 97–126 in *Primate Ecology,* edited by T. Clutton-Brock, New York: Academic Press.

EDGARDO MONDOLFI
Escuela de Biologia
Facultad de Ciencias
Universidad Central de Venezuela
Caracas

JOHN F. EISENBERG
National Zoological Park
Smithsonian Institution
Washington, D.C. 20008

New Records for *Ateles belzebuth hybridus* in Northern Venezuela

ABSTRACT

Records for *Ateles belzebuth hybridus* derived from sightings and collections were analyzed and isolated populations in the north coast range of Venezuela are described for the first time. It is concluded that *Ateles* shows a disjunct distributional pattern in the premontane forests along the foothills of the Andes and into the north coast range.

RESÚMEN

Se analiza información de *Ateles belzebuth hybridus* derivada de observaciónes en el bosque y colecciónes de museos y se describen por primera vez poblaciónes aisladas de la montaña de la costa norte de Venezuela. Se concluye que *Ateles* tiene una distribución descontinuada en los bosques de premontaña a lo largo de las colinas a los pies de los Andes y entre las montanas de la costa del norte.

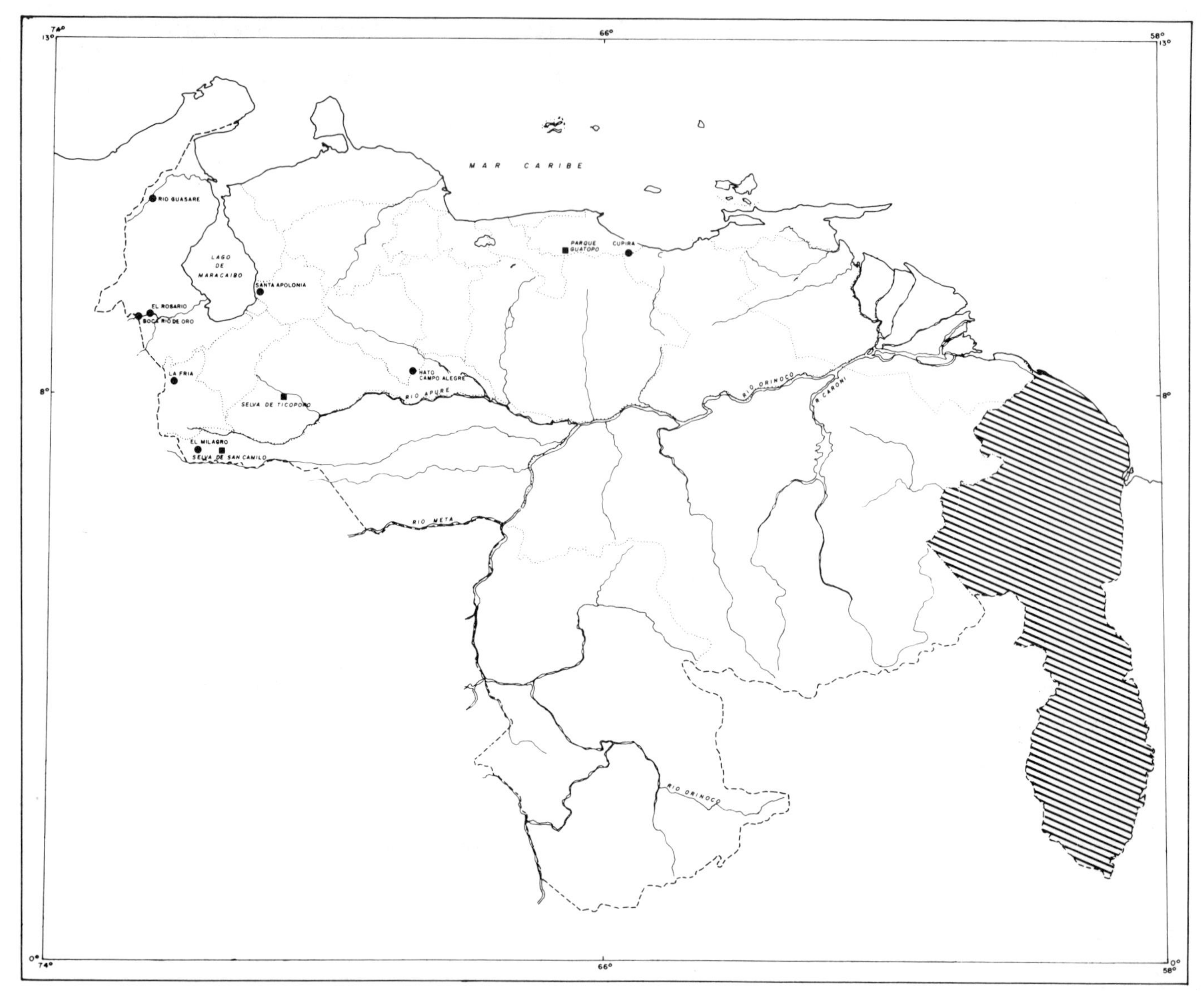

Figure 1. Distribution of *Ateles belzebuth hybridus* in Venezuela. Circles refer to collection localities; squares designate localities of sightings. See text for explanation of numbers and specimens examined and sighted.

The distribution of *Ateles belzebuth hybridus* I. Geoffroy (1829) in Venezuela has been unclear for many years. According to Kellogg and Goldman (1944), *A. b. hybridus* was found in the Serranía de Valledupar of the Sierra Perijá and the upper Río Tarra, a tributary of the Río Catatumbo between Colombia and Venezuela, to the west of Lake Maracaibo. This view was shared by Hershkovitz (1949) who notes its occurrence from the Sierra Perijá to Lake Maracaibo. This subspecies was widely distributed in northeastern Colombia east of the Río Magdalena. The type locality was restricted to La Gloria, Río Magdalena, Colombia (Kellogg and Goldman, 1944).

Recent records for Venezuela indicate the *Ateles belzebuth hybridus* in fact occurs on both the eastern and western sides of Lake Maracaibo and then extends eastward, apparently not continuously, in the coastal range of Venezuela at least to Cúpira in the State of Miranda. These records are plotted in Figure 1. Three females collected in the vicinity of La Fria, State of Táchira, were taken in March of 1958 by E. Mondolfi and A. Bodini. The male specimen from Boca de Caño Norte, Río Guasare in the Montes de Oca, State of Zulia, was collected by I. Rodriguez in 1957.

The specimen from Cúpira in the extreme northeast of the coastal range of the State of Miranda was obtained by A. Barcelona and I. Rodriguez during January 1961 while on a field trip organized by Professor R. Lancini and Dr. M. A. Schön. All these specimens are now deposited in the Collection of the Museo de Biología de la Universidad Central de Venezuela (U.C.V.). One female (U.S.N.M. 372804) from near Santa Apolonia, 49 km. W. of Valera, State of Trujillo; one female (U.S.N.M. 443221) from El Milagro, 4 km NW of Nula, State of Apure; one female (443225) from Boca de Río de Oro, 60 km NW of Encontrados, Zulia; and another female (443223) from El Rosario 45 km NW of Encontrados, are in the Mammal Collection of the U.S. National Museum. Those collection sites are included in Figure 1.

External measurements for specimens collected by Mondolfi, Bodini and Rodriguez and the measurements for the Cúpira skull are included in Tables 1 and 2.

The Cúpira skin differs slightly from the specimens collected in Zulia and Táchira. It is considerably darker on the forearms, arms, and hands; the upper parts of the head, the dorsum and tail are darker than the specimens from La Fria and Río Guasare. This individual has a conspicuous buffy forehead patch. The darker coloration of the Cúpira skin falls among the range variation for *hybridus* (see Hershkovitz, 1949).

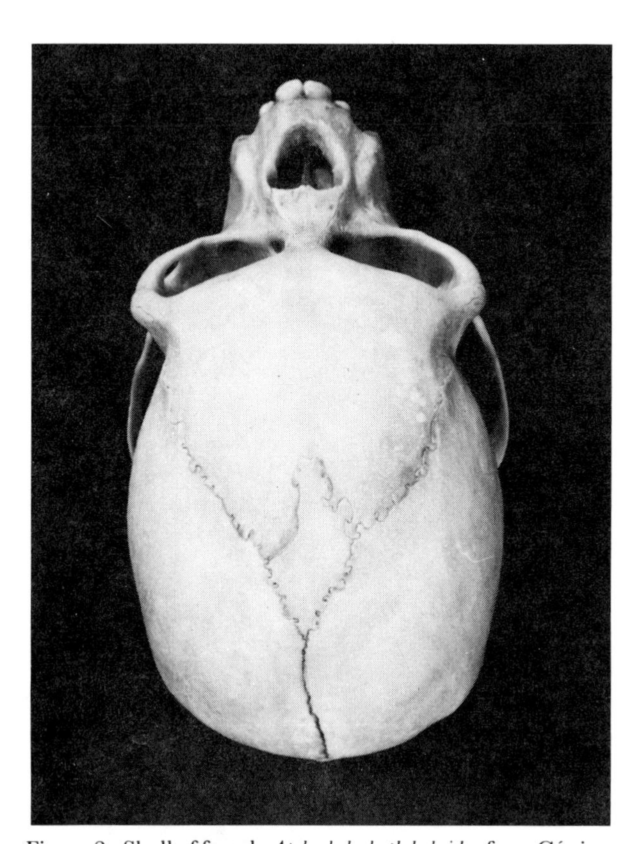

Figure 2. Skull of female *Ateles belzebuth hybridus* from Cúpira, showing an interparietal bone.

Table 1. Measurements for *Ateles belzebuth hybridus*

Locality	*Measurements (mm)*				*Sex*	*Date*
	Tl	*T*	*Hf.*	*E*		
La Fria	1116	707	165	39	♀	III/30/58
La Fria	1273	793	181	42	♀	III/30/58
La Fria	1250	790	184	40	♀	III/30/58
Rio Guasare						
Caño Colorado	1140	860	185	35	♀	VIII/23/57
Caño Norte	1200	800	170		♂	I/4/58

Table 2. Skull of *Ateles belzebuth hybridus* adult ♀ from Cúpira

Greatest length	113.3[1]
Orbital width	60.0
Postorbital constriction	49.1
Width of brain case	60.0
Zygomatic breadth	68.6
Upper maxillary tooth row	28.7

[1] Measurements are in mm.

It is worth mentioning two anatomical variations shown by Venezuelan specimens of *A. belzebuth hydridus.* The Cúpira skull has an interparietal bone (os incae) irregularly rhomboid shaped (Figure 2) that measures 2.8 × 1.35 cm. One individual from La Fria has a vestigial thumb, bearing a nail, on each hand. In the dry skin the vestigial thumb of the right hand measures 1.42 cm., on the left hand, 1.39 cm. Hershkovitz (1949) mentions one individual from Guaimaral, Río Cesar, Colombia, that possessed a vestigial thumb bearing a nail on each hand.

Confirmed sightings of *Ateles belzebuth hybridus* have been recorded for "Campo Alegre" Ranch north of Arismendi in the State of Barinas (Lopez de Ceballos, 1974, p. 124); Selva de San Camilo in the State of Apure; forests between the Río Escalante and Río Catatumbo, State of Zulia; Selva de Mijagual and Selva de Ticoporo in the State of Barinas (information from the Ministry of Sanitation, Yellow Fever Control Program); and Guatopo National Park. These sightings are included in Figure 1.

The presence of *Ateles* in Guatopo National Park was noted by the park rangers and confirmed in July 1974 by Professor C. Rivero Blanco and Dr. E. J. Kormondy who sighted one group. In January 1975, the authors and Dr. D. Marcellini sighted three *Ateles belzebuth* on the same trail. The exact location of the sighting is located about 2 km. south of "Los Alpes" where a jeep trail leaves the main road winding southeast to the narrow valley where the Río Taguaza flows.

It is noteworthy that this group of three spider monkeys was associated with a large troop of *Cebus nigrivittatus,* a habit also mentioned by Lopez de Ceballos (1974). The same type of association has been noted for *Ateles geoffroyi* on Barro Colorado Island (Eisenberg and Kuehn, 1966). Associations between *Ateles* and *Alouatta* are discussed at some length by Klein and Klein (1973).

During a second visit to Guatopo, Eisenberg and R. Rudran located on 11 March 1975 an *Ateles* troop on the banks of the Río Taguaza some 4 km east of the previous sighting. The troop was divided into at least two subgroups about a half km apart. The animals were located by their loud "long calls" which are employed in communication between subgroups. Such long call exchanges were heard at 1120, 1240 and about 1258 hours.

In Western Venezuela, as in eastern Colombia, *A. belzebuth hybridus* is called "marimonda." In the northern coastal range of Venezuela, it seems to be known from Cabo Codera to Cúpira and Guatopo as "mono frontino," probably referring to the characteristic whitish frontal patch.

Capt. W. Robinson (1902) gathered information at San Julian, in the northern Coastal Range, east of La Guaira and west of Cabo Codera, about a species of monkey locally known as "mona frontina," described by the natives as "a white crowned monkey which barks like a dog."

The habitat preferences of *A. belzebuth hybridus* in Venezuela include: tall evergreen very humid tropical forest (La Fria, Encontrados, Catatumbo); tall evergreen very humid tropical montane forest (Guatopo); tall semi-deciduous tropical humid forest, almost evergreen but where some trees may be leafless for a few weeks [this type of forest occurs in the western llanos and extends to the piedmont in Barinas (Ticoporo, Mijagual) and Apure (San Camilo)]; riverine semi-deciduous tropical forest (Rio Guasare); semi-evergreen dense but not very tall, tropical seasonal montane forest (Cupira). The elevation of these habitats extends from 28 to about 600 meters in Guatopo National Park.

The three females from La Fria were shot at about 1700 hours when they were seen in a tall tree in a forest block that was being cut down. Examination of the stomach contents of these three indivduals showed: (1) leaves, two kinds of fruits (one of a palm); (2) flowers of a *Papilionatae,* one seed of a kind of berry, one seed of a kind of drupe and crushed leaves; (3) flowers of an *Euphorbiaceae* and of a *Papilionatae,* and one palm seed.

According to Müller (1973) in his figure 93, Guatopo lies at the edge of Arboreal Dispersal Center #17 (i.e., Venezuelan Coastal Forest Center). This area has strong faunal relationships with the region to the south of Lake Maracaibo (the Catatumbo Center #16). The presence of *Ateles* in Guatopo then is in conformity with other faunal discontinuities as outlined by Müller (1973).

Acknowledgments

The authors thank Professor R. Lancini, Director of the Caracas Natural Science Museum for permission to examine the Cúpira specimen, and Dr. C. O. Handley for permission to examine specimens of Venezuelan *Ateles* in the U.S. National Museum. Sr. A. Mendoza prepared the range map and Sr. P. Meyn made the photograph of the skull.

Literature Cited

Eisenberg, J. F. and R. E. Kuehn.
1966. The behavior of *Ateles geoffroyi* and related species. *Smithsonian Misc. Colls.*, 151(8):1–63.

Hershkovitz, P.
1949. Mammals of northern Colombia. Preliminary Report No. 4: Monkeys (Primates) with taxonomic revisions of some forms. *Proc. U. S. Nat., Hist., Mus.*, 98:323–427.

Kellogg, R. and E. A. Goldman.
1944. Review of the spider monkeys. *Proc. U. S. Nat. Hist. Mus.*, 95:1–45.

Klein, L. and D. Klein.
1973. Observations on two types of neotropical primate intertaxa associations. *Amer. J. Phys. Anthrop.*, 38(2): 649–653.

Lopez de Ceballos, E.
1974. *Fauna de Venezuela y su Conservación.* Caracas: Editorial Arte.

Müller, P.
1973. *The Dispersal Centres of Terrestrial Vertebrates in the Neotropical Realm.* The Hague: W. Junk.

Robinson, W. and M. W. Lyon, Jr.
1902. An annotated list of mammals collected in the vicinity of La Guaira, Venezuela. *Proc. U. S. Nat. Mus.*, 24:135–162.

R. W. THORINGTON, JR.
National Museum of Natural History
Smithsonian Institution
Washington, D. C. 20560

R. RUDRAN and D. MACK
National Zoological Park
Washington, D. C. 20008

Sexual Dimorphism of *Alouatta seniculus* and Observations on Capture Techniques

ABSTRACT

Howler monkeys (*Alouatta seniculus*) were captured in Guarico State, Venezuela. They were measured, marked, and released. The capture and marking techniques are described and evaluated. Sernylan continues to be the best drug for use in darting *Alouatta*. Ear tags are useful for marking the animals, although there are difficulties in coding individuals—either by the position or color of the tag—so that they can be recognized from a distance. Dental casts were made from impressions taken on the anesthetized monkeys. From eruption and wear patterns, the relative ages of the animals were estimated. By analogy with other studies, we estimated that the 36 individuals ranged approximately from 3 to 13 years old. A brief discussion of these aging procedures is included.

Sexual dimorphism in body size of adult *Alouatta seniculus* is greater than that in *A. palliata* and approximately the same as that in *A. caraya*. In *A. seniculus* it is paralleled by the extreme enlargement of the hyoid and by precocious scrotal development in infant and juvenile males. In *A. caraya* the sexual dimorphism in size is paralleled by the development of sexual dichromatism. In *A. palliata* the lesser degree of sexual dimorphism in body size is paralleled by a lesser development of the hyoid, the lack of sexual dichromatism, and by the fact that the genitalia of the males mimic those of the females until puberty. These patterns of sexual dimorphism are discussed.

A few other differences and similarities between the species are discussed. *A. palliata* is more

RESÚMEN

Monos alladores (*Alouatta seniculus*) fueron capturados en el Guarico, Venezuela. Cada animal fue medido, marcado, y soltado. El método de captura y marcado está descrito. El sernylan sigue siendo la mejor droza para adormecer por medio de dardos al Alouatta. Las marcas en las orejas son útiles para marcar a los animales, aunque hay dificultades para la clasificación de diferentes individuos por la posición o por el color de la marca para que puedan ser reconocidos a la distancia. Se hicieron moldes dentales de monos anestesiados y se estimó su edad de acuerdo a la información obtenida con este método. Por analogia con otros estudios, estimamos que la edad de estos individuos variaba entre los 3 y los 13 años de edad.

La diferencia de tamaño por sexo es mayor en el *A. seniculus* adulto que en el *A. palliata* y approximadamente iqual el del *A. caraya*. En *A. seniculus* está acompañado de un gran ensanchamiento del hioides y de desarrollo precoz del escrots en infantes y jóvenes machos. En *A. caraya* se puede encontrar dicromatismo sexual ademas del dimorfismo sexual. En *A. palliata* el menor grado de dimorfismo sexual está acompañado de un menor desarrollo del hioides, falta de dicromatismo sexual, y los genitales de machos y hembras son iguales hasta la pubertad.

Ademas, *A. palliata* es más robusta que las otras dos especies. Pezones supernumerarios son comunes en la población de Guàrico de *A. seniculus*, pero no hay información comparativa de otras poblaciones. *A. seniculus* y *A. caraya* están infectados de piojos que no existen en ningún

robust than the other two species. Supernumerary nipples are common in the Guarico sample of *A. seniculus*, but no comparative data are available from other populations of any species. Both *A. seniculus* and *A. caraya* are heavily infested with lice, which were not found on any specimens of *A. palliata* from Barro Colorado Island.

otro especímen de *A. palliata* de la isla Barro Colorado.

Introduction

As part of a long-term study of red howler monkeys (*Alouatta seniculus*) in Venezuela (see Rudran, p. 107), it became desirable to have a number of male monkeys marked for individual recognition. This paper results from our marking program. We report on the techniques we used to capture and mark these animals. In addition, we provide some comparisons of the notes and measurements we took of these animals with similar data collected by Thorington and others on *Alouatta palliata* from Barro Colorado Island, Panama Canal Zone, and with data reported in the literature for *A. caraya*.

The study site is at Masaguaral, Guarico State, Venezuela. The red howlers have been studied here previously by Neville (1972, 1976). Descriptions of the area, the vegetation, and the highly seasonal climate are given by Neville (1972) and Rudran (see p. 107). This work was conducted between 6 and 18 March 1978 during the dry season, at which time many of the trees were leafless and the vegetation of the llano was brown and dry.

The capture techniques are those developed by Scott and reported by Scott, et al. (1976). They and the marking techniques have been modified and developed further by Thorington, Froehlich, and others on Barro Colorado Island, Panama Canal Zone. Certain modifications were desirable at Masaguaral because of the lack of electricity, the inaccessibility of dry ice and liquid nitrogen, and also because the trees are much lower and the animals much more accessible than on Barro Colorado Island. These last two factors particularly influenced our choice of marking techniques.

Alouatta seniculus is one of six species of howler monkeys presently considered valid. The differences between the species are seldom emphasized (e.g., Napier and Napier, 1967) except for the differences in coat color. Some facts, such as that *A. caraya* is sexually dichromatic, whereas the other species are not, and that there is such a difference in the sizes of the hyoid bones, as emphasized by Hershkovitz (1949), should give us warning that they probably differ dramatically in some aspects of their biology. In this paper we have noted some other differences between the species, especially between *A. seniculus* and *A. palliata* with which we are most familiar, in the hope that eventually the differences will be recognized as a comprehensible pattern, reflecting different evolutionary strategies or different stages in the evolution of howler monkeys.

Methods and Materials

The animals were darted with a CO_2 gun using the Pneu-dart system, described by Scott, et al (1976). The darts hold between 0.6 and 0.7 ml. of drug, which is injected on impact by a small cap which explodes behind the plunger. Three different kinds of drugs were used: Sernylan (Phencyclidine), Ketamine (Ketamine HCl) and Rompun. Sernylan was our drug of choice in the darting and ketamine was the drug we used to maintain the initial anesthesia. We used Sernylan at a concentration of 100 mg/cc, so the darts delivered a dose of 60–70 mg. The animals weighed between 3 and 7 kg, so the dosage varied between 10 and 20 mg/kg. An injection of Ketamine of 20 mg/kg would provide us with an extra hour to work on the animal.

Most of the animals captured fell to the ground and were not hurt. A lesser number were caught in a blanket held by two persons, while a third person carefully plumbed the point where the animal would fall.

All animals were marked with numbered ear tags. One, two, or three tags were used on each animal. The tags were painted with one of four colors of "Flo-Paque" paints: F-54 ultramarine blue, F-92 flo-glo orange, F-95 flo-glo citron, or F-96 flo-glo green. A tag could be placed at any of 6 positions—at the top, halfway down, or at the bottom of the pinna—on the two ears. Two independent codes were used—a color code and a position code—to permit the recognition of each individual from a distance. The two codes were interrelated so that a failure in either system—the loss of a color, the loss of a tag, or a change in the position of a tag—was recognizable.

All animals were given a small freeze-brand on the face. These brands were applied with a jet of freon from a spray can of Realistic© component cooler. The spray was administered for 3 to 5 seconds. By using a mask of cardboard we were able to restrict the brand to an area of ¼ cm^2. The freeze brands were placed at different positions on the face. Animals from the same troop received the same brand.

All animals were given a two-character tattoo with a set of tattoo punches and green tattoo ink. Animals from the same troop were tattooed in the same place on the thorax or upper arm. The location of the tattoo differed for each troop.

Animals were measured with a tape measure and were weighed with a calibrated spring scale. Measurements are defined in the appendix. Tooth casts were made with dental stone from molds of alginate following normal dental procedures.

The anesthetized animals were kept in burlap bags until they had recovered sufficiently to climb trees without falling, at which time they were released beneath their troop. Special care was taken to keep the bagged animals out of the sun and to make certain that the monkeys could breath easily in the bags. Animals that felt overheated were removed from the bags and wetted.

Evaluations of Techniques

The capture procedures resulted in the successful marking and release of 35 monkeys. One animal died as a result of its fall, in spite of the fact that it had one of the shortest falls of all the animals. The only other injuries incurred by falls were some chipped teeth suffered by one animal. Nothing else was detected, although all animals were examined carefully for broken limbs, ribs, and teeth, and for any loss of muscle tone which might indicate spinal injuries undetected by palpation.

Although we tried all three drugs for capturing the animals, only Sernylan was dependably successful. The dosage of 60–70 mg Sernylan in a dart was suitable for animals of 3–7 kg. Using 60 mg of Ketamine in each dart, we were not able to capture immature howlers, even when we darted an animal twice in a short period of time. We had three failures with animals estimated to weigh between 3 and 4 kg. We also experimented with Rompun but had variable results. An old male of 6.6 kg went into a very deep anesthesia for an hour when injected with only 0.3 mg/kg. Other animals required several injections totalling approximately 2 mg/kg before they experienced a light anesthesia. Using Rompun in the darts, we captured one animal with a single injection of 0.8 mg/kg and a second animal with three sequential darts, 15 minutes apart, totalling 1.9 mg/kg. With other animals we injected as much as 4–5 mg/kg without being able to drop them from the trees. Rompun seems to make the animals sleepy and causes them to find a safe and comfortable place to lie down. At the dosages we used, the animals can arouse themselves when pressed to do so. The result is that a drugged howler finds a comfortable and secure fork in a tree, falls asleep and starts to fall out of the tree, arouses himself just enough to pull himself back into the fork, and then falls asleep again. Being unwilling to experiment with potentially lethal doses of the drug, we did not test it further.

It would seem desirable to catch all howler monkeys when they fall from the trees, but this is neither practical nor safe. Ten percent of the monkeys we successfully darted never fell. Either their tails were overlapped around a branch or they became sufficiently tangled in vines that someone had to climb the tree to bring them down. Other monkeys would hang precariously for 15–30 minutes before falling, which made catching them a very tedious and time-consuming practice. The major problem, however, is with animals which do not have a free fall. In these cases it is very difficult to judge exactly where the animal will land and thus it is difficult to catch them. Furthermore there is far more risk that one of the catchers will be hit and injured by a falling monkey than there is a risk that the monkey will be hurt in the fall. We subjectively weighed these risks in each case and decided whether to attempt to catch the animal or not. The one monkey death resulted from a case in which we thought the monkey was unlikely to be hurt in the fall, and we did not attempt to catch him.

The marking procedures met with mixed success. All freeze brands faded away within one month after they were applied. This suggests that five seconds is not a long enough exposure time, and that the problem is analagous to that which Lazarus and Rowe (1975) had in branding *Myocaster.* They suggested an exposure time of 10–15 seconds. On Barro Colorado Island, Thorington obtained good facial brands on *Alouatta palliata* by using the head of a nail at the temperature of liquid nitrogen and using an exposure time of 20–30 seconds. It is probable that exposure time must be adjusted according to the thickness of the skin, so as to assure that the dermal melanocytes are well frozen, if freeze-branding is to be successful.

The tattoos were not visible from the ground, and since none of the animals were recaptured, we could not judge the failure rate of this technique. On *Alouatta palliata* the tattoos were readily visible on animals which were recaptured up to five years later.

Five months after the animals were tagged, Rudran noted the conditions of the ear tags on 33 monkeys. Two monkeys had lost one ear tag each and one monkey had lost both ear tags. This is a failure rate of 3 out of 33 monkeys, or 9 percent, and of 4 out of 54 ear tags, or 7 percent. The position of the ear tags had shifted in 7 monkeys, thus the positional code had a failure rate of 21 percent, in addition to the 9 percent loss. The color came off the ear tags of 5 animals and the color had faded on most of the others, so that Rudran judged the color code to be a success on only 3 of the 33 animals. The colors which lasted best were green and orange. In total, 23 of the 33 animals were still identifiable from the ground on the basis of one of the original codes with the ear tags, and if recaptured 32 of the 33 would be identifiable by the numbers on their ear tags. The number of animals recognizable from the ground by the codes would have been greatly increased if we had used ear tags of different shapes, rather than ear tags of different colors. In fact, however, by using a combination of natural marks and the codes, Rudran was able to unambiguously identify from the ground all 33 of the animals which remained in or near the study area five months after they were marked.

Tooth Eruption, Breakage, and Wear

Among the younger males captured, the upper third molars or upper canines were erupting. The order of the eruption of the upper canines and third molars is variable. In the youngest animal (no. 4), only one permanent canine had started to break through the gum, but the upper third molars were both visible above gum. In contrast, in an older animal (no. 28) the lower canines were completely erupted and the upper canines were approximately two-thirds of the way erupted, but there was no sign of the third upper molars. This animal may completely lack third upper molars, but it seems more likely that they will erupt very late. Such a situation seems to have occurred in a still older animal (no. 31), which had its canines completely erupted and had wear facets on the buccal cusps of the first molars, but its left third upper molar was not yet in the occlusal plane. These last two animals contrast with nine others of the same age span, in which the third molars had reached or would reach the occlusal plane before the canines were fully erupted. Thus we think that the third upper molars erupt later than the canines in approximately 20 percent of the males of this population. The lower third molars seem to erupt invariably before the canines.

Howler monkeys in this population seem to break or lose their incisors frequently. Two old males (no. 9 and no. 33) had lost three and two lower incisors respectively. Another male (no. 36) and a female (no. 16) had each lost one. Two old males (no. 30 and no. 9) had broken their central upper incisors and a young male (no. 15) had lost its left central incisors. In only one of these animals (no. 30) is there any possibility that the loss resulted from the capture technique. This animal had also lost its most posterior upper premolar (prior to capture). It was the only monkey missing a post-incisor tooth. Thus six of 36 animals had lost a total of 13 teeth, mostly incisors.

Tooth wear usually is first noticeable on the buccal cusps of the first molars. Soon thereafter it can be seen as facets on the buccal cusps of the posterior two premolars and the second molar. As these facets become larger with wear, facets appear on the lingual cusps of the first molars, then of the second molar and last premolar. The next distinctive change is the extension of the facets on the middle and last upper premolars along the whole posterior ridge of the cusp, so that one can see two parallel ridges of enamel. In the two oldest animals captured, the facets of the lingual cusps of the first upper molars are confluent and those of the buccal cusps of the same tooth and the lingual cusps of the second upper molars are barely or almost confluent. On the lower molars confluency occurs first across the teeth, between the facets of anterior buccal and lingual cusps.

The honing of the canines leads to interesting wear patterns. The upper canine is honed against the posterior edge of the lower canine and the anterior edge of the most anterior premolar. In the older males the posterior one-third to one-half of the lower canine may be worn away, although a lingual posterior part of the tooth may remain, making the lower canines almost bicuspid. Canine wear doesn't appear to correlate well with molar wear, suggesting that some animals grind their canines more than other animals do.

Using these progressions of eruption and wear, we ranked 22 males from old to young. (They are as follows: 30, 33, 9, 5, 6, 36, 2, 35, 34, 10, 31, 7, 13, 15, 1, 3, 28, 12, 32, 8, 29, 4.) We can not assign chronological ages to these animals, but by analogy with the howlers of B.C.I., we suspect that they ranged from approximately 3 to 13 years of age. The first ten of these animals were considered to be fully adult because all teeth were occluding. The last five were juvenile males with their upper canines just breaking through the gum to approximately half-way erupted. With the addition of two more males of approximately the same weight, these last form an arbitrarily-defined group of juvenile males which may be compared with the adult males and adult females.

A series of four females with teeth fully erupted forms our small series of adult females. All of these were young adults with little wear on the molars. The buccal and lingual cusps of their molars were pitted but not grooved.

Body Size and Sexual Dimorphism

Table 1 present our mensural data for *Alouatta seniculus* and for comparison Table 2 presents similar data for *A. palliata* from Barro Colorado Island. These data suggest that *A. seniculus* is more dimorphic in size than *A. palliata*. The females of *A. seniculus* average 92 percent of the males in linear measurements and 69 percent in weight. The ratio of the cube roots of weights suggests that the females average 88% the size of the males overall in linear dimensions. In comparison, the females of *A. palliata* average 97 percent the size of the males in linear dimensions and 84 percent in weight. The ratio of the cube roots of weights suggests that overall females average 94 percent the size of the males in linear dimensions. Thus we conclude that *Alouatta seniculus* is more dimorphic in size than is *A. palliata* by approximately 5% in linear dimensions. A larger sample of female *A. seniculus* from the Guarico population is needed to test this conclusion, however.

Although the two species of *Alouatta* have approximately the same lengths of head and body, the weights of the *Alouatta palliata* from Panama are significantly higher than those of the *A. seniculus* from Guarico. The magnitude of the difference is demonstrated by the comparisons of the ponderal ratios, formed by dividing the cube root of the weight by the length of the head and body. Within each species, there is no significant difference between the ratios of males and females. Between the two species there is a large and significant difference. If the males of *A. seniculus* were as robust as the *palliata* males, they would average 2.1 kg or 33 percent heavier.

The relative tail length of *A. palliata* is greater than that of *A. seniculus*. For both sexes the difference between the species is 11 percent of the length of head and body. This may be a slight over-estimate, due to the fact that tail length was measured in slightly different ways on the two samples, but we believe the difference between the species is real and biologically significant. In both species the relative tail length is greater for females than it is for males by 4–5 percent.

Supernumerary Nipples

We noticed that male number 23 had supernumerary nipples. We looked closely at all the rest of the animals and found supernumerary nipples in 5 of the remaining 13 animals. Thus our best estimate is that approximately 40 percent of the males have supernumerary nipples, and that we overlooked the condition in 6 animals before we became alert to it. The normal position of the nipple in males is in the axilla just slightly posterior to the posterior border of the pectoralis muscle. In this area they are conspicuous because of the lack of guard hairs in the axilla. The most common location for the extra nipple was anterior to the normal position, among the hair on the skin covering the pectoralis muscle. On some animals the supernumerary nipples were inconspicuous while in others they were obvious and could be seen with binoculars on an animal in the trees. One animal had two nipples in the axilla and one on the pectoral muscle. Most animals were bilaterally symetrical for presence or absence of supernumerary nipples. This suggests that there is a genetic basis for the condition, which suggests in turn that the character might prove useful in helping to determine paternities and geneologies.

Other Observations

Most of the *Alouatta seniculus* were heavily infected with lice. The nits were common and conspicuous in the fringing hairs along the abdomen. The adults were found in the heavy hairs of the beard. The lice were identified as *Pediculus mjöbergi* Farris 1916 (= *P. lobatus* Fehrenholz 1916). In contrast, the *Alouatta* of Barro Colorado Island appear to be free of lice. Thorington and others have carefully examined many of the more than 100 *A. palliata* captured on B.C.I. without finding any nits or adults.

There is a striking difference between *Alouatta seniculus* and *A. palliata* in the secondary sexual characters of juvenile animals. In *A. seniculus*, it is possible to distinguish males from females by visual inspection very shortly after their birth. Neville (1972, 1976) and Rudran (this volume) were able to sex the infants in their study troops by examining them through binoculars or a spotting scope. The testes descend early and the penis is obvious. The only possible problem would be the misidentification of females because the clitoris of the juvenile is protrudant and the labia bear some resemblance to a scrotum. In contrast, it is virtually impossible to distinguish male from female *Alouatta palliata* before puberty without manipulating the genitalia. This was noted by Carpenter (1934) and has been a source of annoyance in all censuses conducted on B.C.I. since then. The genitalia of juvenile males mimic the genitalia of females until the testes descend and the scrotum becomes evident.

Discussion

The most striking thing about *Alouatta seniculus* to anyone who has worked with another species of howler monkey is its golden red color. The significance of interspecific variation in the coat colors of howlers has not yet been studied, but it must have a significant effect on the absorption of radiation from the sun and sky. As with other mammals, there is surely no effect of coat color on the radiative loss of heat from the animal to his environment. Thus it is probable that *A. seniculus* with a pale red coat is better adapted in this respect to areas of high insolation, in which the animals face serious problems of overheating in direct sunlight. In contrast, the black howler monkeys, like *A. pigra, A. palliata* and *A. belzebul*, are better designed for sunning themselves and absorbing solar radiation—either to warm their bodies or dry their hair or both. This argument, together with data on the distributions of the species, suggests that *A. pigra* and *A. palliata* are basically adapted to the rain forest environments of Central America, as is *A. belzebul* to the rainforest of the Amazon. *A. seniculus* is basically adapted to the more savanna habitats of northern South America. *A. fusca*, the brown howler, is an animal of the Brazilian coastal forest, while the dichromatic *A. caraya* seems to be more adapted to the drier habitats of Argentina and southern Brazil. It may be significant that this dichromatic species of howlers is subjected to the greatest seasonal changes of temperature. Thus the relative advantages of being a dark color or a light color vary seasonally. Within each species, the geographic pattern of coat color variation appears also to parallel climatic variables. For example, where *A. seniculus* occurs in heavily forested areas, it is a much darker shade of red than it is on the Venezuelan llano. Thus it seems probable to us that a combination of environmental factors determines whether it is more advantageous for howlers in a particular area to be darker so that they can absorb solar radiation better or to be paler so that they are not subject to overheating in bright sunlight. Where we have studied *Alouatta*, in Panama, Venezuela, Colombia, Brazil, and Argentina, the correlations between coat color and climate have seemed striking. Therefore we contend that the interspecific and intra-specific variations in coat color of howler monkeys are best explained as adaptations to the thermal environments of the animals, rather than by the competing hypothesis of metachromism (Hershkovitz, 1977). Certainly our hypothesis is more easily tested—by comparative studies of the activity budgets of the different species of *Alouatta* under similar thermal circumstances or by experimental introductions—than is the hypothesis of metachromism.

Alouatta seniculus has the largest hyoid of all species of howler monkeys (Hershkovitz, 1949). Contrary to Hershkovitz (1969, p. 20) the major effect of the large hyoid is to make the howls deeper pitched, not louder. In contrast with the vocalizations of *A. palliata*, the roars of *A. seniculus* are lower pitched and more bubbly in quality. The howls of a distant troop of *A. palliata* sound rather like the cheers of a crowd in a football stadium, whereas the howls of *A. seniculus* are much like the sound of surf on a distant shore or a roaring distant wind. As noted and illustrated by Hershkovitz (1949, pp. 395, 397 and 399), the enlargement of the hyoid is most extreme in the males. Thus *A. seniculus* exhibits not only the largest hyoids, but also the most sexual dimorphism in this character. Since the enlarged hyoid is obviously the derived state, it can clearly be argued that this aspect of the sexual dimorphism results from special selection on the males, not the females. Less convincingly, it can be argued that the same is true for the sexual dichromatism of *A. caraya*, since it is the males which moult at maturity and change to dark-coated animals, whereas the females retain the pale pelage of the juveniles. It is probably also true for the dimorphism in body size and other characters, but it is less compelling logically to argue that there has been selection for larger males rather than for smaller females. The existence of combat between males does, however, provide a reasonable mechanism for selection of larger body size.

Estimates of the degree of sexual dimorphism in body size in *Alouatta seniculus* and *A. palliata* are given in Tables 1 and 2. The degree of sexual dimorphism of *A. caraya* can be estimated from the data given by Stahl et al (1968). In their Tables 1 and 2 they list measurements of males and females. However, they define their classes of animals by weight. Class C animals range from 4 to 6.5 kg and class D animals (males only) range from 6.6 to 9.9 kg. They suggest that sexual maturity is reached by both sexes at approximately 4 kg. With some trepidation we combined classes C and D of males and compared them with class C of the females. We suspect that this underestimates the degree of sexual dimorphism, because we think that more immature males are included than are immature females. However, the female to male ratios are 0.74 for body weight, 0.92 for length of head and body, and 0.91 for length of hind foot. The cube root of the weight ratio is 0.91, which provides an overall estimate for the ratio of linear measurements. In Table 11 of Stahl et al (1968), lens weight and body weight are compared for both sexes, and the data are separated into three postpubertal classes. If lens weight correlates well with age, and there are no differences between the sexes in this measure of age, then this table enables us to compare the weights of like-aged males and females. The female/male ratio of postpubertal class 1 is 0.76, which we would anticipate to be an underestimate of the degree of dimorphism. The ratio for class 2 is 0.65 and that for class 3 is 0.71. These are probably our best estimates of the sexual dimorphism of full adult *Alouatta caraya.* They correspond to ratios of linear measurements of approximately 0.88 for a total sample size of 33 males and 19 females. Thus we conclude that in *A. caraya*, adult females weigh approximately 68 percent as much as the males, in *A. seniculus* they weight 69 percent as much as the males, but in *A. palliata* they weight 84 percent as much as the males. Therefore, *A. palliata* is only half as dimorphic in weight as the other two species.

It is perhaps in the context of adult dimorphism that we should view the interspecific differences in juvenile dimorphism. If the adult dimorphism results from competition between males, as seems likely, if we are dealing with age-graded hierarchies among the males, as proposed by Eisenberg et al (1972), and if a male's status in the hierarchy influences his ability to sire offspring, then it is reasonable to hypothesize that the hierarchy established among juvenile males is an

Table 1. Measurements of *Alouatta seniculus*

Measurement	*Means and standard deviations*			*Ratio Female/ male*
	Adult males (N = 10)	*Adult females (N = 4)*	*Juvenile males (N = 7)*	
Weight (kg)	6.5 (0.6)	4.5 (0.2)	4.1 (0.1)	.69
Total length[1]	115.4 (4.4)	105.9 (2.5)	106.6 (0.1)	.92
Tail length	63.0 (3.3)	59.0 (1.1)	60.0 (3.0)	.94
Head and body	52.3 (2.0)	46.8 (2.2)	46.6 (2.2)	.90
Femur length	17.6 (0.6)	16.4 (0.2)	16.3 (0.6)	.93
Knee height 1	19.2 (0.6)	17.3 (0.5)	17.5 (0.4)	.90
Hind foot	14.6 (0.3)	13.7 (0.3)	14.1 (1.6)	.94
Upper arm 1	16.6 (0.4)	15.7 (0.6)	15.0 (0.4)	.94
Ulna	17.0 (0.3)	15.4 (0.4)	15.3 (0.4)	.90
Elbow to finger	27.8 (0.6)	24.5 (0.5)	26.0 (0.5)	.91
Average for linear measurements				.92
$kg^{1/3}$/H+B)	.0356 (.0013)	.0353 (.0016)	.0344 (.0016)	.99
Tail/(H+B)	1.20 (.07)	1.26 (.07)	1.29 (.07)	1.05

[1] All length measurements are in cm.

Table 2. Measurements of *Alouatta palliata*

Measurement	*Means and standard deviations*		*Ratio Female/ male*
	Adult males (N = 15)	*Adult females (N = 15)*	
Weight (kg)	7.8 (0.9)	6.6 (0.7)	.84
Tail length[1]	66.5 (3.2)	65.1 (2.9)	.98
Head and body	50.6 (2.3)	47.5 (1.4)	.94
Femur length	16.4 (0.7)	15.7 (0.4)	.96
Knee height 2	16.0 (0.8)	15.7 (0.5)	.98
Hind foot	14.4 (0.6)	13.7 (0.5)	.95
Upper arm 2	16.6 (0.7)	16.5 (0.6)	.99
Forearm	17.3 (0.5)	16.6 (0.5)	.96
Average for linear measurements			.97
$kg^{1/3}$/(H+B)	.0392 (.0012)	.0394 (.0014)	1.01
Tail/(H+B)	1.31 (.05)	1.37 (.07)	1.04

[1] All length measurements are in cm.

important determinant of subsequent reproductive success. The precocial development of secondary sexual characters could be important in determining a juvenile male's status, and thus it could be subjected to natural selection in howler monkey social systems. This framework provides a possible mechanism for the selection of the conditions found in *Alouatta seniculus*, but of course, it is not relevant to *A. palliata*, in which the genitalia of the juvenile males so closely mimic those of the females, nor to *A. caraya*, in which the juvenile males retain the coat color of the females until they reach maturity. These two cases accord much better with the hypothesis that young males are avoiding the aggressions of the adult males until they reach sexual maturity, by resembling the females. However, we do not have enough information to assess which condition is primitive and which is derived within *Alouatta*, or to determine if both are derived. If we did, it would be obvious that we should focus our attention on the derived condition. Nor do we have observational data that suggest that there are interspecific differences in social structure or dynamics which accord with these ideas. In fact, the observational data available do not test any of these ideas very adequately, and it remains for future investigators to formulate these and other hypotheses sufficiently carefully that they lead to testable predictions.

The demographic structure of *Alouatta* populations is poorly known, for usually investigators have no measure of the ages of the adult animals. In our sample of 36 individuals, there were no very old animals, as best we can judge from tooth wear. The oldest animal had tooth wear fitting it in adult class D-2 of Pope (1968a, p. 20). In Pope's analysis of a sample of *Alouatta caraya*, fewer than 5% of the adult males had more tooth wear than class D-2, thus it is not surprising or significant that none of our sample of 32 males was classified as D-3 or D-4. Rosenthal (1968) used strontium-90 levels to determine the age of the animals in the same sample of *A. caraya*, and reported that no animals exceeded 12.5 years of age. The average of the estimated ages of three males of *Alouatta caraya* in the D-2 class was 8 years, and this might be taken as an estimate of the age of the oldest *A. seniculus* in our sample. We have difficulty believing this however. Froehlich et al (m.s.) have scored the tooth wear of 126 *Alouatta palliata* captured on Barro Colorado Island. Some of the animals were recaptured after 1 to 5 years and the tooth wear was scored a second time. According to their calculations, the average animal with tooth wear in the D-4 class will be on the order of 20 years of age, and an animal in tooth-wear class D-2 like the oldest in our sample of *A. seniculus* would be 12 or 13 years of age. The differences between these estimates of ages will only be resolved by more work with marked animals. Not till then will we be able to discern differences in the demographic structures of populations in different habitats, and perhaps demographic differences between species.

The difference in the robustness of *Alouatta palliata* from Barro Colorado Island and this sample of *A. seniculus* from Guarico is striking. From the data provided by Stahl et al (1968), several values of our ponderal index can be calculated for *A. caraya*. From the regression equations given in their Table 7, the average ratio for males is 0.0362 and that for females is 0.0356, while the value calculated for adult non-pregnant females from their Table 4 is 0.0358. These averages are not significantly different from those for *A. seniculus*. The obvious question raised by these data is whether this is a consistent difference between *A. palliata* on the one hand and *A. seniculus* and *A. caraya* on the other, or whether there is a seasonal difference in the weights of the last two species. Our data on *A. seniculus* were collected in the middle of the dry season, the *A. caraya* specimens were collected at the beginning of the dry season in Argentina, and the data on *A.* palliata were collected both at the beginning and the end of the dry season. Seemingly, there is comparability in the time of collection, but the dry season on Barro Colorado Island is scarcely comparable with the dry seasons of the Venezuelan llano or those of Northern Argentina. Thus the critical question cannot be answered without data on the robustness of *A. caraya* or *A. seniculus* in the wet season. We suspect that there is some seasonal change in the weight of these animals, but we doubt that they ever average as heavy as do *A. palliata* of the same body length on Barrow Colorado Island.

A small part of the difference probably lies in the relative tail lengths of the different animals. *A. palliata* has the longest relative tail length, *A. seniculus* has a relatively shorter tail, and *A. caraya* has the shortest. (Stahl et al, 1968, p. 66, list means ranging from 106 percent to 110 percent of body length for males and from 110 percent to 118 percent for females.) Hoping that the measurements are reasonably comparable, we note that the longest relative tail length is found in a sample of *A. palliata* which is found in continuous forest. These animals rarely come to the ground, in contrast with the *A. seniculus* of Guarico and probably the *A. caraya* of Argentina. For animals which always transfer from tree to tree across overlapping branches, frequently 30 m above the ground, there may be a special significance in having a slightly longer prehensile tail. Relative tail length may prove to be a good measure of the degree to which the different species of *Alouatta* are adapted to continuous rain forest or to a savanna type of environment.

It is also notable that in all species the females have relatively longer tails than do the males. The significance of the difference is not always overwhelming, but it is consistent in the five samples of adults of these three species and it is definitely significant in some of them. Two explanations come readily to mind. First, the females frequently allow their infants to cross over them while they hold the branches of one tree with their hands and the branches of the tree behind with their feet and tail, or sometimes just their tail alone. Such behavior could reasonably be considered as leading to more selection for long tails among females than among males. The second explanation is that the relatively longer tail of a female compensates for sexual dimorphism. Both the males and the females cross between trees at the same points. Since the males are larger, their bodies are slightly longer than are those of the females. The relatively longer tail of the female does not completely compensate so as to make the total length of the animal the same for both sexes, but one might guess that it would help to make the females as adept at spanning the distance between trees or as efficient at foraging as are the larger males. Of course, these two explanations are no in conflict with one another, nor are they posed here as hypotheses with testable predictions. However, we think they can be, and in view of the morphological data, we think they should be in the future.

The contrast between the louse infestations of *A. seniculus* and the absence of lice on *A. palliata* is surprising. Pope (1968*b*) noted that there was a high incidence of louse parasitism in *A. caraya* as well. It would be interesting to know if this is a contrast between the species, if louse infestations are characteristic of all howlers in drier habitats, or if there is some other explanation of the absence of lice on the howlers of Barro Colorado Island. Our sampling was not adequate to test whether fewer females than males have lice in *A. seniculus* as Pope found in *A. caraya.*

Appendix

Definition of external measurements taken on Alouatta

Total length was measured from the tip of tail to the most anterior point on head. It approximates but does not equal the sum of the next two measurements.

Tail length was measured from the base, as close as could be estimated to the joint between the sacrum and the first caudal vertebrae, to the tip. The tail was extended straight out behind the animal, not perpendicular to the axis of the body as is usually done with smaller mammals and which yields a slighlty shorter length measurement. On *A. palliata* the tail was measured along its dorsum without the tail being completely straightened out which gives a slightly longer measurement than when curl of the tail is completely straightened out as was done for *A. seniculus*.

Length of head and body was measured from the most anterior point on the head to the most posterior part of the body over the ischium.

Femur length was measured laterally from the greater trochanter to the lateral condyle. This measurement will excede the actual length of the femur by the thicknesses of muscle and skin overlying the bony landmarks.

Knee height-1 was measured from the top of the knee, including the distal end of the femur, to the heel.

Knee height-2 was measured from the proximal end of the fibula, excluding the distal end of the femur, to the heel.

Hind foot was measured from the heel to the end of the longest toe, excluding the nail.

Upper arm 1 was measured from the top of the shoulder, i.e. including the acromion, to the posterior surface of the ulna directly distal to the end of the humerus.

Upper arm 2 was measured from the proximal end of the humerus, i.e. excluding the acromion, to the same point on the ulna as in 1.

Ulna was measured from olecranon to ulna styloid process.

Forearm was measured from olecranon to radial styloid process.

Elbow to finger was measured from olecranon to end of longest finger, excluding the nail.

Literature Cited

Carpenter, C. R.

1934. A field study of the behavior and social relations of howling monkehs (*Alouatta palliata*). *Comp. Psychol. Monogr.*, 10:1–168.

Eisenberg, J. F., N. A. Muckenhirn and R. Rudran.

1972. The relation between ecology and social structure in primates. *Science*, 176:863–874.

Froehlich, J. W., R. W. Thorington, Jr. and J. Otis.

Ms. The demography of *Alouatta palliata* on Barro Colorado Island, Panama Canal Zone.

Hershkovitz, P.

1949. Mammals of northern Colombia. Preliminary report No. 4: Monkeys (Primates), with taxonomic revisions of some forms. *Proc. U. S. Nat. Hist. Mus.*, 98:323–427.

1969. The evolution of mammals on southern continents. VI. The recent mammals of the neotropical region: A zoogeographic and ecological review. *Quart. Rev. Biol.*, 44:1–70.

1977. *Living New World Monkeys (Platyrrhini), with an Introcution to Primates, Vol. 1.* Chicago: University of Chicago Press.

Lazarus, A. B. and F. P. Rowe.

1975. Freeze-marking rodents with a pressurized refrigerant. *Mammal Rev.*, 5:31–34.

Napier, J. R. and P. H. Napier.

1967. *A Handbook of Living Primates.* London: Academic Press.

Neville, M. K.

1972. The population structure of red howler monkeys (*Alouatta seniculus*) in Trinidad and Venezuela. *Folia Primat.*, 17:56–86.

1976. The population and conservation of howler monkeys in Venezuela and Trinidad. Pages 101–109 in *Neotropical Primates, Field Studies and Conservation*, edited by R. W. Thorington, Jr. and P. G. Heltne. Washington, D.C.: National Academy of Sciences.

Pope, B. L.

1968a. Population characteristics. Pages 13–20 in *Biology of the Howler Monkey (Alouatta caraya)*, edited by M. R. Malinow. Basel/New York: S. Karger.

1968b. Parasites. 204–208 in: *Biology of the Howler Monkey (Alouatta caraya)*, edited by M. R. Malinow. Basel/New York: S. Karger.

Rosenthal, H. L.

1968. Chronological age determination as estimated from Strontium-90 content of teeth and bone. Pages 48–58 in *Biology of the Howler Monkey (Alouatta caraya)*, edited by M. R. Malinow. Basel/New York: S. Karger.

Rudran, R.

1979. The demography and social mobility of a red howler (*Alouatta seniculus*) population in Venezuela. Pages 107–126 in *Vertebrate Ecology in the Northern Neotropics*, edited by John F. Eisenberg. Washington, D.C.: Smithsonian Institution Press.

Scott, N. J., A. F. Scott and L. A. Malmgren.

1976. Capturing and marking howler monkeys for field behavioral studies. *Primates*, 17:527–533.

Stahl, W. R., M. R. Malinow, C. A. Maruffo, B. L. Pope and R. Depaoli.

1968. Growth and age estimation of howler monkeys. Pages 59–80 in *Biology of the Howler Monkey (Alouatta caraya)*, edited by M. R. Malinow. Basel/New York: S. Karger.

R. RUDRAN
National Zoological Park
Washington, D. C. 20008

The Demography and Social Mobility of a Red Howler (*Alouatta seniculus*) Population in Venezuela

ABSTRACT

Twenty bisexual groups of red howlers comprising between 169 and 186 individuals were regularly censused over a period of 18 months. Within this population, immature individuals slightly outnumbered adults and there were more or less equal numbers of both sexes. However, the sex ratio favored females in the adult age class, and males in the immature age class. This was probably due to the faster rate of maturation of young females combined with increased mortality of adult males resulting from inter-male aggression.

The mean group size was 8.9 (range 4–17). The composition of groups changed frequently due to births, deaths, immigrations, emigrations and maturation of individuals. Births and immigrations were slightly greater than deaths and emigrations. Therefore, the population increased at the rate of 1.5 percent per annum during the study. In this population, the most frequent cause of mortality was infanticide. While the primary function of infanticide was probably the elimination of food competitors who do not benefit the infanticidal individual or its offspring in any way, this behavior pattern plays an important secondary role in regulating the growth of the population. It is predicted that the gradually increasing population will reach maximal numbers and then face a period of decline, largely due to the effects of infanticide. This prediction of a decline is supported by evidence collected from a population living a few kilometers east of the study area.

Males as well as females which belong to

RESÚMEN

Se hizo un censo por 18 meses en veinte grupos de ambos sexo de monos aulladores rojos. Había 169 a 186 individuos en esta poblacion. Había más individuos jóvenes que adultos en un número más o menos igual de ambas sexos. Sin embargo, en adultos, había más hembras que machos, y había más machos entre los jóvenes. La razón para esta diferencia es el resultado de un desarrollo más rapido en hembras jóvenes y una mortalidad mayor de machos adultos resutantes de agresiones de macho a macho.

El tamanõ medio de los grupos era de 8.9 (rango 4–17). La compasición de los grupos cambió frecuentamente a causa de nacimientos, muertes, inmigraciónes, emigraciónes y desarrollo de los individuos. Los nacimientos y inmigraciones fue un poco mas grande que las muertes y las emigraciones. Por eso, la población creció una velocidad de 1.5 percent anual durante este estudio. En esta población, la mayor causa de mortalidad fué el infanticidio. La función primaria de infanticidio fué, problemente, la eliminación de competidores de alimentos, que no benefician al individuos que asesin o a la cría. Además este comportamiento tiene una importante función en la regulación del crecimiento de la población. Se predice que la población crecerá hasta alcanzar cifra máxima y después declina, principalmente debido al asesinato de infantes. Una población observada a pocos kilometros al oriente de esta area de estudio muestra este fenómeno y apoya esta predicción.

Machos y hembras jóvenes o viejos que viviran

medium sized juvenile or older age classes lived outside established bisexual groups. They were contacted as solitaries or in associations of two, three or four individuals. Associations of three animals or less and solitary life appeared to be more or less temporary modes of existence. Associations of four animals seemed to be somewhat more permanent and it is possible that under certain conditions they may form stable bisexual groups. Some extra group individuals of both sexes entered bisexual groups as adults or when nearing sexual maturity.

The ecological density within the study area was 150 animals/km^2. This yielded a biomass estimate of 622 kg/km^2. Adults of both sexes contributed nearly equal proportions to the biomass and their combined weights accounted for nearly two-thirds of the total biomass.

afuera establecieron grupos de ambos sexos y existen como solitarias o en grupos de dos, tres or cuatro. Estos grupos, con excepción del última, eran más a menos transitorios. Los grupos de cuatro eran más permanentes y es posible que este haya sido un grupo estable en ambos sexos. También había individuos adultos o cercanos a la madurez sexual que se integraron a grupos de ambos sexo.

La densidad ecológica de este sitio fue de 150 animales/km^2 con una biomasa de 622 kg/km^2. Los adultos de ambos sexos contribuyeron en una proporción más o menos equivalente a la biomasa y sus pesos combinados son cerca de dos tercios de la biomasa total.

Introduction

In this paper I present demographic data collected between June 1976 and April 1978 from a population of red howlers (*Alouatta seniculus*) in Hato Masaguaral, Guarico State, Venezuela. Neville (1972, 1976) collected similar data from the same population in 1969–1970 and in 1972. Comparisons are made with Neville's findings as well as with demographic data collected from other *Alouatta* species. I also report the first observed case of infanticide in any New World Primate species and describe certain aspects of inter-group migrations (or Social Mobility) of male as well as female red howlers.

The study area was located in the seasonally inundated "bajo llano" or low plains (Blydenstein, 1962; Troth, this volume) about 45 km to the south of Calabozo town in Guarico State. The habitat comprised of a mosaic of savannas which were either treeless or included palm-fig associations, and patches of short statured trees (matas) growing to about 6 m. Lianas sometimes inter-connected these matas and the larger matas included some trees which were taller than 6 m and infrequently reached heights of about 20 m. The lianas served as arboreal pathways between patches of woody vegetation but the monkeys also came down to the ground during group movements. In this rather "open" habitat, visibility of monkeys was good particularly during the dry season, and several groups could be approached to within 10 m.

The dry season from December to May alternated with a wet season from June to November when extensive areas of the study site became flooded. The flooding, however, was less than about 60 cm and all parts of the study site were accessible on foot throughout the year. The seasonal distribution of rain had a dramatic effect on the vegetation. More detailed descriptions of the study area can be found in Neville (1972) and Troth (this volume).

Methods

Usually during the first week of each month of the study, I conducted a census of the bisexual groups comprising the study population. I located these groups either by their howls or by searching out different parts of the study area. In these censuses I also contacted individuals living outside bisexual groups and groups living adjacent to the study area. The size of a group was noted after three to five consistent counts during each period of observation. Group composition was recorded by classifying individuals into infants, small, medium or large-sized juveniles, and into subadults and adults. Categories intermediate to the above major classes were also used whenever necessary. Physical characteristics such as body size, color of pelage and naked parts of infants, size and shape of nipples and genitalia of females, size of throat and beard of males and behavioral criteria, such as locomotor abilities, propensity to play and join howling bouts were important in the age classification of animals.

I considered individuals who were 10 months old or less as infants. Females between 10 months and 3 years, and males up to 3½ or 4 years of age were classified as juveniles. My study was not conducted over a sufficiently long period to enable reasonable estimates of the duration of older age classes.

During the initial stages of the study, it was possible to recognize a few individuals through natural scars and deformities. Later 36 animals (about 20 percent of the study population) were marked with ear tags which were coded according to their position and color (see Thorington, Rudran and Mack, this volume). This made individual recognition easier and greatly enhanced the collection of unambiguous data from the groups. Individual recognition, the size of groups, their composition and the location of sightings were all used in group identification.

Results

Much of the data presented below were collected during 18 monthly censuses conducted in June and July 1976 and from January 1977 to April 1978. Some data collected between May and October 1978 were also used.

The Census and the Study Population

During the monthly censuses I obtained accurate group counts from 28 bisexual groups living within and adjacent to the study area. Twenty of these groups were regularly censused and formed the study population. Ten study groups were observed during all 18 monthly censuses mentioned above. Seven study groups were censused through 16 months and one group each was censused for 13, 14 and 15 months respectively. Only two study groups were not contacted on more than two consecutive censuses after their first month of observations.

Between June 1976 and April 1978, the study population fluctuated between 186 and 169 individuals ($\bar{X}$ = 176.9 individuals). The population was relatively stable during the study and I present certain population parameters from the terminal census of April 1978 in Table 1. Immatures slightly outnumbered adults, and there were more or less equal numbers of males and females. However, in the adult class, the sex ratio was biased in favor of females, while in

Table 1. Population parameters of red howlers derived from twenty bisexual groups regularly censused over eighteen months

From terminal census (April '78) of bisexual groups	
Total population size	178
a. Adult to immatures	82:96 (1:1.17)
b. Males to females	86:92 (1:1.07)
c. Adult males to females	34:50 (1:1.47)
d. Subadult males to females	9:3 (1:0.33)
e. Large juvenile males to females	14:4 (1:0.29)
f. Medium juvenile males to females	13:10 (1:0.77)
g. Small juvenile males to females	13:13 (1:1)
h. Infant males to females	3:12 (1:4)
i. One male to multi-male groups	8:12
Mean (range) bisexual group size	8.90 (5–14)
% adult males:adult females:subadult:juveniles: infants	19:28:7:38:8
Other statistics from bisexual groups	
Population size	
a. Monthly fluctuation	169.0–186.0
b. Monthly mean	176.9
Population density	
a. Ecological density No./sq.km.	150.0
Biomass	
a. Biomass based on above kg./sq.km.	622.0
b. % contribution of adults	69.1
c. % contribution of immatures	30.9
d. Monthly fluctuations in kg.	665.0–703.9
Population increase	
a. % net rate (per annum)	1.5
b. % Immigration (18 months)	4.5
c. No. of births (19 months)	40.0
Infant mortality	
a. % mortality (19 months)	25.0
b. % of mortality due to infanticide (19 months)	40.0
Statistics from extra groups individuals (social mobility)	
Extra groups individuals	
a. Mean no./months	2.6
b. % of mean total population	1.4
c. Mean monthly contribution biomass/kg	14.0
d. Ratio of males:females	1:0.58

the subadult class, it favored males. Neville (1972) found the same trend in his earlier study. Among juveniles there were equal numbers of both sexes in the small juvenile class, but the sex ratio became progressively more biased in favor of males in the two older juvenile classes.

As in Neville's study (1972), infant females greatly outnumbered infant males (Table 1). In both studies, however, the unequal sex ratio at birth was probably the result of small sample sizes. A larger sample of 40 infants born into the study population from January 1977 through April 1978 included 17 males, 16 females, and 7 infants which disappeared before they were sexed. Also, among the infants seen in all groups that I contacted, males and females numbered 23 and 27 respectively, giving a birth sex-ratio of more or less 1:1.

Bisexual Group Sizes and Compositions

The mean group size computed from terminal counts of the 20 bisexual groups comprising the study population was 8.90 (Table 1). On the basis of initial and terminal counts of these groups, the mean group size was 9.07 (Table 2). On the same basis, Neville (1972) arrived at a mean group size of 8.46. Using Neville's home range maps and my knowledge of red howler home ranges, I was able to unambiguously identify 14 bisexual groups observed by Neville (Table 3). Comparison of terminal counts of these groups in the two studies showed no significant difference in group size (Wilcoxon signed ranks test $p < 0.05$, Snedecor and Cochran, 1969). Thus, group sizes in this population appear to have been stable during the eight years between the two studies.

On the average, there were 1.65 adult males to 2.68 adult females in each group (Table 2). This is in good agreement with Neville's data (1972). However, our findings differ in the mean number of individuals in each immature age class. Apparently I classified fewer numbers of individuals as infants and subadults and a larger number as juveniles than Neville did. This was probably because, according to my age classification, the duration of the infant stage was shorter, and that of the juvenile stage longer than according to Neville's classification (see Neville, 1972, 1976).

Terminal counts of regularly censused groups show that eight bisexual groups included one adult male each, while 12 groups contained two or three adult males. Thus in this population there was a slight preponderance of multi-male social groups. During the terminal census, eight groups contained subadult males and three included subadult females. All groups containing juveniles and infants were found in ten groups (Table 2).

Table 2. Initial and terminal counts and compositions of bisexual groups of red howlers*

Study population group	*Adults*		*Subadults*		*Juveniles 3*		*Juveniles 2*		*Juveniles 1*		*Infants*		*Total*
	M	*F*	*M*	*F*	*M*	*F*	*M*	*F*	*M*	*F*	*M*	*F*	
I	2	3	—	—	—	—	—	—	—	1	—	—	6
	2	3	—	—	—	—	—	1	2	1	—	—	9
III	3	3	1	—	1	1	—	—	—	1	—	—	10
	1	2	1	1	1	—	—	1	—	—	—	2	9
IV	1	2	—	—	1	—	1	1	—	—	—	—	6
	2	2	1	—	—	—	1	1	1	—	—	—	8
V	3	2	1	—	—	1	—	—	1	—	—	—	8
	2	3	—	—	1	—	1	—	—	—	—	3	10
Tag	2	2	2	—	1	—	—	1	1	—	—	—	9
	2	2	—	—	—	—	1	—	—	—	—	1	6
NC	1	4	—	—	2	—	2	—	—	—	2	1	12
	2	4	1	—	2	—	2	—	1	1	—	—	13
Win	2	2	1	—	—	—	2	—	—	—	—	—	7
	3	2	—	—	1	—	1	—	1	—	—	2	10
Albizia A	1	3	—	—	—	—	—	1	—	2	—	—	7
	2	3	—	—	—	—	1	1	—	—	—	1	8
Albizia B	1	3	—	1	1	—	2	1	—	—	—	2	11
	1	3	1	—	2	1	—	1	—	1	1	—	11
Albizia C	2	3	—	—	—	—	—	—	2	—	—	—	7
	2	2	—	—	—	—	2	—	—	—	—	—	6
Guasimo C	1	2	—	—	1	1	1	—	2	—	—	1	9
	1	1	1	1	1	—	—	1	—	1	1	—	8
Guasimo D	1	2	—	1	1	1	1	—	1	—	1	—	9
	1	2	1	1	1	—	2	—	—	—	1	—	9
S	1	3	—	—	1	—	—	—	—	—	—	—	5
	2	3	—	—	1	—	—	—	—	—	—	1	7
L	2	5	1	—	1	—	1	1	1	1	—	1	14
	3	4	—	—	1	1	—	1	1	2	—	—	13
BNS	2	1	1	—	—	1	1	—	1	1	—	—	9
	1	2	—	—	1	—	—	1	—	1	—	1	7
NBNS	2	2	1	1	1	—	1	—	—	—	1	1	10
	2	2	—	—	1	—	—	—	1	1	—	—	7
Trailer	1	2	1	1	1	—	—	—	1	1	—	—	8
	1	2	—	—	1	—	—	1	1	1	—	—	7
TIF	2	3	—	—	—	—	—	—	—	—	1	lu 1	8
	1	2	—	—	—	—	—	—	1	1	—	—	5
Stiff-toe	1	3	—	—	1	—	1	—	1	2	—	—	9
	1	3	2	—	—	1	1	1	1	1	—	—	11

* Terminal counts are as of 30 April, 1978.
M = male, F = female, u = unsexed, Juvenile 3 = large juvenile, Juvenile 2 = medium sized juvenile, Juvenile 1 = small juvenile

Study population group	Adults		Subadults		Juveniles						Infants		Total
					3		2		1				
	M	F	M	F	M	F	M	F	M	F	M	F	
Power line A	2	4	2	—	1	—	1	2	1	—	1	1	15
	2	3	1			1	1	—	3	2	—	1	14
X̄	1.65	2.68	0.53	0.18	0.70	0.23	0.68	0.48	0.63	0.56	0.25	0.50	9.07
Other groups													
NNC	2	3	1	—	1	—	1	—	—	1	2	—	11
	2	3	1	1	—	—	2	1	—	—	1	1	12
GNF	2	2	—	1	—	—	1	1	—	—	—	—	7
	2	3	—	—	1	1	—	—	—	—	—	2	9
SNG	2	2	—	—	—	—	—	—	—	—	—	—	4
	2	2	—	—	—	—	—	—	—	—	—	1	5
Guasimo A	1	3	1	—	—	—	1	1	3u	—	—	—	10
Guasimo B	2	3	1	—	2	—	2	1	1u	—	—	1u	14
Guasimo E	1	2	—	—	2	—	—	2	1	—	—	—	8
Pumphouse	1	3	—	—	1	—	1	—	—	—	—	1u	7
MR	1	2	2	—	—	—	—	—	1	1	—	—	7

Table 3. Size comparison of groups unambiguously identified as included in Neville's (1972) study and the present study

Group name		*Group count*			
Present study	*Neville's ('72)*	*Neville's ('72)*		*Present study*	
		Initial	*Terminal*	*Initial*	*Terminal*
Guasimo D	1	9	6	9	9
SNG	2a	9	7	4	5
Guasimo C	2b	9	10	9	8
Albizia A	3	7	7	7	8
Albizia B	4a	6	7	11	11
Albizia C	4b	7	8	7	6
I	7	6	6	7	9
III	8	12	12	10	9
Win	9	7	8	8	10
Tag	10	8	10	10	6
Stiff-toe	11	8	9	9	11
NBNS	13	8	9	10	7
V	18b	10	10	7	10
NC	21	8	8	12	13
Total	14	114	117	120	122

Changes in Group Compositions

All study groups underwent changes in composition due to maturation of individuals as well as to births, deaths, emigration and immigration. In computing the number of changes in composition (other than due to maturation) I regarded the individual rather than the observation as the unit of change. Thus, the disappearance of two individuals recorded during the same observation was scored as two changes. In computing the total duration during which these changes took place, I summed the months of study of individual groups *after* their first month of observation. For example, three groups observed over six months yield 15 group-months of observations when changes in group composition could have been detected. The number of changes in group composition are minimal estimates because some changes, such as the birth of an infant and its death, may have occurred between censuses and thus gone unrecorded.

The 20 study groups underwent a total of 111 changes in composition over a period of 314 group-months (Table 4). The number of changes in each study group varied from 2 to 17. Only one group, Group I, underwent changes solely through births (i.e.,

Table 4. Summary of changes in group composition (other than due to maturity) in 20 bisexual groups observed over 314 group-months from June 1976 to April 1978

Group	*Initial count*	*Change and age/sex class*														*Terminal count*	*No. of changes*
		Adult				*Subadult*				*Juvenile*				*Infant*			
		M		*F*		*M*		*F*		*M*		*F*					
		+	−	+	−	+	−	+	−	+	−	+	−	+	−		
I	6	—	—	—	—	—	—	—	—	—	—	—	—	3	—	9	3
III[1]	10	1	3	—	1	3	2	—	—	1	2	—	—	3	1	9	17
IV[1]	6	2	1	—	—	—	—	—	—	—	—	—	—	1	—	8	4
V[1]	8	1	2	—	—	—	—	—	—	—	—	—	—	3	—	10	6
Tag	9	—	—	—	—	—	3	—	1	—	1	—	—	2	—	6	7
NC[1]	12	1	—	—	—	—	—	—	—	—	—	—	—	1	1	13	3
Win	7	—	—	—	—	—	—	—	—	—	1	—	—	4	—	10	5
Albizia A[1]	7	1	—	—	—	—	—	—	—	—	—	—	1	1	—	8	3
Albizia B	11	—	—	—	—	—	—	—	1	—	—	—	—	1	—	11	2
Albizia C	7	—	—	—	1	—	—	—	—	—	—	—	—	2	2	6	5
Guasimo C	9	—	—	—	1	—	—	—	—	—	—	—	1	1	—	8	3
Guasimo D	9	—	—	—	—	—	—	—	—	—	1	—	—	1	—	9	2
S[1]	5	1	—	—	—	—	—	—	—	—	—	—	—	2	1	7	4
L[1]	14	—	—	—	1	2	2	—	—	—	2	—	1	4	1	13	13
BNS	9	—	**2**	—	—	—	—	—	1	—	—	—	1	2	—	7	6
NBNS[1]	10	**2**	**1**	—	1	—	**1 1**	—	—	—	1	—	—	1	1	7	9
Trailer	8	—	—	—	—	—	2	—	1	—	—	—	—	2	—	7	5
TIF	8	—	1	—	1	—	—	—	—	—	—	—	—	—	1	5	3
Stiff-toe	9	—	—	—	—	—	—	—	—	—	—	—	—	3	1	11	4
Power line A	15	—	—	—	1	—	1			—	2	—	—	3	—	14	7
																Total	111

M = male, F = female.

+ Indicates addition in group size.

− Indicates reduction in group size.

[1] These groups underwent male incursions during the study. However, male incursion into NBNS group was due to the death of this group's adult male during ear-tagging operations. Changes related to this death are given in bold type.

disregarding maturation of individuals). TIF Group lost an infant that was born before this group was contacted. Four changes in NBNS Group and two in BNS Group were unwittingly induced by the death of the adult male of NBNS Group during ear tagging operations (see Thorington, Rudran and Mack, this volume). Thus 105 natural changes occurred during the study, which yields a mean of 5.25 changes per group and a rate of one change per three group-months of observations.

Births accounted for a large proportion of the natural changes in group composition (Table 5). As emigrations out of the study area could not be monitored, they, together with presumed deaths, were classified as disappearances. These accounted for 38.1 percent of the changes. Emigrations within the study population and immigrations from outside the study area contributed 7.6 percent each, while immigration within the population and known deaths accounted for small, and more or less equal proportions to the changes in composition.

All infants which disappeared were dependent offspring and therefore must have died. Among juveniles and older individuals which disappeared, there were 20 males and 13 females. However, in the juvenile and subadult age classes, male disappearances out-num-

Table 5. The frequency of different types of natural changes in group composition (in the 20 Study Groups) between June 1976 and April 1978 inclusive

Nature of change	*Adult*		*Sub-adult*		*Juve-nile*		*Infant*		*Unsexed*	*Total*	*% Frequency*
	M	F	M	F	M	F	M	F			
Births	—	—	—	—	—	—	17	16	7	40	38.1
Immigration from outside study areas	3	—	5	—	—	—	—	—	—	8	7.6
Immigration within population	4	—	—	—	1[1]	—	—	—	—	5	4.8
Deaths	1	—	—	1	—	—	—	2	—	4	3.8
Disappearance	4	7	7	2	9	4	1	1	5	40	38.1
Emigration within study population	2	—	4	1	1[2]	—	—	—	—	8	7.6

[1] This immigration was preceded by a temporary emigration[2].
M = male, F = female.

bered female disappearances, while among adults the trend was reversed. These data do not allow a discrimination between rates of death and emigration, but one or both of these factors may have had differential effects on the two sexes in the juvenile and older age classes.

Immigration and Emigration

Between June 1976 and the terminal census of April 1978, a total of 13 individuals immigrated into my study groups. Eight of these animals originated from outside the study area. Immigration into the study population was therefore 4.5 percent of the average population. Immigrations and births accounted for an increase of 48 individuals, while deaths and disappearances amounted to 44 losses. In other words, the population increased by 2.26 percent of their average numbers during the study, or at the rate of 1.5 percent per annum.

All of the above mentioned immigrants were males. Immigrations of two females nearing sexual maturity were also recorded by 30 June 1978. Twelve of the immigrant males were either subadults or adults and one was a juvenile. The juvenile male's immigration actually resulted from a temporary emigration followed by a return to the original group. The 12 older males were responsible for seven incursions into bisexual groups (Table 4). Four incursions involved single adult males and the other three were due to simultaneous invasions by two adult males, two subadult males, and one adult with three subadult males, respectively. These incursions took place over 299 group-months of observations when these changes could have been detected. Thus, on the average, bisexual groups experienced male incursions once every 43.7 group-months of observations. In this computation, I excluded the duration of observation, and male incursions of NBNS Group because it took place immediately after the death of the dominant male of this group during ear tagging operations. At the time of the natural incursions, six groups contained an adult male each, while the seventh included two adult males. Therefore, one-male groups appeared to be more vulnerable to male incursions than multi-male groups.

The eight emigrations within the study area involved two adult males, four subadult males, a subadult female, and the juvenile male whose temporary emigration was described before. Of the other seven emigrants, two adult males and two subadult males eventually immigrated into bisexual study groups. The other two subadult males and the subadult female joined a solitary female in an association which in some ways resembled bisexual groups. More details of this association can be found below.

Individuals Living Outside Heterosexual Groups

Individuals which lived outside bisexual groups were located as solitaries, or in associations of two to four animals. The quartet associations, unlike the other associations, were stable over several months. For this reason and because quartet associations may have a special significance in the social system of red howlers, I have discussed them separately from the other types of associations.

I contacted some solitaries, pairs and trios only once, but others were seen several times. In the latter case, I occasionally located the same animals in different parts of the study area, and these animals usually stayed within the study area only for short periods. Pairs and trios frequently disbanded within a few months and sometimes individuals of these associations later joined other extra group animals. Thus, although I did not actively search and follow these individuals, they appeared to be extremely mobile transients. During the study I recorded 65 sightings of solitaries and 20 and 6 sightings of pairs and trios, respectively. These observations probably represented a minimum of 23 solitaries, 10 pairs and 3 sets of triplets and included individuals of both sexes from medium-sized juvenile and older age classes (Table 6). Large juvenile and older age classes were found as solitaries and represented in pairs and triplets as well. Medium-sized juveniles, however, were seen only in pairs. This suggests that the tendency to leave a bisexual group may occur at an earlier age, if immediate pair association is possible.

All extra group individuals designated as adult females were nulliparous and in some instances may not have reached full sexual maturity. Fully adult females or females with infants were not seen living outside bisexual groups. Among adult males, there were visibly old and sometimes scarred individuals as well as males in their prime. All three triplets were heterosexual and each contained an adult male, an adult or subadult female, and one other individual. Among the 10 pairs, however, only 3 were heterosexual. Also, pair associations contained several combinations of age/sex classes, but pairs of subadult males predominated. Three like-sex pairs (Nos. 1, 3 and 4) originated from my study groups and possibly the members of these pairs were related. Medium-sized juveniles were found only in two like-sex pairs of immature individuals. Thus, these were not parent-offspring pairs but they may have been related.

Table 6. Age/sex class of solitary, pairs, and triplets of red howlers seen between June 1976 and April 1978 inclusive

		Adult		*Subadult*		*Juveniles*						*Unsexed*
		M	F	M	F	3		2		1		
						M	F	M	F	M	F	
Solitary		6	2	7	5	1	1	—	—	—	—	1
Pairs	1	2	—	—	—	—	—	—	—	—	—	—
	2	1	1	—	—	—	—	—	—	—	—	—
	3	1	—	—	—	1	—	—	—	—	—	—
	4	—	—	2	—	—	—	—	—	—	—	—
	5	—	—	2	—	—	—	—	—	—	—	—
	6	—	—	2	—	—	—	—	—	—	—	—
	7	—	—	1	1	—	—	—	—	—	—	—
	8	—	—	1	1	—	—	—	—	—	—	—
	9	—	—	—	1	—	—	—	1	—	—	—
	10	—	—	—	—	1	—	1	—	—	—	—
Triplets	1	1	1	—	—	—	1	—	—	—	—	—
	2	1	2	—	—	—	—	—	—	—	—	—
	3	2	—	—	1	—	—	—	—	—	—	—

M = male, F = female.

Juveniles 1, 2, and 3 = small, medium, and large sized respectively.

Quartet Associations

I first contacted the quartet association (2 males and 2 females) mentioned previously on 14 August 1977. At this time all four individuals were nearing sexual maturity. Prior to the formation of this association, the two males (NBF and BF) were members of Trailer Group and one female (BRM) probably belonged to BNS Group (Table 4). The other female (YLU) was a solitary, first contacted on 26 March 1977. The members of the quartet used a well defined home range and both males copulated with YLU on several occasions until 26 January 1978. After this date female YLU began associating with the neighboring NBNS Group and by June 1978 she became a peripheral member of this group. From February 1978 through April 1978, the two males stayed together and frequently associated with female BRM. By early May 1978, however, their association began to disintegrate. On May 4, 1978, I saw NBF for the last time, and on this day he had nine or more deep gashes on his body and scrotum. It is possible that NBF succumbed to these injuries, and this may have precipitated the break-up of the association. BF was alone on the two occasions that I saw him in May 1978. BRM was once seen in a temporary association with an unidentified female nearing maturity and the adult male of Tag Group. Later (on 26 May 1978) BRM was found as a solitary about 1 km from the area which she had regularly used during the previous eight months.

Another association of four animals that I initially contacted on 5 March 1978 consisted of an old adult male (LM), two adult females and a large-sized juvenile female. By 16 May 1978, an adult male (RRU), who was previously ear-tagged as a solitary, entered this association after evicting male LM. The size and shape of the nipples and genitalia of the two older females in this association indicated that they were nulliparous. Hence, neither could have been the juvenile female's parent. Also these individuals ranged in a locality within the study area where I had not seen bisexual groups of red howlers before. These observations suggest that the four individuals belonged to a recently formed quartet rather than to a stable bisexual group. Their association persisted through October 1978.

Social Mobility

As indicated above, individuals of both sexes were socially mobile (i.e., migrated between bisexual groups), and when these individuals lived outside bisexual groups, their social status sometimes changed because of their associations with variable combinations of extra group individuals. For example, male BF left a bisexual group and spent different amounts of time as a member of a quartet and trio and later became a solitary individual. Female YLU, who was solitary during the early stages of my study, joined a quartet association and later became a peripheral member of a bisexual group.

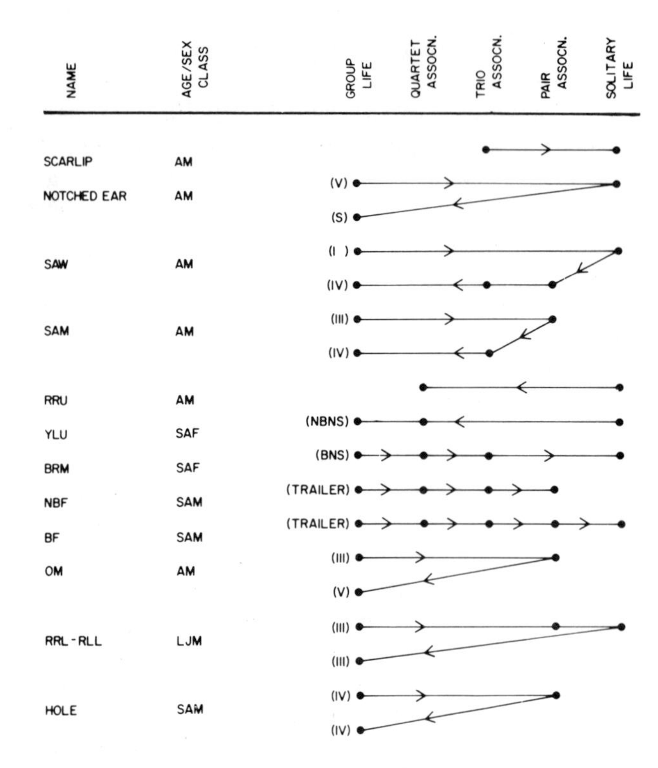

Figure 1. Recent histories of certain identifiable individuals.—➤— indicates direction of change in status. Groups within parentheses. ● indicates status of the individual during the period of observation. AM = adult male, SAF = subadult female, SAM = subadult male, LJM = large juvenile male.

A composite picture of social mobility may be obtained by considering identifiable individuals who lived outside bisexual groups during some period of my study (Figure 1). Many of these individuals were males. Some of them show the entire sequence of group desertion and re-entry into bisexual groups (e.g., adult males, Notched Ear, SAW, Sam and OM), and this cycle repeated itself during the life of some males. For example, SAW and Sam and two other extra group individuals entered Group III by evicting OM, another adult male and the juvenile male RRL-RLL. Later SAW and Sam left Group III at different times but joined to form a triplet association with a young adult female. Subsequently, they left this female and together invaded and took over Group IV.

SAW spent about seven months as an extra group individual between the time he left Group III and joined Group IV, while Sam spent only two months outside these two groups. OM, who was evicted from Group III, formed a pair association with his juvenile group-mate RRL-RLL for a few weeks, but successfully entered Group V within two months of eviction. On the other hand, Scar-lip was never known as a bisexual group member and lived outside these groups for at least 16 months during the study. This shows that males spend variable amounts of time as extra group individuals.

Over a period of 18 months, I estimated that I saw at least 46 different individuals living outside bisexual groups; on the average 2.6 extra group individuals per month. Therefore, a minimum of 1.4 percent of the average total study population was socially mobile during any particular month. The 46 extra group individuals included 19 adults (12 males, 7 females), 21 subadults (14 males, 7 females) and 6 large and medium sized juveniles (3 males, 3 females). This shows that adults and subadults left bisexual groups more frequently than juveniles, and males have a greater tendency to be socially mobile than females. In the above sample, the ratio of males to females was 1:0.58. In a limited sample of 15 extra group individuals who were identifiable through natural marks or ear tags, the male to female ratio was 1:0.50. In other words, the social mobility of males was about twice as great as that of females. Nevertheless, female red howlers appear to be more socially mobile than females of many other primate species.

Mortality Factors

Between June 1976 and May 1978 inclusive, I recovered the dead bodies of an adult male, a subadult female and three infants.

The infants, two from NC Group and one from S Group, died shortly after males invaded these groups. On 23 May 1978, I saw the invader of NC Group bite and fatally wound one of the infants. The other two infants died of fractured skulls, on which teeth marks particularly canine puncture wounds were clearly visible. A fourth infant disappeared from Group III shortly after male invasions. Infant deaths under similar circumstances have been reported in certain Old World species and there is little reason to doubt that the infants mentioned above were killed by invading males. Between June 1976 and May 1978, I recorded a total of 40 births and 10 infant deaths or disappearances. Thus, infanticide accounts for a minimum of 40 percent of the infant losses. Furthermore, infant losses amounted to 25 percent of the births, and infanticide contributed at least 10 percent toward these losses.

Infants which disappeared probably died, but older individuals that disappeared either died or emigrated. Some visibly senile adults which disappeared, particularly multiparous females who do not appear to emigrate, probably died of old age. Among immature individuals, I was less certain about the cause of death. The subadult female whose body I recovered appeared to have been in good physical condition at the time of her death. External and internal examinations did not reveal any obvious cause of mortality.

My observations indicate that injuries sustained through aggression between males especially during male invasions also probably led to some deaths. Resident as well as invading males involved in four of the seven incursions sustained serious injuries during the period of invasions. The adult male whose dead body I found on 5 March 1978 was the dominant male of Group IV, and in prime condition at the time of his death. Shortly before this male died, I saw him chasing SAW and Sam who took over his group immediately after his death. Between 7 and 10 June 1978, Yama, who invaded NC Group, was badly wounded in several parts of his body and his movements were seriously hampered. At the same time, the resident male of NC Group had a wound on his throat, and prior to these observations I had seen these two males involved in chases. On 4 July 1978 most of the gashes on Yama's body were still open and the wound on his left foot had become infected. The foot was swollen and it dangled uselessly exposing many bones. Bothered by insects, he climbed down unsteadily, slaked his thirst at a pool of water and lay on his back for several hours in the open grassland. By the following evening, he had lost his left foot and seemed extremely feeble. He slipped many times as he climbed and appeared to prefer the ground. By 9 July he disappeared from the study area.

NBF, whose wounds I referred to earlier, was the most seriously injured animal that I saw during this study. Except for his left arm and tail, all parts of his body were covered with serious injuries. My efforts to relocate him after he was injured were unsuccessful.

There is no doubt that chases similar to those seen in Groups IV and NC sometimes result in body contact and serious wounding. On 19 May 1977, I saw OM who initially belonged to Group III being chased and bitten several times by three of the four individuals which invaded his group. While the invaders were unscathed, OM bled profusely, but he survived the attack and later joined Group V. Although males do recover even from serious wounds, some probably die as a result of injuries sustained during inter-male aggression.

On 25 May 1978 I located an adult female which had a gaping hole in her lower abdomen through which her ruptured intestines protruded and fecal matter was periodically ejected. This female was greatly emaciated and disappeared from the study area the day after she was seen. The lower abdomen is generally well protected by the hind legs and is not likely to be injured during fights. In the few instances of fights involving females, they sustained injuries on the hands and tail. Therefore, I suspect that the abdominal injury was sustained accidentally, perhaps during a fall.

In summary then, infanticide, inter-male aggression and also probably senility appeared to be the most frequent causes of mortality. Accidental deaths probably occur infrequently. I did not find any convincing evidence to suggest that deaths occurred as a result of parasitic infestation, disease and predation, although deaths due to these factors probably also occur infrequently. Almost the entire population carried pin worms and periodically a few individuals were infected with bot fly (*Dermatobia homonis*) larvae. None, however, appeared to be overwhelmed by these infestations. On one occasion each, I saw a jagouaroundi (*Felis yagouaroundi*) and an ocelot (*Felis pardalis*) walk under trees with red howlers in them. Neither the cats nor the monkeys showed any response to each other. On another occasion, I saw an ocelot resting in a tree for several hours within 75 m of a group of red howlers, who may or may not have been aware of the cat's presence. Observations such as these suggest that predation is not an important mortality factor in this population of red howlers.

Population Density

In estimating the population density, I considered the sizes of the first 23 of the 28 groups listed in Table 2. The sizes of five groups were omitted from the calculations of population density because they were either incursive or had home ranges outside the study area. Eighteen of the groups considered had their entire home ranges and the other five (Ablizia C, Guasimo D, NNC, GNF and SNG) had part of their home ranges within my study area.

Following Neville (1972), I used mean group sizes obtained from initial and terminal counts to compute the total population size. Groups whose home ranges were partially included within the study area were assumed to contribute half their mean group sizes toward the total numbers. The area was also calculated according to Neville "by exluding the space surrounding the house compound out to where forest began and the Laguna de los Guasimos and by including all other parts of the work area including pastures, streams, cattle runs and even the extensive treeless strip between forest proper and the highway." According to this method, I estimated my study site to be 154.8 hectares in extent and obtained a population density of 118 red howlers/km^2. Neville (1972) computed an area of 190.6 hectares and estimated a density of 87 red howlers/km^2, within the same locality.

The above population estimates were not true ecological densities because extensive treeless expanses where I have not seen monkeys were included in the computation of the area of the study site. In order to obtain an accurate estimate of ecological density, I used an area of 42 hectares which was surveyed and marked by grid lines. Although this grid also contained open areas, all parts of it were used by red howlers. The grid included the entire home ranges of five study groups and partial home ranges of four. From this area, I estimated an ecological density of 150 red howlers/km^2.

Biomass Estimates

During the study I obtained the weights of 31 males and 9 females of different age/sex classes (Table 7). This sample included 10 socially dominant males of which the lightest individual was 6.2 kg. Therefore, I assumed this weight as the minimum weight for classifying a male as a full sized adult. My size classifications of the other age/sex classes conformed with the categories of known weights. The above sample does not include weights of small and medium sized juveniles and infants over three months of age. Thus, in estimating the biomass of infants and juveniles, I assumed the following mean weights by inspection of known weights: infants 0.63 kg (range 0.25–1.0 kg), juvenile females 2.25 kg (range 1.0–3.5 kg), and juvenile males 2.75 kg (range 1.0–4.5 kg). For biomass estimates of adults and subadults of both sexes, I used the mean weights of these age/sex classes presented in Table 7. The sample size of weights of females was small but so was the variation in these weights.

Using the above mean weights, the ecological density within the study grid translated into 261.3 kg (i.e., 622 kg/km^2). Similarly, the total red howler biomass of the 23 groups which used the study area amounted to 728.0 kg (i.e., 470 kg/km^2). Adults of both sexes contributed equal proportions and their combined weights accounted for nearly two-thirds of the total biomass (Table 8). Juvenile males contributed 15 percent to the total biomass, while each of the other age/sex classes accounted for less than this amount. On the average extra group individuals contributed an additional 14 kg to the red howler biomass within the study area.

During my investigations, the study population fluctuated between 169 individuals in October 1977 and 186 individuals in March 1977. The total biomass for these months were 665.0 kg and 703.9 kg, respectively (Table 9). The average total biomass of these months and the terminal census of April 1977 was 688.1 kg. Monthly fluctuations in total biomass were small, and less than the variations in total population size. Coefficients of variations of total biomass and total population size for the months mentioned above were 2.96 percent and 4.78 percent, respectively. The infant age class had the greatest coefficient of variation. This was due to the fact that, in March 1977 and a few months prior to it, many more births occurred than during other months. Coefficients of variation of subadults of both sexes were also relatively large and probably reflect the greater tendency of these individuals to leave bisexual groups.

Discussion

Group Size and Composition

Comparisons within the genus *Alouatta* show that sizes of *A. seniculus* and *A. caraya* groups were similar and usually about half that of *A. palliata* groups (Table 10). The adult sex ratio was less disparate in *A. seniculus* and *A. caraya* than in *A. palliata*, and in groups of all three species adult females out-numbered adult males. However, within the immature age classes of *A. seniculus* males out-numbered females (Table 1, Neville, 1972). Pope (1968) also found a similar situation in *A. caraya*. My observations indicate that adult males

Table 7. Mean weights (in kilograms) of certain age/sex classes

	N	*Mean*	*Range*	*± Std. deviation*
Adult males	14	6.70	6.200–7.300	0.33
Adult females	4	4.50	4.250–4.650	0.17
Subadult males	9	5.70	5.000–6.050	0.38
Subadult females	2	3.70	3.650–3.800	0.11
Large juvenile males	7	4.10	3.930–4.300	0.14
Large juvenile females	1	3.30	—	—
Infants	3[1]	0.36	0.230–0.539	0.16

[1] All infants were under three months of age and recently dead when weighed.

Table 8. Percent contribution of different age/sex classes towards biomass (in kilograms) within study grid and study area

	Study grid (42 ha.)		*Study area (154.8 ha)*	
	Biomass	*%*	*Biomass*	*%*
Adult males	92.13	35.27	234.50	32.21
Adult females	88.25	33.78	243.00	33.38
Subadult males	21.38	8.18	64.63	8.88
Subadult females	3.70	1.42	12.95	1.78
Juvenile males	38.50	14.74	105.88	14.54
Juvenile females	13.51	5.17	56.82	7.80
Infants	3.78	1.44	10.24	1.41
Total	261.25	100.00	728.02	100.00

Table 9. Comparison of monthly fluctuations of total biomass of study population

	Month			$\bar{X}$	*± Standard deviation*	*% Coefficient of variation*
	Mar. '77	Oct. '77	Apr. '78			
Adult male	201.00	201.00	227.80	209.93	15.47	7.37
Adult female	238.50	216.00	225.00	226.50	11.32	5.00
Subadult male	91.20	79.80	51.30	74.10	20.55	27.73
Subadult female	14.80	18.50	11.10	14.80	3.70	25.00
Juvenile male	93.50	93.50	110.00	99.00	9.53	9.63
Juvenile female	47.25	45.00	60.75	51.00	8.52	16.71
Infant	17.64	11.34	9.45	12.81	4.29	33.49
Total biomass (kg.)	703.89	665.14	695.40	688.14	20.37	2.96
Total population size	186.00	169.00	178.00	177.67	8.50	4.78

Table 10. Comparison of some population parameters of *Alouatta* species in different localities

Species—Location and Date of Study	*Percent group composition*					*No. of groups studied*	*Mean group size*
	Adult male	*Adult female*	*Sub-adult*	*Juve-nile*	*Infant*		
Alouatta seniculus							
Hato Masaguaral, Venezuela '70	19.2	30.1	13.0	21.9	15.8	26.00	7.7[1] ± 2.3
" " " '72	18.4	28.8	11.7	25.1	15.9	19.00	8.6 ± 2.5
" " " '76	19.1	28.1	6.8	37.6	8.4	28.00	8.9 ± 2.5
La Macarena Colombia '68	28.0	40.0	—	22.0	10.0	6–15.00	~ 5.0
Alouatta palliata							
Barro Colorado Island, Panama '32	16.0	43.0	—	23.0	18.0	23.00	17.3 ± 7.1
" " " " '33	17.0	39.0	—	24.0	20.0	28.00	17.4 ± 7.0
" " " " '35	18.0	38.0	—	30.0	14.0	15.00	18.2 ± 7.1
" " " " '51	15.0	57.0	—	13.0	15.0	29.00	8.1 ± 3.1
" " " " '59	18.0	49.0	—	17.0	16.0	44.00	18.5 ± 9.4
" " " " '67	23.0	41.0	—	19.0	17.0	12.00	14.7 ± 2.5
" " " " '72	18.0	36.0	—	22.0	24.0	12.00	15.2 ± 4.2
Chiriqui, Panama '68	21.0	42.0	—	20.0	17.0	8.00	18.9 ± 6.6
Taboga, Costa Rica '66–'71	21.0	48.0	—	21.0	10.0	11.33[2]	11.5[3]
La Pacifica, Costa Rica '67–'70	20.0	47.0	—	20.0	13.0	5.00[2]	11.9[3]
Santa Rosa, Costa Rica '71–'72	20.0	44.0	—	25.0	11.0	8.00	8.1 ± 7.1
Alouatta caraya							
Tragadero Sur Argentina '64	40.0[4]	36.0	—	20.0	4.0	17.00	7.9 ± 3.0

[1] Mean group size obtained from 19 groups.
[2] Mean number of groups located during different censuses.
[3] Mean of average group sizes in different censuses.
[4] Pope (1968) included peripheral males in her group counts.
[5] Quoted in Heltne Turner & Scott 1976.

sometimes died during male-male aggression. This mortality combined with the more rapid maturation of females after the age of small juveniles probably explains the sex ratio favoring males in the immature age classes and females in the adult age class.

Extra Group Individuals

In *A. seniculus*, I encountered solitary individuals of both sexes, pairs of one or both sexes, as well as triplets and quartets which contained both sexes (Table 6). Sightings of solitary males have also been reported in *A. palliata* and *A. caraya* (e.g., Carpenter, 1934; Pope, 1968), but observations of solitary females are rare. Reliable reports indicate sightings of solitary females in Costa Rican populations of *A. palliata*. Neville (1972) contacted a pair of *A. seniculus* females and Heltne, Turner and Scott (1976) reported an association of two isolated *A. palliata* juveniles. Many investigators have mentioned sightings of groups of three or four individuals (Carpenter, 1934; Collias and Southwick, 1951; Pope, 1968; Neville, 1972; Klein and Klein, 1976; Freese, 1976).

In my study, associations of three animals or less and solitary life seemed to be temporary modes of existence. Pair associations and trios were not stable for more than a few months. Quartet associations, however, were stable for several months, but they too sometimes disbanded. It appears then, that in addition to factors, such as the successful location of a suitable home range, a critical lower limit in numbers is important for the stability of an association, and at the numerical minimum a specific composition may also be vital.

Reference
Neville (1976)
Neville (1976)
Present study
Klein & Klein (1976)
Carpenter (1934)
Carpenter (1934)
Carpenter (1964)
Collias & Southwick (1952)
Carpenter (1964)
Chivers (1967)
Thorington[5]
Baldwin & Baldwin (1976)
Heltne Turner & Scott (1976)
Heltne Turner & Scott (1976)
Freese (1976)
Pope (1968)

Although quartet associations sometimes fragmented, they were numerically similar to SNG Group which was the smallest entity that I designated as a stable bisexual group (Table 2). In fact, I referred to this group as an association until an infant was born into it and gave it an important characteristic of stable bisexual groups. Moreover, in certain behavioral aspects, such as patterns of ranging and rates of grooming, quartet associations resembled bisexual groups (Rudran, in prep.). While it is not unlikely that stable bisexual groups may sometimes decrease in numbers and thus resemble different types of associations, there is the possibility that extra group individuals may form associations which subsequently result in stable bisexual groups. Carpenter (1934) mentioned that new groups of *A. palliata* were formed either by the division of existing bisexual groups or the separation of a subgroup which later may or may not become associated with a peripheral male. In *A. seniculus* I did not see any of these processes taking place, but Neville (1972) reported that the largest *A. seniculus* group that he contacted showed signs of instability with subgroups detaching frequently from the main body of animals.

During my investigations some extra group individuals joined bisexual groups. The successful entry of an extra group individual into a bisexual group appeared to be dependent on several factors. Immature individuals RRL-RLL and Hole entered groups which they had left temporarily when males invaded these groups (Figure 1). The re-entry of these individuals may have been facilitated by previous affiliations with some of the group members. Female YLU's attempts to enter NBNS Group was quickly accepted by the resident males of this group who copulated with this female when she was sexually receptive, but still peripheral to the group. However, the resident females and juveniles initially resisted female YLU's entry into the group. The resident females particularly, often chased female YLU and sometimes bit her. After about four months, aggression waned and YLU was less peripheral to the group than before.

Extra group males that entered bisexual groups were usually in good physical condition and in their prime. However, OM was visibly old when he joined Group V, and Notched-ear who entered S Group was sexually mature but smaller than the resident male of this group. Therefore, it appears that size and age were less important than good physical condition for gaining entry into bisexual groups.

SAW reached his prime after he left Group III, but he lived as a solitary and did not enter Group IV until Sam joined him (Figure 1). Also, earlier when these two individuals invaded Group III, they were accompanied by an adult male and a subadult male. Of the seven natural male incursions reported previously, only in these two incursions did invaders successfully evict all resident males. Among the groups where resident adult males remained after the incursions, four (Groups S, NC, Albizia A, and V) were invaded by solitary adult males and one (Group L) was invaded by a pair of subadult males. This shows that, while single adult males enter bisexual groups, cooperation between invading males may be necessary for the successful eviction of resident males. Conversely, cooperation between resident males probably prevent incursions by extra group males. Of nine bisexual groups which were multi-male during the initial stages of my study, only one (Group III) was invaded by males. Two bisexual groups that were initially multi-male became one-male groups (V and L) before they were invaded by males. Of nine groups which initially included single adult males, four were invaded by males. The greater stability of multi-male groups may explain why resident males of one-male groups tolerate incursions by solitary extra group males and sometimes accept old and small sized adult males as group members.

Social Mobility

One of the most interesting aspects of social mobility in *A. seniculus* is the conspicuous migratory behavior of young females. Social mobility of young females due to factors operating within a bisexual group appears to be rare in primates, and has been previously documented only in the red colobus and chimpanzees (Struhsaker and Leland, in press; Wrangham quoted in Struhsaker and Leland, in press). These female migrations are different from those in hamadryas baboons and gorillas where young bisexual group females are apparently kidnapped by extra group males (Kummer 1968; Harcourt, et al, 1976), or from those in the purple-faced langurs where young females leave bisexual groups subsequent to male invasions (Rudran, 1973).

In *A. seniculus,* where females are nearly half as mobile as males, it is interesting that the female external genitalia closely resemble those of males. In the infant and juvenile stages, the external genitalia of females resemble the scrotum in size and shape, and the extensible and erectile clitoris is very similar to the male penis. As females mature, the clitoris becomes relatively inconspicuous but the external genitalia enlarge and continue to resemble the scrotum. In *A. palliata,* Carpenter (1934) mentioned the strong resemblance between the external genitalia of the two sexes and the difficulty in distinguishing young males from adult females of this species.

I suggest that the female mimicry of male genitalia in *A. seniculus* is mainly the result of the social mobility of young females. This suggestion derives from certain observations mentioned previously. They are:

(1) Resident females resist the entry of migratory females into bisexual groups, e.g., see YLU's entry into NBNS Group.

(2) Resident males frequently howl at and chase extra group males (see Mortality Factors).

(3) Resident males leave their bisexual groups for short periods to associate with migratory females nearing maturity, e.g., association between Tag Group's adult male and female BRM and another female.

(4) Resident males copulate with extra group females, e.g., copulations between males of NBNS Group and female YLU.

(5) Extra group adult and subadult males do not appear to form associations with juvenile females unless these associations include females nearing sexual maturity (Table 6).

It should also be noted that (1) subadult and older females can exhibit piloerection and increase their apparent body size, (2) immature extra group females are sometimes subjected to sexual harassment by adult males, and (3) extra group males copulate with migratory adult females (Rudran, unpublished data).

Female mimicry of male genitalia probably functions best as a visual signal within the medium to long distance range. At these distances, body size differences between migratory young females and females nearing maturity are relatively indistinguishable. Therefore, young females mimic males with their prominent and frequent clitoral erection in order to minimize sexual harassment by resident and extra group males and to reduce associations with extra group males. As these young females mature, they attempt to enter bisexual groups but their entry is resisted by the resident females. Thus they simulate extra group males through genital mimicry as well as piloerection in order to induce the resident males of bisexual groups to leave the confines of their closely spaced groups and investigate a "male." This results in the resident male either associating with the migratory adult female, or copulating with her if she is sexually receptive. These copulations are probably crucial in gaining the tacit support of the resident male in joining his bisexual group. Mimicry by a migratory adult female also attracts extra group males to an individual who may aid in group incursions, but turns out to be a possible sexual partner and an ally in the formation of a social group. Hence the associations between these two age/sex classes (Table 6). While female mimicry minimizes sexual harassment of young migratory females and later facilitates their entry into bisexual groups, within the bisexual group, female mimicry may increase the apparent number of males and hence discourage male incursions.

Population Density

Although my method of computation of population density was similar to that of Neville, our most comparable estimates differed considerably. The difference, however, is probably more apparent than real and may have been caused by the non-identical areas that we used in our computations. This I believe was due to a difference in judgement as to whether or not open savannas, particularly at the boundaries of the study area without identifiable landmarks, should be included in the calculations. Nevertheless, there appears to have been a true increase in population density since the time of Neville's major study. Neville (1976) recensused the population nearly two years after his initial investigations and documented an increase in one part of the study area. Also by April 1978, the total size of the 14 groups that I was able to identify as having been observed during both studies showed an increase of 7 percent of their original numbers (Table 3). This amount probably reflects the minimal increase in population density during the period between the two studies. The increase, however, has not significantly affected the size of groups. In view of this and considering the time elapsed between Neville's study and mine, the growth of the population seems to have been gradual. Increases due to births and immigrations from outside the study population are only slightly greater than the losses due to deaths and disappearances (Table 5).

One probable reason for the gradual population increase is the progressive increase of howler habitat. For nearly the last two decades, Sr. Tomas Blohm, the owner of the ranch on which my study site is located, has prohibited all hunting, destruction of forest and firing of grassland. This has led to a gradual increase in the regeneration of woody plants, and in the size of existing forest patches (Troth, in prep.). The study area included some parts which I considered suitable howler habitat but did not contain resident groups. In the future these areas may become occupied by either the expansion of the home ranges of nearby groups or the establishment of new social units.

Mortality Factors

Mortality factors operating within this population included infanticide, and male-to-male aggression during incursions of bisexual groups. Infanticide appeared to be the most frequent cause of mortality and it contributed a minimum of 40 percent toward the mortality of infants. In the absence of infanticide, the annual rate of population growth would have been 2.64 percent, or 76 percent greater than the present rate of population increase. Therefore, the reduction in numbers through infanticide must be considered significant, particularly because the red howler is a relatively long-lived species. To a lesser extent mortality during inter-male aggression also probably serves to retard the growth of the population.

Evidence of infanticide in Old World Primates was documented by several investigators (Sugiyama, 1967; Mohnot, 1971; Rudran, 1973; Hrdy, 1975, 1977a and b; and Struhsaker, 1977), and Rudran (1973) presented a model suggesting the population consequences of infanticide in the langurs of southeast Asia. The model predicts that an infanticidal population will reach maximal numbers in relation to carrying capacity and then gradually decline through the effects of infanticide. At high densities, the frequency of male invasions of bisexual groups and hence infanticide will be high. Therefore, population size will decrease largely through reduced rates of recruitment into adult age classes. This in turn will decrease infant production and the number of extra group males in the population during the period of decline. With the reduction in the number of extra group males and therefore male invasions, the frequency of infanticide will decline and the population will again increase. Thus, this cyclic process will repeat itself. When a population undergoing such changes begins to decline, the ratio of adult to immatures will progressively be biased in favor of adults. Also some groups may completely lack immature individuals and group sizes would tend to be small. These population characteristics, however, are also shown by primate species experiencing declines due to natural reduction of food resources (Struhsaker, 1973, 1975).

In view of infanticide in red howlers, I suggest that the above model has a wider application among primate species, and predict that my present study population will undergo cyclic fluctuations in population size. Furthermore, infanticide in this red howler population occurred at a time when it was gradually increasing. Therefore, I reiterate that infanticide per se is neither pathological behavior nor is it a result of high population densities. This behavior occurs at all densities but, within a given infanticidal population, it is the frequency of infanticide that will vary with density (see Rudran, 1973). At very high population densities, however, physiological stress may lead to an unusually high frequency of infanticide and depress the population well below carrying capacity. It is only under these circumstances that infanticide could be considered pathological.

Long-term observations are necessary to test the applicability of the above model. Nevertheless, periodic short-term observations may also show population trends and fluctuations predicted by the model. The best known case of population fluctuations in howler monkeys is of course the one documented by the black howler (*Alouatta palliata*) on Barro Colorado Island, Panama (Table 10). During the three censuses in the early 1930s, there was an apparent increase in the number of groups and the population inhabiting this island (Carpenter, 1934, 1964). However, by the 1951 census, the population size had crashed to an all time low but the number of groups had increased slightly (Collias and Southwick, 1952). During this census, mean group size was less than half the previous values and some groups did not include immature individuals. In addition, there were relatively fewer adult males than before and the adult to young ratio was 1:0.4. Later censuses in 1959, 1967 and 1972 showed a gradual build-up in the number of groups and population size (Carpenter, 1964; Chivers, 1969; and Thorington quoted in Heltne, Turner and Scott, 1976).

Collias and Southwick (1952) suggest that an epidemic of yellow fever was possibly an important cause of the population decline. If this was the case, then variability in group composition should have been greater than was reported during the 1951 census, and groups without adults of one or both sexes should have been seen. Furthermore, on the basis of a yellow fever epidemic the apparently greater mortality of adult males and juveniles cannot be easily explained. On the other hand, Collias and Southwick (1952) report an observation where an adult male bit the tail of an infant in half and threw the infant to the ground. Unfortunately, they do not mention the consequences of this injury, but the observation resembles the manner in which one of the infants in my study population was fatally wounded. Therefore, I suggest that infanticide as well as male mortality during invasions may have played an important role in the reduction of the size of this population. This suggestion is supported by the population statistics collected during the period of decline. That these statistics reflect a population undergoing decline due to reduction in food resources is improbable, because during the period of decline continued growth of secondary forest provided progressively larger areas suited to howlers (Collias and Southwick, 1952).

The decline of two populations of *Alouatta palliata* in Costa Rica was reported by Heltne, Turner and Scott (1976). Deforestation and insecticide poisoning were suggested as possible causes of the decline. Red howlers, living sympatrically with the wedge-capped capuchin (*Cebus nigrivittatus*) about two kilometers east of my study site, have also apparently declined in recent years. In this area, Neville (1972) observed five groups of red howlers and noted a mean group size of 9.6 (range 5–16) and an adult to immature ratio of 1:1. Robinson (pers. comm.) who is now working in the same area contacted twenty groups and found a mean group size of 4.6 (range 2–7) and an adult to immature ratio of 1:0.54. Thus, within a period of about eight years, the mean group size in this area has been drastically reduced and the adult to immature ratio has been nearly halved. This decline was probably not caused by epidemics, insecticide toxicity or habitat clearing. The area has remained undisturbed for several years and is sufficiently far from agricultural areas to be free of insecticide contamination. Yellow fever epidemics have not occurred in this locality for several decades. It is possible that in recent times red howlers may have faced greater food competition from the *Cebus* living in the same area and therefore declined in numbers (Eisenberg, in press). It is more likely, however, that the decrease was mainly due to a high frequency of infanticide in this area. This locality is near my study site and relatively continuous with it. Thus, individuals presumably migrate between these two areas. I have seen groups moving across a highway which is probably the major barrier to movements between the two localities. Therefore, behavioral traits found in my study population are also probably typical of the population located near my study site.

Undoubtedly factors, such as epidemics, insecticide toxicity and deforestation can bring about the decline of primate populations, but these factors are temporally unpredictable and their effects are only local. Their effects, however, may be superimposed on the effects of predictable natural processes of population regulation, such as infanticide (Rudran, 1973; Kummer, et al., 1974; see also Dittus, 1974). Although this behavior pattern serves to regulate population growth and thereby maintain numbers at or below the carrying capacity and food levels of the environment, the primary function of infanticide is probably the elimination of unrelated food competitors who do not benefit the infanticidal individual or its offspring in any way (Rudran, in prep.). In other words, I propose that food is the basis of the primary function as well as an important secondary consequence of infanticide. The profound effects of food as one of the principal factors influencing the social organization of primates cannot be over-emphasized (Crook, 1970; Eisenberg, et al., 1972; Struhsaker and Leland, in press).

Acknowledgments

This research was funded by the Smithsonian Institution and the National Institute of Mental Health, Grant No. MH 28840-01 awarded to J. F. Eisenberg and R. Rudran. I wish to thank Tomas and Cecilia Blohm for giving me permission to work on Hato Masaguaral and for their unfailing support during every stage of my investigations and stay in Venezuela. I also wish to thank Edgardo and Ruth Mondolfi for their warm hospitality and contribution toward my study, and Migel Caravallho and his family for their assistance and friendship. I am grateful to J. F. Eisenberg, J. G. Robinson, and C. M. Crockett for constructive criticism of the manuscript; to F. Boccardo, N. A. Muckenhirn and D. S. Mack for short term field assistance; and to Wy Holden for secretarial help. Finally, I wish to record my gratitude to my wife, Ranji, whose forbearance and quiet courage have always been the mainstay of my field endeavors.

Literature Cited

Baldwin, J. D. and J. I. Baldwin.
1976. Primate populations in Chiriqui, Panama. Pages 20–31 in *Neotropical Primates, Field Studies and Conservation,* edited by R. W. Thorington, Jr., and P. G. Heltne. Washington, D. C.: National Academy of Sciences.

Blydenstein, J.
1962. La Sabana de trachypogon del alto llano. *Bol. Soc. Venez. Cienc. Nat.*, 23:139–206.

Carpenter, C. R.
1934. A field study of the behavior and social relations of howling monkeys (*Alouatta palliata*). *Comp. Psychol. Monogr.*, 10:1–168.
1964. *Naturalistic Behavior of Non-human Primates.* University Park: Pennsylvania State University Press.

Chivers, D. J.
1969. On the daily behavior and spacing in howling monkey groups. *Folia Primat.*, 10:48–102.

Collias, N. and C. Southwick.
1952. A field study of population density and social organization in howling monkeys. *Proc. Amer. Phil. Soc.*, 96:143–156.

Crook, J. H.
1970. The socio-ecology of primates. Pages 103–168 in *Social Behavior in Birds and Mammals,* edited by J. H. Crook. London/New York: Academic Press.

Dittus, W. P. J.
1974. The ecology and behavior of the toque monkey, *Macaca sinica.* Ph.D. Dissertation, University of Maryland, College Park.

Eisenberg, J. F.
In press. Habitat economy and society: Correlations and hypotheses for neotropical primates. In *Ecological Influences on Social Organization: Evolution and Adaptation,* edited by I. Bernstein and E. Smith. New York: Garland Press.

Eisenberg, J. F., N. A. Muckenhirn and R. Rudran.
1972. The relation between ecology and social structure in primates. *Science,* 176:863–874.

Freese, C.
1976. Censusing *Alouatta palliata, Ateles geoffroyi,* and *Cebus capucinus* in the Costa Rican dry forest. Pp. 4–9 in: *Neotropical Primates, Field Studies and Conservation,* edited by R. W. Thorington, Jr. and P. G. Heltne. Washington, D. C.: National Academy of Sciences.

Harcourt, A. H., K. J. Stewart and D. Fossey.
1976. Male emigration and female transfer in wild mountain gorilla. Nature, 263:226–227.

Heltne, P. G., D. C. Turner and N. J. Scott, Jr.
1976. Comparison of census data on *Alouatta palliata* from Costa Rica and Panama. Pages 10–19 in *Neotropical Primates, Field Studies and Conservation,* edited by R. W. Thorington, Jr. and P. G. Heltne. Washington, D. C.: National Academy of Sciences.

Hrdy, S. B.
1975. Male-male competition and infanticide among the langurs (*Presbytis entellus*) of Abu, Rajasthan. *Folia Primat.*, 22:19–58.
1977a. Infanticide as a primate reproductive strategy. *Amer. Scientist,* 65:40–49.
1977b. *The Langurs of Abu.* London/Cambridge: Harvard University Press. Klein, L. L. and D. J. Klein.

Klein, L. L. and D. J. Klein.
1976. Neotropical primates: Aspects of habitat usage, population density and regional distribution in La Macarena, Colombia. Pages 70–78 in *Neotropical Primates, Field Studies and Conservation,* edited by R. W. Thorington, Jr. and P. G. Heltne. Washington, D. C.: National Academy of Sciences.

Kummer, H.
1968. *Social Organization of Hamadryas Baboons.* Chicago: University of Chicago Press.

Kummer, H., W. Goetz, and W. August.
1974. Triadic differentiation: An inhibitory process protecting pair bonds in baboons. *Behaviour,* 49:62.

Mohnot, M.
1971. Some aspects of social changes and infant killing in the Hanuman langur, *Presbytis entellus* (Primates: Cercopithecidae) in Western India. *Mammalia,* 35: 175–198.

Neville, M. K.
1972. The population structure of red howler monkeys (*Alouatta seniculus*) in Trinidad and Venezuela. *Folia Primat.*, 17:56–86.
1976. The population and conservation of howler monkeys in Venezuela and Trinidad. Pages 101–109 in *Neotropical Primates, Field Studies and Conservation,* edited by R. W. Thorington, Jr. and P. G. Heltne. Washington, D. C.: National Academy of Sciences.

Pope, B. L.
1968. Biology of the howler monkey (*Alouatta caraya*). Population characteristics. *Bibl. Primat.*, 7:13–20.

Rowell, T. E.
1969. Long term changes in a population of Ugandan baboons. *Folia Primat.*, 11:241–254.

Rudran, R.
1973. Adult male replacement in one-male troops of purple-faced langurs (*Presbytis senex senex*) and its effect on population structure. *Folia Primat.*, 19:166–192.

Snedecor, G. W. and W. G. Cochran.
1967. *Statistical Methods.* Ames: Iowa State University Press.

Struhsaker, T. T.
1973. A recensus of vervet monkeys in the Masai-Amboseli Game Reserve, Kenya. *Ecology,* 54:930–932.

1976. A further decline in numbers of Amboseli vervet monkeys. *Biotropica,* 8(3):211–214.
1978. Infanticide and social organization in the red-tail monkey (*Cercopithecus ascanius schmidti*) in the Kibale Forest, Uganda. *Zeit. f. Tierpsychol.*, 45:

Struhsaker, T. T. and L. Leland.
In press Socio-ecology of five sympatric monkey species in the Kibale forest, Uganda.

Sugiyama, Y.
1965. On the social change of Hanuman langurs (*Presbytis entellus*) in their natural condition. *Primates,* 6:381–429.

DAVID MACK
National Zoological Park
Washington, D. C. 20008

Growth and Development of Infant Red Howling Monkeys (*Alouatta seniculus*) in a Free Ranging Population

ABSTRACT

Quantitative data on behavioral growth and development of eleven infant howling monkeys (*Alouatta seniculus*) during the first nine months were collected from six groups in a free ranging population. Through the first four months, infants were carried over 50 percent of the time, usually by the mother. Infants nursed frequently throughout the study and mothers often assumed positions to facilitate the nursing. Infants began to move off the mother during the third month and generally spent most of their "off" time locomoting or hanging by their tails. Feeding and playing increased over time, but occurred at relatively low levels in comparison to other behaviors. Infant interactions with other group members occurred while the mother was nearby or carrying the infant. Adult females spent more time carrying and huddling with the infant in comparison to juvenile females and juvenile, subadult, and adult males.

RESÚMEN

Se recogió datos de crecimiento y desarrollo, en once monos aulladores infantes (*Alouatta seniculus*) durante los primeros nueve meses, en seis grupos en una población de diferentes edades en el Hato Masaguaral. Durante los primeros cuatro meses, los infantes son acarreados mas del 50 porciento del tiempo generalmente por la madre. Los infantes mamaron con mucha frecuencia durante el estudio y las madres frecuentemente cambiaban las posiciones de sus cuerpos para facilitarles la tarea. Infantes comenzaron a separarse de su madre durante el tercer mes y generalmente pasaban la mayoria de su tiempo "libra" moviéndose o colgando de sus colas. Con el tiempo aumentaron sus periodos de alimentación y juegos, pero a niveles muy bajos comparados con otros comportamientos. La interacción de los infantes con otros miembros del grupo ocurría cuando la madre se encontraba cerca o acarreando al infante. Las hembras adultas pasaban mas tiempo cargando y abrazando al infante comparadas con el resto de los miembros del grupo.

Introduction

Although behavioral growth and development of infants have been studied in many primate species, both in captivity and in the wild, few comparisons among species have been made. Problems include the use of different measurements of behavioral growth and development (e.g., infant riding, infant nurses, time spent off mother) by various investigators, and the possibility that different indices vary in importance according to species. Some studies are short-term or focus on one specific behavior, such as carrying, while others, including field studies, look solely at mother-infant relationships or infant interactions with other group members.

Clearly no one measurement can describe growth and development and interactions with group members in all species of monkeys since the significance of particular behaviors varies according to species. For example, decrease in the percent of time the mother carries her infant does not reflect infant independence in all species of monkeys; in *Presbytis* spp., *Cercopithecus aethiops, Colobus* spp., callitrichids, and many small New World cebids, other animals in addition to the mother frequently carry the infant.

A few studies compare infant growth and development in individual species of monkeys in captive and free (or semi-free) ranging conditions (*Papio anubis,* Nash, 1978; *Lemur* spp., Sussman, 1977; *Saimiri sciureus,* Baldwin, 1969). The results show minor differences between the two living conditions.

In field studies the context in which each behavior is typically performed can be assessed more accurately. One measure of an infant's independence from the mother is represented by the infant traveling "off" the mother during group movement, even though the infant may still be carried while nursing or when the mother rests. These contexts are more difficult to observe in captive studies.

This behavioral study of howling monkey infants (*Alouatta seniculus*) in a free ranging population provides quantitative data on many aspects of infant behavioral growth and development and interactions with group members for the first nine months. A discussion is provided comparing these aspects in many New World monkeys.

Methods

Hato Masaguaral, a ranch in the llanos of Venezuela, is the study site of red howling monkeys (*Alouatta seniculus*) described by Neville (1972, 1976), and more recently by Rudran (p. 107). During Rudran's research, I collected data on 11 infants in his study groups to evaluate infant development during the first nine months (Table 1). Dates of birth were determined by Rudran (pers. comm.) during group censuses performed at the beginning of each month. In total, over 400 hours were spent observing groups with infants from March through July 1978.

In an effort to observe infants in as many activity states as possible, half-hour focal animal samples were taken for each infant at different times between 0700 and 1200 hours and between 1600 and 1830 hours. During mid-day the animals commonly had long rest periods, sometimes out of sight of the observer. Two half-hour observations of a single focal infant were separated by at least 15 minutes to avoid continuation of long bouts of similar behaviors. In groups with two or more infants, observations were alternated from one infant to the other. Observations could be performed within 10 to 30 meters of groups since all groups had previously been habituated to human presence by Rudran (p. 107).

Focal animal sampling entailed recording the time each infant spent in different behavioral categories: carrying, either ventral or dorsal; resting which includes sitting, lying, quadrupedal standing and huddling (body contact or within six inches of mother or other group members); active which includes generalized locomotion (quadrupedal walk, run, jump, ascent or descent), hanging by tail or in combination with hands and/or feet, exploring (manipulation and biting of objects), feeding, and play. Most play took the form of grappling where two or more animals hang by their tails with arms embraced around one another. Other pulling and pushing interactions were also placed in the play category. Nursing was recorded whenever the infant had its mouth attached or in the area of the mother's nipple.

Durations of each activity were recorded to the nearest five seconds. All behaviors were for the most part mutually exclusive, i.e., an infant would not be recorded sitting at the same time he was feeding. In this case, the active behavior, i.e., feeding, was recorded. Huddling with the mother or being carried while nursing was the only situation where durations of behaviors were recorded simultaneously. The relative frequency of each behavior was analyzed by summing the time spent performing a behavior for all infants during a given month, and then dividing by the number of half-hour observations.

A separate set of data was collected on the two infants, ages 1–4 months and 3–6 months, in Group 3. This systematic sampling involved recording the infants' behavior at the beginning of every 15-minute period during four 6-hour periods, on two mornings (0600 to 1200 hours) and two afternoons (1200 to 1800 hours). The first behavior that was sustained for five

Table 1. Infants and their ages during study

Animal	*Group*	*Sex*	*Age of infant during study (mos)*	*No. of half-hour observations*	*Group composition*[1]			
					A	*S*	*J*[2]	*I*
1	3	F	1–3	13	1♂	1♂	1♂	2
					3♀	—	1♀	—
2	5	F	1–3	22	2♂	—	2♂	3
					3♀	—	—	—
3	Win	M	2–4	18	2♂	2♂	2♂	2
					2♀	—	—	—
4	3	F	2–5	15	1♂	—	1♂	2
					3♀	—	1♀	—
5	5	F	2–5	21	2♂	—	2♂	3
					3♀	—	—	—
6	GNF	F	3–6	13	2♂	—	1♂	2
					3♀	—	1♀	—
7	Small	F	4–6	22	2♂	—	1♂	1
					3♀	—	—	—
8	GNF	F	4–7	18	2♂	—	1♂	2
					3♀	—	1♀	—
9	5	F	4–7	24	2♂	—	1♂	1
					3♀	—	2♀	—
10	BNS	F	5–9	32	1♂	—	1♂	1
					2♀	—	2♀	—
11	Win	F	6–9	22	2♂	2♂	2♂	2
					2♀	—	—	—

[1] A = adult, S = subadult, J = juvenile, I = infant.

[2] Analysis of data of juvenile females was calculated only from groups with juvenile females.

seconds was recorded. Categories of infant behavior were the same as those in the focal animal sampling. Each month I calculated the percentage of time each infant was carried, resting, and acitve. A comparison of these behavior categories (carry, rest, and active) between the two sampling techniques showed no differences (Figure 1). Therefore, only the focal animal sample data were used for the analysis of infant activity.

Results

Carrying

The newborn infant is usually carried ventrally on the mother for the first two months (Figure 2). Thereafter, the infant rides on the mother's back as she travels, but is still frequently carried ventrally until the fourth month during group movements.

During the fifth and sixth months, the infant is

carried while nursing or during group movement, but leaves the mother upon reaching a resting or feeding tree. The infant is carried less than 25 percent of the time during the seventh through the ninth month (Figure 2), ventrally while nursing or dorsally when transported over gaps in the canopy greater than one meter. If the mother is ahead of the infant at these difficult crossings, the infant whines until either the mother returns to carry the infant across or other group members provide assistance, e.g., using the body to bridge gaps between branches or sitting on a flexible branch, and lowering it toward the infant (see also Carpenter, 1934; Baldwin and Baldwin, 1973). Animals other than the mother were only observed to carry an infant twice during group movements.

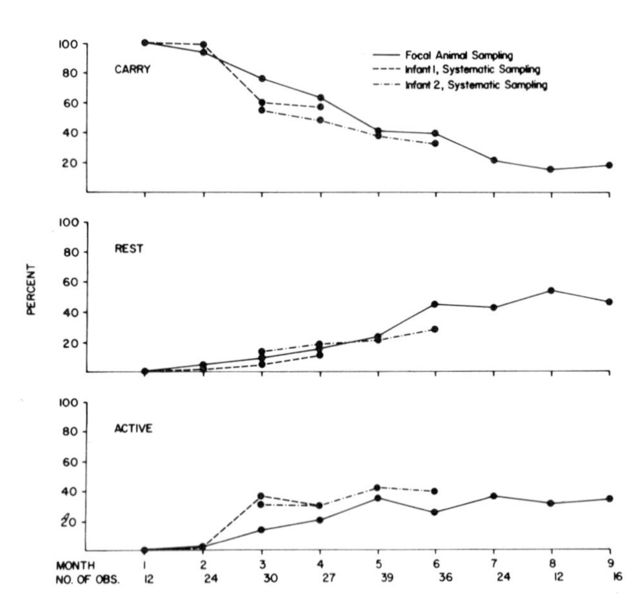

Figure 1. Comparison of the percent of carrying, resting, and active behaviors between the two sampling techniques.

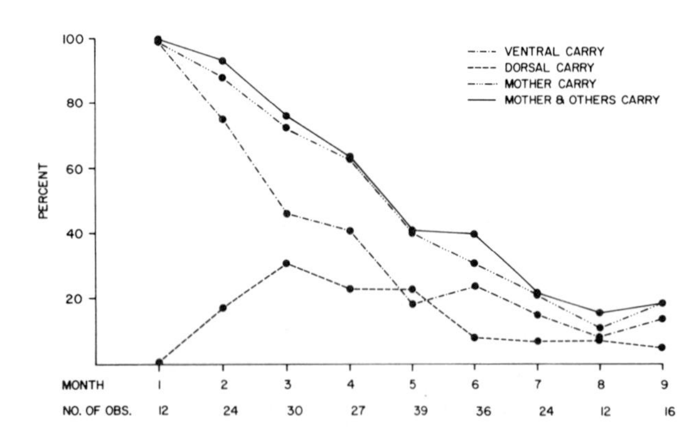

Figure 2. Percentage of time infant howling monkeys are carried through the first nine months.

Mother-Infant Interactions

For the first four months, the activity periods of infant and mother are asynchronous. The infant tends to crawl over the mother's body or move around nearby while the mother rests, but generally remains immobile as the mother travels or feeds. The infant is rarely more than 2 meters away from the mother, and immediately jumps on her back at the slightest movement.

The infant of five to six months is often away from the mother as she rests or feeds, and the mother usually retrieves the infant as the group begins to travel. By the seventh month, a mother tends to ignore the young unless the infant whines.

Except for carrying and nursing, the mother rarely interacts affiliatively with the infant. She also exhibits little overt punitive behavior, such as biting or pushing the infant from her body or nipple. Yet she approaches the infant whenever it whines and accommodates its attempts to nurse by sitting up and/or lifting her arm. The mother rarely interferes as others touch, pull, or remove the infant from her, but she usually remains near other group members interacting with her young through the fourth month. The mother frequently retrieves her infant by presenting her chest to it, which often results in a nursing bout. These behaviors have also been described in *A. palliata* by Carpenter (1934), Altmann (1959), and Baldwin and Baldwin (1973, 1978).

The infant was observed in a nursing position at least once in 70 percent of all observations from the fourth through the ninth month (Figure 3). Prior to this age, the infant was usually carried ventrally, and it was difficult to determine when the infant was nursing. An infant nurses throughout the day, but especially following group movements or foraging. The frequency and duration of nursing did not decline through the ninth month (Figure 3), even though an infant begins eating solid foods as early as the second month, and feeding increases through the ninth month (Figure 4). The average nursing bout lasts 431 seconds (range 50–1755 seconds).

An inverse relationship exists between the amount of time the infant spends on or huddling with the mother and independent rest away from the mother (Figure 5). During the seventh through the ninth month, an infant nurses 52.5 percent of the time it spends with the mother compared to 27.1 percent during months four through six. Thus, as the infant gets older, the amount of time it spends with the mother for purposes other than nursing decreases. Juveniles as old as 17 to 18 months of age were still observed to nurse regularly for long periods of time when resting next to their mothers.

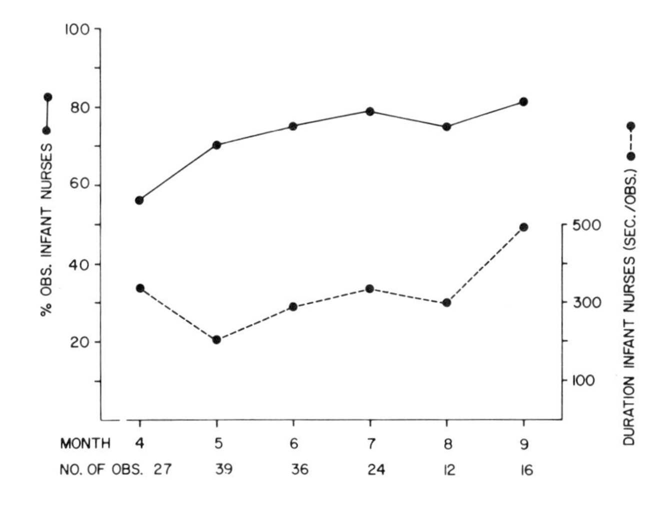

Figure 3. Percent of observations and average duration (seconds) per observation of infant howling monkeys nursing.

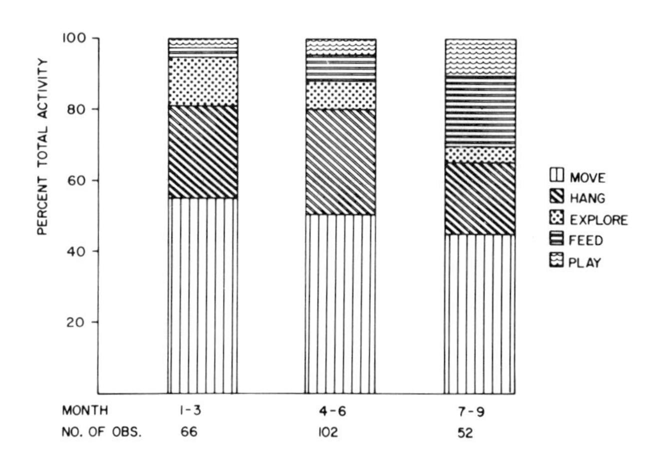

Figure 4. The distribution of behaviors during activity of howling monkey infants.

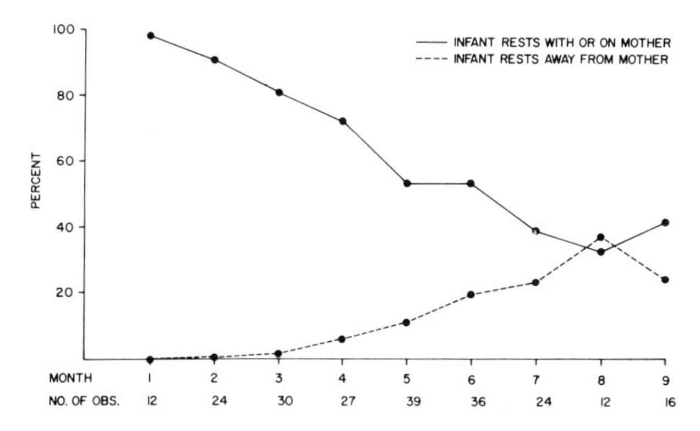

Figure 5. Percent of time infant howling monkeys spend with their mothers (huddling and carried) and resting away from mothers.

Locomotion and Active Behaviors

A newborn infant clings to the ventral surface of the mother by its hands and feet. During the second month, it begins to crawl about the mother's body and grasp objects with its hands. Quadrupedal locomotion on branches begins by the end of the second month, but movements are slow and awkward.

Alouatta have prehensile tails with a grasping pad at the end of the tail which enables an individual to hang by the tail alone. An infant's tail does not wrap around objects until the end of the second momth. Prior to this the tail is usually curled into a tight spiral, curled loosely around the mother's body, or hanging freely by the mother's side. Grasping with the tail begins during the second month when infants switch from the ventral to dorsal carrying position. The infant's tail always curls around the mother's tail base during group movements. By the end of the third month, the infant is able to hang by the tail alone, and employ it in the crossing of branches. An infant attaches the tail pad to the hind branch, either leaps towards or grasps the forward limb, and pulls its feet over before releasing the tail in a manner similar to that used by adults.

Jumping across branches begins during the fourth month and small leaps frequently accompany quadrupedal movements. By the end of the fourth month, the infant has acquired the full repertoire of adult motor patterns except for long leaps into thick brush or between trees.

Infants spend approximately 30 percent of time active from the fifth through ninth months (Figure 1). Walking, running, and jumping exceed other types of active behaviors (Figure 4). An infant often hangs solely by its tail, an activity not observed in adults. Feeding, playing, and exploring occur at lower frequencies than moving or hanging (Figure 4). While exploratory behavior exhibits a decline over time, both play and feeding increase. Play is observed as early as the third month, but the most common type of play, grappling, which involves hanging by the tail, begins during the fourth month.

Prior to six months old, the infant generally explores its environment, using behaviors displayed by adults for feeding. However, the manipulation and biting of limbs and leaves rarely leads to food consumption. By the sixth month, an infant spends more time actually feeding and less time exploring (Figure 4).

Interaction with Group Members Excluding Mother

The newborn is a source of great interest to group members, mainly adult females. They generally sit next to the mother-infant pair and occasionally sniff or pull at the infant. An infant is sometimes carried by other group members during resting bouts. By the second month, the infant may crawl onto other animals or be pulled off the mother's body by a group member. The infant is carried by group members less than 6 percent of the total carrying time for any month. Adult females carry the infant more than juvenile or subadult males and females, and adult males rarely carry infants (Figure 6a). The infant usually initiates interactions with adult males.

By the end of the third month, the infant tends to crawl off animals that pull it toward them. However, group members do not grab or try to initiate play with the infant as he locomotes awkwardly during the third and fourth months.

Most interaction with the infant by group members occurs near the mother. From the fourth through ninth months, an infant spends less than 3 percent of time huddling with other group members when away from its mother. Adult females' huddling rate with the infant or mother-infant pair exceeds that of juvenile females and adult, subadult and juvenile males (Figure 6b and c).

Most play occurs with the other infants and juvenile males and females (Figure 6d). Adult males and mothers rarely, if ever, play with the infant, and other adult females sometimes gently push an infant that approaches and hangs down by its tail in front of them.

Discussion

Even through 10 out of 11 infants studied were females, I suspect there are no sex differences in the growth and development or in maternal care and interactions with other group members during the first year of life. Size differences between males and females were not evident for the first year, and both sexes nurses and spent considerable amounts of time with their mothers until they were weaned. A few studies of *A. palliata* provide information on infant growth and development and describe similar behaviors and interactions with mother and other group members in comparison to this study (Carpenter, 1934; Altmann, 1959; Baldwin and Baldwin, 1973 1978). From his behavioral observations, Carpenter (1934, 1965) presents three age classes for infants: I_1, age 0 to 5–6 mos; I_2, age 5–6 to 10–12 mos; I_3, age 10–12 to 18–20 mos. These age classes are estimates, and based on many of the same behaviors observed in both studies, Carpenter (1934) and this study, I would define: I_1, age 1–4 mos; I_2, age 5–7 mos; I_3, age 8–18 mos or until the infant is weaned.

Comparisons of reproductive and behavioral growth and development indices in many New World cebids and callitrichids are listed in Table 2. The four large

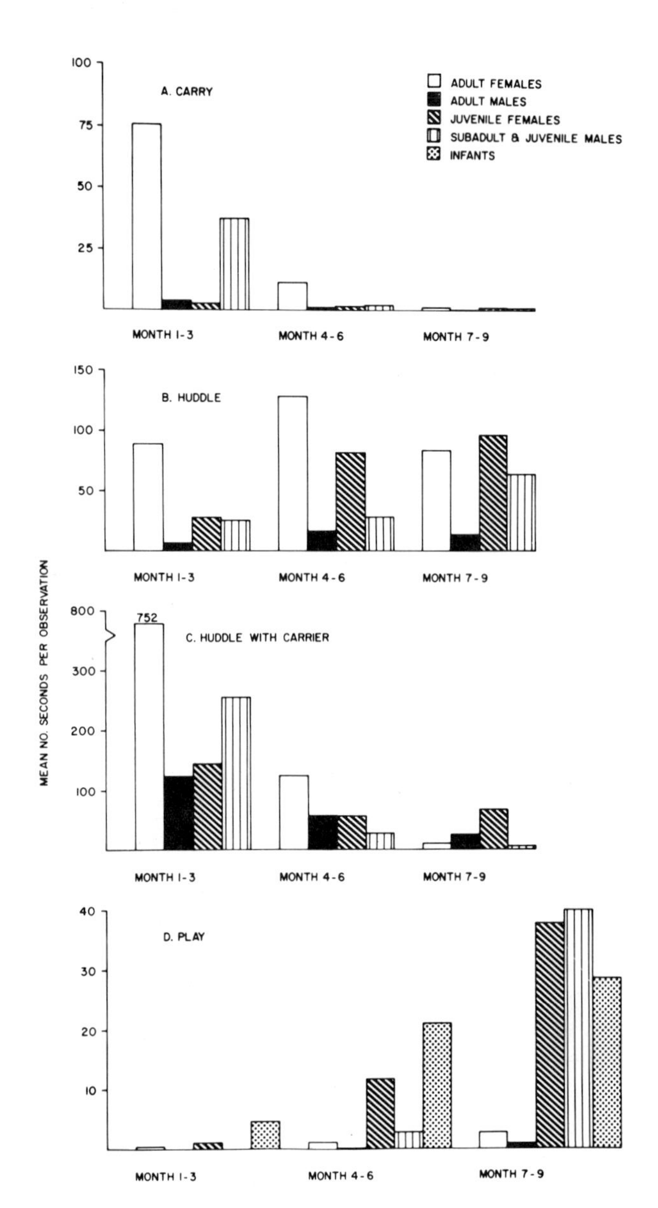

Figure 6. The rate per observation period that:
(a) infant howling monkeys are carried by group members (excluding mother);
(b) infant howling monkeys huddle with group members (excluding mother and other infants);
(c) group members, excluding mother and other infants, huddle with the carried infant;
(d) infant howling monkeys huddle with group members (excluding mother and other infants).

New World monkeys, *Ateles, Lagothrix, Alouatta,* and *Cebus,* exhibit similar patterns of infant behavioral growth and development. Infants begin to get off the mother and move about the environment at the end of the second month. Throughout the first six months, they are frequently carried as the mother travels, and nurse for well over one year.

One major difference among these four species is that *Cebus* infants are carried dorsally at birth while *Ateles, Lagothrix* and *Alouatta* infants are ventrally carried one to two months before assuming the dorsal riding position. *Cebus,* the callitrichids, *Saimiri,* and *Callicebus* all exhibit dorsal carrying at birth. These species all have axial nipples and thus the carrying position at birth appears to facilitate access to the mother's nipple. In callitrichids twinning may also account for dorsal carrying at birth.

Infant *Ateles, Lagothrix* and *Alouatta* are seldom carried by other group members during group movements. Adult females frequently huddle with the mother and newborn, and at times infants are observed crawling over the body of other group members during rest periods.

From the third month, infant *Cebus* monkeys are carried by other group members, especially adult females (Robinson, pers. comm.), and travel on these females occasionally as the group moves. *Saimiri* are similar to *Cebus* in that by the second month infants are carried "40% of daylight hours" by adult females other than the mother (Baldwin, 1969). The high levels of activity and traveling by *Saimiri* and *Cebus* (Terborg, pers. comm.) and the relatively high proportion of infant-mother body weight (Leutenegger, 1973) in comparison to *Ateles, Lagothrix* and *Alouatta* may produce excessive amounts of strain for the *Saimiri* and *Cebus* mothers, and thus the need for other helpers in carrying the young.

Weaning closely precedes the mother's next birth in all species of New World monkeys. In captivity, most groups of *Leontopithecus rosalia* give birth twice yearly (Kleiman, 1977; pers. observation), and infants are weaned by the end of four months when the mother is pregnant again (Hoage, 1978). *Saimiri* are yearly breeders, and juveniles are observed still nursing one month prior to the next birth (Baldwin, 1969). *Alouatta, Ateles, Lagothrix* and *Cebus* have interbirth intervals greater than one year and probably closer to two years (Table 2). In these large New World species, young frequently nurse for over one year. Callitrichids and *Saimiri* lactate until mid to late pregnancy, and it is probable that spider, woolly, howling, and capuchin monkeys also lactate during early gestation. An adult female howling monkey mated with an adult male one month prior to the weaning of her 19-month old juvenile female (pers. observation). In these four species with interbirth intervals greater than one year, nursing behavior probably accounts for the close association between mother and young for over one year, as observed in howling monkeys.

Carrying behavior terminates prior to weaning in New World monkeys having interbirth intervals greater than one year (*Ateles, Lagothrix, Alouatta* and *Cebus*). Callitrichid and *Saimiri* mothers simultaneously wean and stop carrying infants which appears to be induced by late pregnancy.

Punitive behavior (nipping or pushing the infant) by the mother is most intense in callitrichids. This may be due to mothers becoming pregnant early in lactation as they generally produce twice a year in captivity. The mother's energy expenditure following birth (lactation, gestation, and carrying behavior) may be excessive, and thus produce the need for other group members to participate in carrying the infants. In captive studies of *Leontopithecus,* fathers carry infants as much as do mothers, and while mothers terminate carrying at 6 to 8 weeks, fathers continue to carry infants (Hoage, 1977, 1978).

Summary

In this study, behavioral growth and development and interactions with mother and other group members were observed. Quantitative data were obtained from 11 infant howling monkeys (*Alouatta seniculus*) in a free ranging population. The infant is usually carried by the mother, ventrally at birth and dorsally by the end of the second month. By the third month, the infant begins to move about the environment, but remains close to the mother. An infant spends most of its time off the mother by the fifth month, and travels independently by the seventh month. The infant nurses throughout the first nine months, and the mother exhibits no punitive behavior in this context.

Interest in the newborn is high, and adult females spend the most time near the mother-infant pair. By the fifth month, infants and juveniles interact in play. Adult males generally ignore the infant and spend the least amount of time with it.

Acknowledgments

This project could not have been conducted without the generosity and time provided by Dr. John Eisenberg and Dr. R. Rudran. My thanks to Dr. Devra Kleiman, Dr. John Robinson, and Dorothy Gracey for critiques and editing the manuscript, and also to Dr. Doree Fragaszy, Marydele Donnelly and Dr. Robinson who provided me with comparative information on *Callicebus* and *Cebus.*

Table 2. Comparison of growth and development of infant new world monkeys

Species	*Gestation (mos)*	*Interbirth interval (yr)*	*Weaned*	*Carrying position during group movement*	*Most carrying terminated (mos)*	*Helpers involved in carrying during group movement*
Ateles fusciceps and *A. geoffroyi*	7–7½	2–3	>1 yr	Ventral (1–4 mos) Dorsal (5–12 mos)	~6	None
Lagothrix lagotricha	7½	1½–2	>1 yr	Ventral (1–3 mos) Dorsal (3–6 mos)	6	None
Alouatta seniculus, *A. villosa* and *A. palliata*	6–6½	>1	>1 yr	Ventral (1–4 mos) Dorsal (4–9 mos)	7	None
Cebus nigrivittatus	6	2	>1 yr	Dorsal (from birth)	7–8	Juvenile and adult females (from 3rd month)
Callicebus molloch	~5½	~1	5–10 mos	Dorsal (from birth)	5	Adult male and other family group members
Saimiri sciureus	5½	1	8–11 mos	Dorsal (from birth)	4	Juvenile and adult females (after 1st month)
Leontopithecus rosalia	4½	½–1	3–4 mos	Dorsal (from birth)	4	Adult male and other group members (after 1st week)
Callithrix jacchus	5½	½–1	3–4 mos	Dorsal (from birth)	4	Adult males and other group members (after 1st day)

Literature Cited

Altmann, S. A.

1959. Field observations on a howling monkey society. *J. Mammal.*, 40:317–330.

Baldwin, J. D.

1969. The ontogeny of social behavior of squirrel monkeys (*Saimiri sciureus*) in a seminatural environment. *Folia Primat.*, 11:35–79.

Baldwin, J. D. and J. I. Baldwin

1973. Interactions between adult female and infant howling monkeys (*Alouatta palliata*). *Folia Primat.*, 20:27–71.

1978. Exploration and play in howler monkeys (*Alouatta palliata*). *Primates*, 18(3):411–422.

Carpenter, C. R.

1934. A field study of the behavior and social relations of howling monkeys. *Comp. Psychol. Monogr.*, 19(2):1–168.

1965. The howlers of Barro Colorado Island. Pages 250–291 in *Primate Behavior: Field Studies of Monkeys and Apes*, edited by I. DeVore. New York: Holt, Rinehart and Winston.

Eisenberg, J. F.

1973. Reproduction in two species of spider monkeys, *Ateles fusciceps* and *Ateles geoffroyi*. *J. Mammal.*, 54:955–957.

1976. Communication mechanisms and social integration in the black spider monkey, *Ateles fusciceps robustus*, and related species. *Smithsonian Contribs. Zool.*, 213:1–108.

Punitive behavior by mother	Reference
Mild	Eisenberg 1976, 1973 Eisenberg and Kuehn 1966 Pers. observation
Mild	Williams 1967, 1968 Mack and Kafka 1978
Mild	Carpenter 1934 Glander 1975 This study
Mild	Robinson pers. comm.
Rare	Fragaszy pers. comm. Donnelly pers. comm.
Mild	Baldwin 1969 Rosenblum 1968 Wiswell 1965
Intense	Hoage 1977, 1978 Kleiman, in press Pers. observation
Intense	Ingram 1977a, b Hampton and Hampton 1965 Hearn and Lunn 1975 Hearn 1977

Eisenberg, J. F. and R. E. Kuehn
1966. The behavior of *Ateles geoffroyi* and related species. *Smithsonian Misc. Coll.,* 151(8):1–63.

Glander, K. E.
1975. Habitat and resources utilization: An ecological view of social organization in mantled howling monkeys. Ph.D. Dissertation, University of Chicago, Chicago.

Hampton, J. K., Jr., and S. H. Hampton
1965. Marmosets (Hapalidae): Breeding seasons, twinning, and sex of offspring. *Science,* 150:915–917.

Hearn, J. P.
1977. The endocrinology of reproduction in the common marmoset, *Callithrix jacchus.* Pages 163–171 in *The Biology and Conservation of the Callitrichidae,* edited by D. G. Kleiman. Washington, D. C.: Smithsonian Institution Press.

Hearn, J. P. and S. F. Lunn
1975. The reproductive biology of the marmoset monkey, *Callithrix jacchus. Lab. Anim. Handbooks,* 6:191–202.

Hoage, R. J.
1977. Parental care in *Leontopithecus rosalia rosalia*: Sex and age differences in carrying behavior and the role of prior experience. Pages 293–305 in *The Biology and Conservation of the Callitrichidae,* edited by D. G. Kleiman. Washington, D. C.: Smithsonian Institution Press.
1978. Biosexual development in the golden lion tamarin (*Leontopithecus rosalia rosalia*). Ph.D. Dissertation, University of Pittsburgh.

Ingram, J. C.
1977a. Parent-infant interactions in the common marmoset (*Callithrix jacchus*). Pages 281–291 in *The Biology and Conservation of the Callitrichidae,* edited by D. G. Kleiman. Washington, D. C.: Smithsonian Institution Press.
1977b. Interactions between parents and infants, and the development of independence in the common marmoset (*Callithrix jacchus*). *Anim. Behav.,* 25:811–827.

Kleiman, D. G.
In press. Sociobiology of captive propagation. In *Conservation Biology,* edited by M. Soulé and B. Wilcox. La Jolla: Sinauer Assoc.
1977. Characteristics of reproduction and sociosexual interactions in pairs of lion tamarins (*Leontopithecus rosalia*) during the reproductive cycle. Pages 181–190 in *The Biology and Conservation of the Callitrichidae,* edited by D. G. Kleiman. Washington, D.C.: Smithsonian Institution Press.

Leutenneger, W.
1973. Maternal-fetal weight relationships in primates. *Folia Primat.,* 20:280–293.

Mack, D. and H. Kafka
1978. Breeding and rearing of woolly monkeys (*Lagothrix lagotricha*) at the National Zoological Park, Washington. *Internat. Zoo Yb.,* 18:117–122.

Nash, L. T.
1978. The development of the mother-infant relationship in wild baboons (*Papio anubis*). *Anim. Behav.,* 26:746–759.

Neville, M. K.
1972. The population structure of red howler monkeys (*Alouatta seniculus*) in Trinidad and Venezuela. *Folia Primat.,* 17:56–86.
1976. The population and conservation of howler monkeys in Venezuela and Trinidad. Pages 101–109 in *Neotropical Primates, Field Studies and Conservation,* edited by R. W. Thorington, Jr. and P. G. Heltne. Washington, D. C.: National Academy of Science.

Rosenblum, L.
1968. Some aspects of female reproductive physiology in the squirrel monkey. Pages 147–169 in *The Squirrel Monkey,* edited by L. A. Rosenblum and R. W. Cooper. New York: Academic Press.

Rudran, R.
1979. The demography and social mobility of a red howler (*Alouatta seniculus*) population in Venezuela. Pages 107–126 in *Vertebrate Ecology in the Northern Neotropics,* edited by John F. Eisenberg. Washington, D. C.: Smithsonian Institution Press.

Sussman, R. W.
1977. Socialization, social structure, and ecology of two sympatric species of lemur. Pages 515–528 in *Primate Biosocial Development,* edited by S. Chevalier and F. Poirier. New York/London: Garland Publications, Inc.

Williams, L.
1967. Breeding Humboldt's woolly monkey *Lagothrix lagotricha* at Murrayton Woolly Monkey Sanctuary. *Internat. Zoo Yb.,* 7:86–88.
1968. *Man and Monkey.* Philadelphia/New York: J. B. Lippincott.

Wiswell, O. B.
1965. Gestation, parturition and maturation in the *Saimiri sciureus* (squirrel monkey). *Excerpta Med. Intern. Congr. Ser.,* 99, E-52 (Abst.).

JOHN G. ROBINSON
National Zoological Park
Washington, D.C. 20008

Correlates of Urine Washing in the Wedge-Capped Capuchin *Cebus nigrivittatus*

ABSTRACT

In *Cebus nigrivittatus,* urine may be caught in the hand and then rubbed on the hind foot and the hind limbs. The frequency of urine-washing is highest during the dry season when humidity is low and the temperature is high. There is no sexual difference in the rate and frequency of urine washing. Urine washing may occur during social interactions, especially between adult males of neighboring troops during an encounter. Frequent urine washing may be shown by some estrous females. It is concluded that urine washing may have several functions including social signalling and thermoregulation.

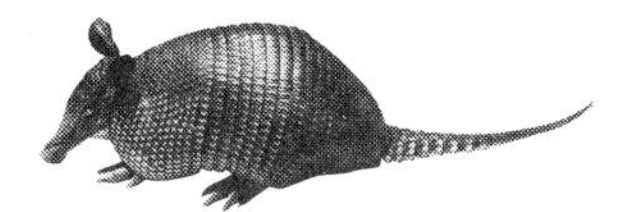

RESÚMEN

En el mono *Cebus nigrivittatus,* la orina puede ser tomada por éste en su mano y luego frotada sobre la pata trasera y los miembros traseros. La frecuencia de éste lavado con orina es más alto durante la época seca cuando la humedad es baja y la tempatura alta. No hay differencia sexual en la velocidad y frecuencia del lavado con orina. Se observa el uso del lavado con orina en situaciones de intercambio social, especialmente cuando los machos adultos de grupos vecinos se encuentran. Además, algunas hembras en estro también se hacen frecuentemente con orina lavados. Se concluye que el lavado con orina puede tener varias funciones incluyendo las de comunicación social y termoregulación.

Introduction

Urine washing in capuchins *Cebus nigrivittatus* consists of an initial urination over the raised ipsolateral fore and hind limbs (Fig.1). The hand forms a receptacle for the urine which is then rubbed over the foot and hind limb. The foot rubs the hand and forelimb. The tail often passes under the body and entwines with the other limbs. This is often repeated symmetrically so the other limbs are washed. In the adult male this is sometimes followed by a stereotyped rubbing of the head and shoulders with the forelimbs while the tail and hind limbs pass over other areas of the body.

Among primates, urine washing has been described in prosimians (Petter, 1962; Charles-Dominique, 1976), and in cebids (Hill, 1938; Oppenheimer, 1968; Thorington, 1968; Milton, 1975). Oppenheimer (1968) has described urine rubbing in *C. capucinus*. Yet the behavioral and physiological consequences of this complex behavior are still unclear. Charles-Dominique (1977) summarizes numerous hypotheses concerning the possible effects of urine washing in primates, and a number of reviews (Ralls, 1971; Johnson, 1973; Eisenberg and Kleiman, 1973) deal with the larger question of scent marking in mammals.

With the exception of Milton's (1975) observations on the howler *Alouatta palliata* and Charles-Dominique's (1977) elegant study of *Galago alleni*, studies have relied on captive animals. The stimuli eliciting urine washing and the responses of conspecifics are most usefully examined in a laboratory setting, but understanding the adaptive significance of behaviors depends on observations of the natural contexts associated with a behavior. In this field study of the wedge-capped capuchin *Cebus nigrivittatus*, I examined the variation in urine washing frequency among individuals and age-sex classes, and between seasons. I noted contexts associated with urine washing, and responses of other group members.

Methods

This report is based on 544 examples of urine washing recorded between June 1977 and July 1978 from a regularly followed, well-habituated group of 18 capuchins. The study site is an extensive gallery forest bordering the Rio Guarico in the *llano bajo* of Estado Guarico, Venezuela. The area is characterized by pronounced dry and wet seasons (Blydenstein, 1967), with almost all rainfall occurring between May and November.

The study group was followed continuously during the first five days of each month, and thereafter on a more irregular basis. I noted each instance of urine washing seen during the course of the study and recorded the context associated with each urine wash and the response, if any, of nearby animals. In a sample of 131 urine washes collected during systematic sampling, I also noted the identity of the nearest neighbor and its distance from the washing animal.

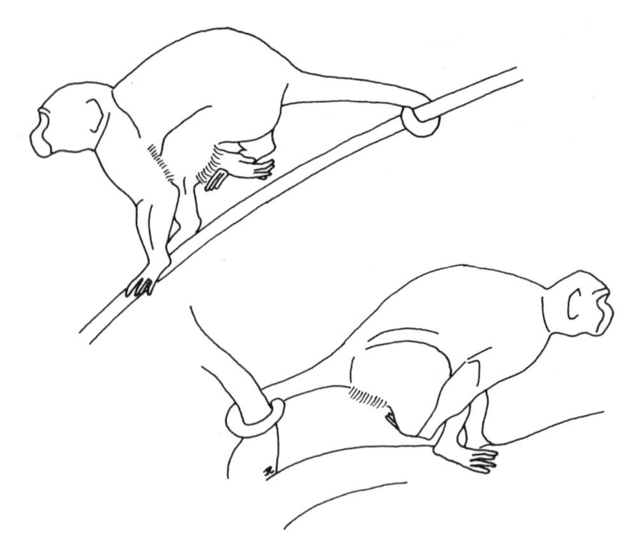

Figure 1. Urine washing in capuchin monkeys (drawn from photographs).

Results

Table 1 outlines the circumstances associated with urine washing. In the majority of cases (80%) I could discern no eliciting stimulus. Prior to the wash, animals were usually moving or foraging. Animals did not urine wash on particular surfaces or trees or in particular locations in the home range. Nor were they highly aroused or unusually active.

The remaining 109 cases were placed into contextual categories. However, no context was invariably associated with urine washing. A non-social disturbance, such as the close proximity of an observer or the barking of dogs from a nearby ranch house, often evoked urine washing. Many of the examples noted by Oppenheimer and Oppenheimer (1973) probably can be attributed to these circumstances. I noted two cases in which noxious stimuli might have elicited urine washing. Once the large adult male BM urine washed after taking Hemiptera (Pentatomidae) nymphs, and once after attempting to take an unidentified vertebrate from a tree bole. Thorington (1967) reports *Cebus apella* urine washing while extracting ants from dead branches.

However, most cases of urine washing attributable to a specific context involved social interactions. On 7 occasions, during 9 intergroup encounters, males, es-

Table 1. Contexts associated with urine washing

Context	*Occurrences*
No evident stimuli	435 (80.0%)
Disturbance	5 (1.0%)
Noxious stimuli to hands and feet	2 (0.4%)
Intergroup encounter	7 (1.0%)
Intragroup social interaction (total)	90 (17.0%)
Following decrease in interindividual distance	36 (7.0%)
Following agonistic interaction	7 (1.0%)
During solicitation by an estrus female	25 (5.0%)
Response to this solicitation	9 (2.0%)
Response to another animal's urine wash	13 (2.0%)
After sniffing substrate	5 (1.0%)

Table 2. Monthly variation in frequency of urine washing

Month	*Observations*	*Contact hours*	*Washes/ Contact hour*
June 1977	0	2.0	—
July 1977	2	42.5	0.05
August 1977	0	27.0	—
September 1977	0	—	—
October 1977	6	56.0	0.11
November 1977	10	75.0	0.13
December 1977	25	64.0	0.39
January 1978	65	85.0	0.76
February 1978	138	103.0	1.34
March 1978	146	100.0	1.46
April 1978	63	85.5	0.74
May 1978	48	79.0	0.61
June 1978	3	85.5	0.04
July 1978	5	60.0	0.08

pecially the large, fully adult male BM, urine washed. This urine washing was of the most complete form, which spreads urine over the whole body. On 90 occasions animals urine washed immediately following or preceding a social interaction within the group, though urine washing was not generally associated with these circumstances. In 43 of these cases the animal had just approached or been approached by another animal. In all but one of these cases the pair involved were of opposite sexes. In 12 cases grooming or grooming solicitation followed. Animals also urine washed 7 times following agonistic interactions. On 25 occasions estrous females approached the large male BM and, in addition to presenting, thrusting against him and grooming him, urine washed next to or above him. While there was often no obvious response from the male, sometimes he would sniff the branch beneath the female or lick the urine. On 9 occasions BM urine washed in turn. Urine washing in response to another's wash was evident an additional 12 times.

The most striking variation in the level of urine washing was seasonal. Table 2 illustrates the large differences in urine washing frequency between months, and Figure 2 plots frequency of urine washing by month and allows comparison with monthly levels of rainfall, relative humidity and temperature. (Rainfall and temperature interact to affect relative humidity. Relative humidity is the ratio of the observed vapor pressure to the saturation vapor pressure for the observed temperature. Thus for a given observed vapor pressure, which depends on accessible water in the environment, an increase in temperature lowers the value of the relative humidity.) Animals continue urine washing at high levels during April and May following the start of the rains, suggesting that high temperature or low relative humidity, but not rainfall, affect this behavior. If these are the principal determinants of urine washing, one would expect more washing in the middle of the day, as "the daily minimum of relative humidity occurs at, or shortly before, the time of day when temperature reaches its daily maximum" (Richards, 1952). Using only days when I followed animals throughout the day, thus avoiding bias resulting from hour of observation, I plotted urine washing across hours (Fig. 3). The level of urine washing remains high between 0900 and 1400 and peaks between 1100 and 1200. Capuchin activity patterns do not generate this distribution: Animals urine wash less when they are resting, an activity more common at midday. The increased frequency of urine washing in the early evening (between 1700 and 1800) appears to be real, as animals continue washing at high rates even after 1800. I offer no explanation for this phenomenon.

The seasonality of urine washing is not determined by seasonal differences in the frequency of specific social interactions. Intergroup encounters were rare during the dry season. Of 9 encounters that I witnessed involving this group, only one occurred in December

Figure 2. Seasonal variation in frequency of urine washing. Temperature and relative humidity data collected at the MARNR meteorological station Los Llanos, which is approximately 40 km from the study site. Monthly maximum mean, mean, and minimum mean from June 1977 to June 1978. Rainfall was calculated by averaging the rainfall for each month from the Hato Masaguaral ranch house, 5 km from the study site, and the MARNR meteorological station Corozopando, which is 8 km away. Averages from June 1977 to July 1978. Urine washes per contact hours from June 1977 to July 1978.

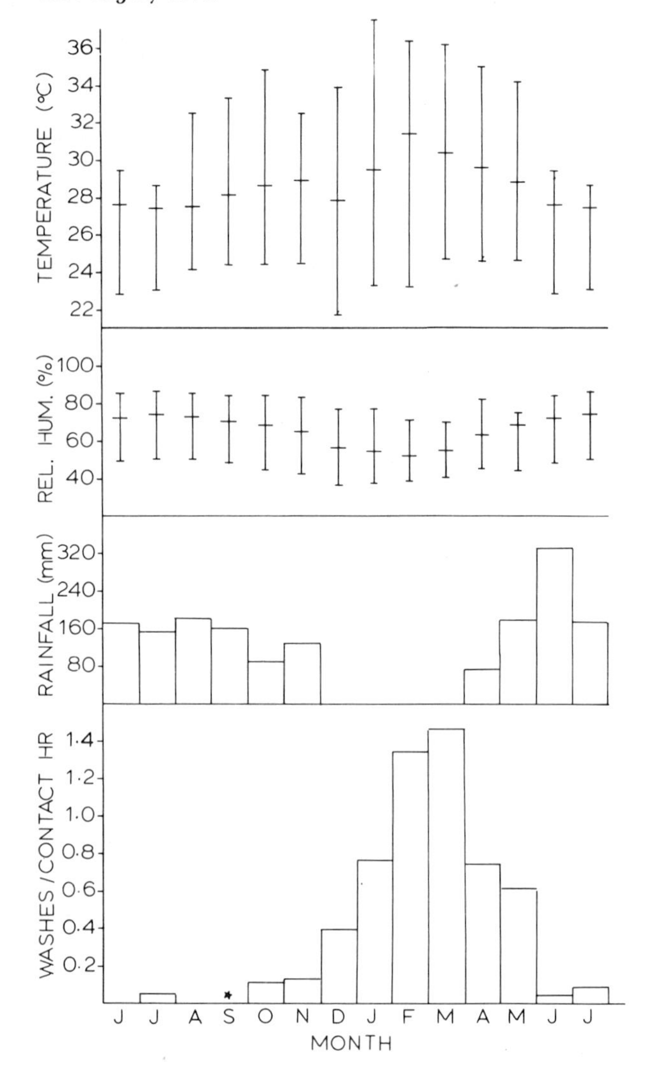

Figure 3. Daily variation in the frequency of urine washing.

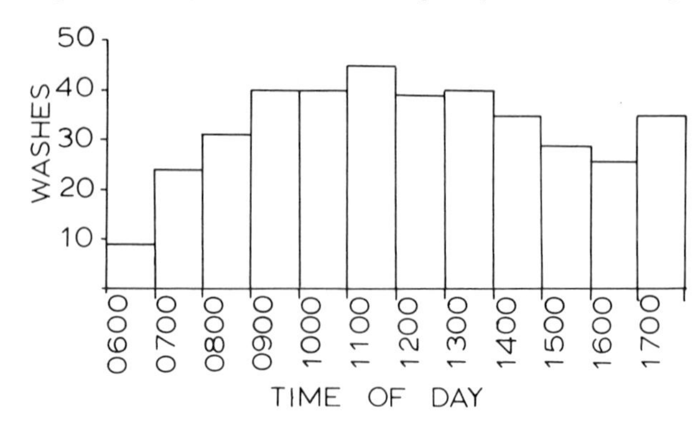

Table 3. Individual variation in frequency of urine washing

Age/sex class	*Individual*	*Occurrences*	*Rate*	*Rank*
Adule male	BM	152	0.54	1
Primiparous female	PN[1]	17	0.31	15
Primiparous female	MO[1]	38	0.26	12
Subadult female	MF	44 (16)	0.24	8
Adult female	HI[1]	59	0.23	2
2nd yr. juv. male	SN	40	0.22	11
Adult female	BU	19	0.16	14
3rd yr. juv. male	MM	25	0.14	6
1st yr. infant female	CR	16	0.13	18
Subadult male	BL	17	0.12	3
4th yr. juv. male	PG	13	0.09	5
2nd yr. juv. female	WH	16	0.08	7
Adult female	BE	12	0.08	10
Subadult male	WO	10	0.08	4
2nd yr. juv. female	PF	9	0.08	16
Adult female	AM	11	0.08	9
1st yr. infant female	RW	6	0.06	17
Adult female	FF	2 (9)	0.03	13

[1] Pregnant January–June 1978.

Numbers in parentheses are the additional number of times these females urine washed while in estrus. Number not included in rate calculation.

through April. Levels of intragroup interactions did not vary seasonally, with the exception of the increase in estrus behavior during November and December. These months were not characterized by high levels of urine washing.

There is considerable variation between individuals in levels of urine washing (Table 3). For all comparisons between individuals, I used a urine washing rate which is the urine washing frequency normalized for each individual's visibility. The adult male BM urine washed considerably more than any other animal. High rates of urine washing were exhibted by three adult females, all of which were pregnant during the study, and by a subadult female who frequently consorted with the male BM. However, I could discern no further pattern based on age-sex class. I calculated an index of rank by lumping all one-on-one displacements, threats and chases. A rank hierarchy was clearly evident, but I found no correlation between urine wash rate and rank ($p > 0.05$, Spearman rank correlation). Neither is urine wash rate correlated with age ($p > 0.05$, Spearman rank correlation). I detected no differences between males and females in levels of urine washing ($p > 0.05$, t test). This contrasts with Oppenheimer's (1968) observations that in *C. capucinus* only males "from about 3 years of age up" urine rub and in *C. nigrivittatus* (Oppenheimer and Oppenheimer, 1973) urine rubbing "was primarily performed by adult males."

The sex of the nearest neighbor, if less than 5 m away, has an effect on the probability of urine washing (Table 4). Urine washing in males is associated with the close proximity of females, and in females by the close proximity of males ($p < 0.005$, G log-likelihood test). This suggests that urine has a role in olfactory communication between sexes.

This interpretation is supported by an analysis of which animals respond by approaching and investigating the site of a urine wash (Table 5). Animals usually sniff or lick the branch or leaf onto which urine has fallen, and sometimes urine wash in turn. Males investigate female urine washing sites, while females investigate those of the big male BM ($p < 0.005$, G log-likelihood test). No animal ever responded to urine washing by any other male. BM was responsible for 65 percent of the reactions to female urine washes.

In a small sample in which I noted the animal investigating the site where another had been sitting (Table 6), the same pattern of cross-sex investigation holds ($p < 0.005$, Fisher exact test). 75 percent of female investigations were directed to sites where BM had been sitting. Of responses to female sitting sites, BM was responsible for 45 percent.

Table 4. Sex of nearest neighbor (if under 5 m) to animal urine washing

Urine washer	*Nearest neighbor*		*Total*
	Male	*Female*	
Male	12	28	40
Female	39	19	58
Total	51	47	98

Table 5. Sex of animal responding to urine washer

Sex washing	*Sex responding*		*Total*
	Male	*Female*	
Male	0	15	15
Female	31	2	33
Total	31	17	48

Table 6. Sex of animal investigating a sitting site

Sex sitting	*Sex investigating*		*Total*
	Male	*Female*	
Male	1	4	5
Female	12	0	12
Total	13	4	17

Discussion

A major portion of the variation in urine washing frequency can be attributed to season. Animals urine wash at higher rates during the dry season, a time of year characterized by little rainfall, low relative humidity and high temperatures. In addition, animals urine wash at times during the day when relative humidity is lowest and temperature highest. As both temperature and relative humidity affect evaporative sweat loss, this suggests that urine washing, by wetting the bare, heat labile, plantar surfaces of the hands and feet, has thermoregulatory effects. Schmidt and Seitz (1967) advanced this hypothesis following experimental work with *Saimiri,* and Milton (1975) considered but finally rejected this possibility for *Alouatta palliata.* While the physiological consequences of urine washing have never been examined, self grooming and the resultant wetting of the body with saliva have been correlated with body temperature drops in rats (Hainsworth and Stricker, 1970; Stricker *et al.*, 1963) and mongolian gerbils (Thiessen *et al.*, 1977). Thermoregulation has been suggested as the effect of saliva "self anointing" in hedgehogs (Burton, 1957). And I have been told of a custom in the Venezuelan llanos to encourage young boys to cool themselves by urinating on their feet.

Another possible effect of urine washing is to limit water loss. The importance of water conservation in *Cebus nigrivittatus* during the dry season is indicated by the frequent (up to 4 times daily) drinking from streams at this time, and the rare observation of drinking during the wet season. High temperatures and low relative humidity increase evaporative sweat loss, and low relative humidity increases insensible perspiration (passive water diffusion through the skin). Davies *et al.* (1976) found that sponging down athletes after prolonged exercise reduced evaporative sweat loss by up to 10%, and wetting the skin surface would limit water loss through insensible perspiration as this depends directly on relative humidity at the skin surface (Ladell,1965).

Urine washing has, in addition, social corelates and behavioral consequences. The fully adult male BM urine washed more frequently than any other animal. Following close approach of neighboring groups, the adult males reacted by approach to one another, sometimes urine washed, and were most active during these agonistic interactions. These observations suggest that the frequent urine washing of the adult males might mark group locations, allowing groups to avoid one another, and thus regulate intergroup spacing. Probability of urine washing was not correlated with location as Charles-Dominique (1977) found in *Galago,* but since group ranges overlap completely this would not be expected. The alterantive is that the frequent urine washing of these males is associated with their rank within the group. More frequent marking by dominants has been reported in duikers (Ralls, 1971), marmosets (Epple, 1970, 1974), rabbits (Mykytowycz, 1965), and muntjacs (Barrette, 1977). But while the *dominant* marks more in capuchins, *dominance* and urine washing are not correlated.

Males mark more than females in most species examined, including rabbits (Mykytowycz, 1962), canids (Kleiman, 1966), hamsters (Payne and Swanson, 1970), and muntjacs (Barrette, 1977). However I found no difference in washing frequency between males and females. The large adult male, as noted, frequently urine washed, but the two smaller, sexually active subadult males urine washed infrequently.

Capuchins are more likely to urine wash when proximate to animals of the opposite sex. This has also been noted in female hamsters (Johnston, 1977), which vaginal mark more frequently in the presence of males, and in hedgehogs (Brockie, 1976), which self-anoint with saliva in the presence of the opposite sex. These observations suggest that urine might give information on reproductive state as has been demonstrated in macaques (Michael and Keverne, 1968; Michael *et al.,* 1971). The frequent urine washing by some estrous capuchin females, and the investigation of urine marks by animals of the opposite sex, support this speculation. Eisenberg (1976) reports that spider monkey males sniff the urine marks of females, and Charles-Dominique (1977) noted that *Galago* males rub their cheeks on the urine marks of females.

Other effects of urine washing that have been suggested are not appropriate to capuchins. Seitz (1969), studying *Saimiri,* and Milton (1975) with *Alouatta* suggested the potential of urine marks as orientation cues during group movement. But group spread in capuchins is considerable, often over 100 m, and animals do not use common pathways. Moistening or cleaning the plantar surfaces has no apparent value.

It is clear, however, that urine washing in *Cebus nigrivittatus* has no one function. Rather some of the behavioral and physiological consequences of this complex behavior are beneficial, and thus have selective value. The high incidence of urine washing during the dry season suggests that this behavior affects the animals' thermoregulation and water conservation. In addition, the contexts and responses to urine washing indicate olfactory signaling, particularly between the sexes within the group, and possibly also between males of different groups.

Acknowledgments

I give my sincere thanks to John Eisenberg and R. Rudran for their continual support and enthusiasm for this work. David Mack kindly read and commented on the manuscript. Linda Cox was a constant help with the observations and critically evaluated the results. Funds for this research were provided by an IESP-Smithsonian grant to J. F. Eisenberg and an NIMH grant to J. F. Eisenberg and R. Rudran.

Literature Cited

Barrette, C.
1977. Scent marking in captive muntjacs, *Muntiacus reevesi. Anim. Behav.,* 25:536–541.

Blydenstein, J.
1962. La sabana de Trachypogon del alto llano. *Bol. Soc. Venez. Cienc. Natur.,* 102:139–206.

Brockie, R.
1976. Self-anointing by wild hedgehogs, *Erinaceus europaeus,* in New Zealand. *Anim. Behav.,* 24:68–71.

Burton, M.
1957. Hedgehog self-anointing. *Proc. Zool. Soc. London,* 129: 452–453.

Charles-Dominique, P.
1976. *Ecology and Behaviour of Nocturnal Primates.* London: Duckworth.
1977. Urine marking and territoriality in *Galago alleni* (Waterhouse 1837—Lorisoidea, Primates)—A field study by radio-telemetry. *Zeit. f. Tierpsychol.*, 43:113–138.

Davies, C. T. M., J. R. Brotherhood and E. Zeidifard.
1976. Temperature regulation during severe exercise with some observations on effects of skin wetting. *J. Appl. Physiol.*, 41:772–776.

Eisenberg, J. F.
1976. Communication mechanisms and social integration in the black spider monkey, *Ateles fusciceps robustus*, and related species. *Smithsonian Contribs. Zool.*, 213:1–108.

Eisenberg, J. F. and D. G. Kleiman.
1973. Olfactory communication in mammals. *Ann. Rev. Ecol. Syst.*, 3:1–32.

Epple, G.
1970. Quantitative studies on scent marking in the marmoset (*Callithrix jacchus*). *Folia Primat.*, 13:48–62.
1974. Olfactory communication in South American primates. *Ann. N. Y. Acad. Sci.*, 237:261–278.

Hainsworth, F. R. and E. M. Stricker.
1970. Salivary cooling by rats in the heat. Pages 611-626 in *Physiological and Behavioral Temperature Regulation*, edited by J. D. Hardy, A. P. Gagge and J. A. J. Stolwijk. Springfield, Ill.: C. C. Thomas.

Hill, W. C. O.
1938. A curious habit common to lorisoids and Platyrrhine monkeys. *Spolia Zeylanica*, B:21–65.

Johnson, R. P.
1973. Scent marking in mammals. *Anim. Behav.*, 21:521–535.

Johnston, R.
1977. The causation of two scent-marking behaviour patterns in female hamsters (*Misocricetus auratus*). *Anim. Behav.*, 25:317–327.

Kleiman, D. G.
1966. Scent marking in the Canidae. *Symp. Zool. Soc., London*, 18:167–177.

Ladell, W. S. S.
1965. Water and salt (sodium chloride) intakes. Pages 235-300 in *The Physiology of Human Survival*, edited by O. G. Edholm and A. L. Bacharach. London: Academic Press.

Michael, R. P. and E. B. Keverne
1968. Pheromones in the communication of sexual status in primates. *Nature*, 218:746–749.

Michael, R. P., E. B. Keverne and R. W. Bonsall.
1971. Pheromones: isolation of male sex attractants from a female primate. *Science* 172:964–966.

Milton, K.
1975. Urine-rubbing behavior in the mantled howler monkey, *Alouatta palliata. Folia Primat.*, 23:105–112.

Mykytowycz, R.
1962. Territorial function of chin gland secretion in the rabbit, *Oryctolagus cuniculus* (L.). *Nature*, 193:799.
1965. Further observations on the territorial function and histology of the sub-mandibular cutaneous (chin) glands in the rabbit (*Oryctolagus cuniculus* L.). *Anim. Behav.*, 13:400–412.

Oppenheimer, J. R.
1968. Behavior and ecology of the white-faced monkey, *Cebus capucinus*, on Barro Colorado Island. Ph. D. dissertation, University of Illinois, Urbana.

Oppenheimer, J. R. and E. C. Oppenheimer.
1973. Preliminary observations of *Cebus nigrivittatus* (Primates: Cebidae) on the Venezuelan llanos. *Folia Primat.*, 19:409–436.

Payne, A. P. and H. H. Swanson.
1970. Agonistic behaviour between pairs of hamsters of the same and opposite sex in a neutral observation area. *Behaviour*, 36:259–269.

Petter, J. J.
1962. Recherches sur l'ecologie et l'ethologie des Lemuriens malgaches. *Mem. Mus. Hist. Nat. A.*, 27(1):146.

Ralls, K.
1971. Mammalian scent marking. *Science*, 171:443–449.

Richards, P. W.
1952. *The Tropical Rain Forest.* Cambridge: Cambridge University Press.

Schmidt, U. and E. Seitz.
1967. Waschen mit Harn zum Zweck der Thermoregulation bei Totenkopfaffen (*Saimiri sciureus* L.). *Anthrop. Anz.*, 30:162–165.

Seitz, E.
1969. Die Bedeutung geruchlicher Orientierung beim Plumplori *Nycticebus coucang* Boddaert 1785 (Prosimii, Lorisidae). *Zeit. f. Tierpsychol.*, 26:73–103.

Stricker, E. M., J. C. Everett and R. E. A. Porter.
1963. The regulation of body temperature by rats and mice in the heat: Effects of desalivation and the presence of a waterbath. *Commun. Behav. Biol.*, 2(part A): 113–119.

Thiessen, D. D., M. Graham, J. Perkins and S. Marcks.
1977. Temperature regulation and social grooming in the mongolian gerbil (*Meriones unguiculatus*). *Behav. Biol.*, 19:279–288.

Thorington, R. W., Jr.
1967. Feeding and activity of *Cebus* and *Saimiri* in a Colombian forest. Pages 180–184 in *Progress in Primatology*, edited by D. Starck, R. Schneider and H. H. Kuhn. Stuttgart: Fischer.
1968. Observations of squirrel monkeys in a Colombian forest. Pages 235-253 in *The Squirrel Monkey*, edited by L. A. Rosenblum and R. W. Cooper. New York: Academic Press.

SECTION 4:

Bats, Carnivores, and Rodents

Cerdocyon thous, the crab-eating fox. This medium sized canid is one of the more important small predators adapted to the llanos ecosystem. A versatile feeder, it lives in monogamous pairs.

Introduction

When the numbers of mammalian species are compared for areas in the tropics showing comparable climate and topographic diversity, the neotropics stands out by consistently exhibiting a high percentage of bat species. Handley (1966) offers data demonstrating that 51 percent of the mammalian species found in the Republic of Panama are members of the order Chiroptera. In his publication of 1976, Handley lists 307 species of terrestrial mammals for Venezuela of which 55 percent are bats. The numbers of bat species in the neotropics presents a formidable challenge to the mammalian ecologist. Studies of behavior have only just begun. The pioneering efforts by Bradbury and Vehrencamp (1976a, 1976b, 1977a, 1977b) with the family Emballonuridae in Central America set a standard for years to come. The vast majority of Chiropteran species in the neotropics belong to the family Phyllostomatidae. Our knowledge concerning the ecology and behavior of this group has been increasing exponentially since 1960s. The volumes edited by R. J. Baker, J. Knox-Jones and D. Carter (1976, 1977, 1978) serve to summarize what we know concerning the morphology, physiology and ecology of the phyllostomatids.

With over 150 species of bats known to exist in Venezuela, the mammalian ecologist must attempt to understand not only the behavior patterns of individual species but the manner in which such a diverse group of sympatric species partitions the available energy in the ecosystem. The task is made doubly challenging because of the nocturnal habits of bats and the great difficulties in obtaining data by direct observation.

In the contribution by August, the problem of communication and group activities is analyzed by means of play-back experiments to *Artibeus jamaicensis* when foraging. His experiments suggest that the distress calls of a captured *Artibeus* serve to attract large numbers of *Artibeus* to the site. They respond by swooping over the source of the disturbance. The calling bat may in fact be attempting to startle the predator with its loud cries. Secondarily, other bats are attracted and their activities may serve to drive the predator away from the vicinity of high foraging activities. The system itself may not be considered a form of altruistic behavior and may be partially analogous to the mobbing behavior shown by birds toward potential avian predators.

Brady has conducted an in-depth study of crab-eating fox behavior in captivity (Brady, 1978, 1979). His field observations were conducted in two successive years during both the dry and the rainy season. He was able to individually identify six adult foxes and rather adequately map home ranges. What emerges here is a typical canid pattern of social organization and habitat utilization. Apparently the crab-eating fox forms pairs on a rather permanent basis. Although the male and female do not necessarily hunt together, through their activities they repel like-sexed invaders and cooperate in the rearing of the young. Hunting techniques vary with the seasons reflecting different strategies for the most abundant prey items. At certain times of the year, the crab-eating fox is a significant predator on land crabs, but at other times of the year the fox avails itself of the abundant terrestrial rodent fauna. *Zygodontomys brevicauda* is the most abundant terrestrial rodent at the close of the dry season and, with the transition to the rainy season, lowland areas which had been used as refugia by *Zygodontomys* are flooded. As a result, the density of the rodents in the available high ground becomes vastly increased. It is at this time that *Cerdocyon* avails itself of the relative rodent density. In Section 5, *Zygodontomys* is considered together with the other members of the rodent fauna in the llanos.

The caviomorph rodents represent an important mammalian radiation in the neotropics. Caviomorphs have been well established in South America since the Miocene and in the course of their long evolutionary history have occupied a wide variety of ecological niches. The agouti, *Dasyprocta aguti*, the paca, *Agouti paca*, and the spiny rat, *Proechimys semispinosus*, are the three most abundant caviomorph rodents in Guatopo National Park. In the llanos, the agoutis and pacas are generally confined to gallery forest and *Proechimys* is replaced by the arboreal *Echimys semivillosus*. *Proechimys* has been the subject of a field study by Fleming (1971) in Panama. Smythe (1978) has summarized the natural history of the Central American agouti, *Dasyprocta punctata*, again in Panama.

The article by Kleiman, et al., explores the reproduction of caviomorph rodents and points up the diversity in litter size, interbirth interval, and growth and maturation, when data are pooled from a variety of sources. The large *Hydrochaeris hydrochaeris*, or capybara, of the Venezuelan llanos produces large litters, given its size, and although the growth of the young is rather slow compared with the smaller forms, the potential productivity of the capybara in the llanos is tremendous (Ojasti, 1973). The basic data set compared by Kleiman provides useful fundamentals for the interpretation of productivity by other caviomorph rodents. Most of the data are derived from captive studies, many unpublished, which were conducted here at the National Zoological Park. Field data from Smythe, Ojasti, Jon Rood (1972) and Peter Dalby (1975) is incorporated to give a balanced picture of trends in reproductive adaptation as exemplified by this suborder of rodents.

The rodents were an important part of our small mammal studies and considerable effort was expended by Peter August and M. A. O'Connell in a trap, mark and release program. Some of their results are presented in Section 5. Other aspects of their studies will be published separately. The caviomorph data are here presented as a convenient unit synopsizing some 14 years of research at the National Zoological Park.

J.F.E.

Literature Cited

Baker, R. J., J. Knox-Jones and D. Carter (editors).

1976. *Biology of Bats of the New World Family Phyllostomatidae. Part I.* Special Publications of the Museum, No. 10. Lubbock: Texas Tech Press.

1977. *Biology of Bats of the New World Family Phyllostomatidae. Part II.* Special Publications of the Museum, No. 13. Lubbock: Texas Tech Press.

1978. *Biology of Bats of the New World Family Phyllostomatidae. Part III.* Lubbock:Texas Tech Press.

Bradbury, J. W. and S. L.Vehrencamp.

1976a. Social organization and foraging in emballonurid bats. I. Field studies. *Behav. Ecol. and Sociobiol.*, 1:337–381.

1976b. Social organization and foraging in emballonurid bats. II. A model for the determination of group size. *Behav. Ecol. and Sociobiol.*, 1:383–404.

1977a. Social organization and foraging in emballonurid bats. III. Mating systems. *Behav. Ecol. and Sociobiol.*, 2(1):1–19.

1977b. Social organization and foraging in emballonurid bats. IV. Parental investment patterns. *Behav. Ecol. and Sociobiol.*, 2(1):19–31.

Brady, C. A.

1978. Reproduction, growth and parental care in crab-eating foxes, *Cerdocyon thous*, at the National Zoological Park, Washington. *Int. Zoo Yb.*, 18:130–134.

1979. Mechanisms of communication in the crab-eating fox (*Cerdocyn thous*), the maned wolf (*Chrysocyon brachyurus*), and bush dog (*Speothos venaticies*). Ph.D. thesis, Ohio University, Athens.

Dalby, P. L.
1975. Biology of pampa rodents, Balcarce Area, Argentina. *Pubs. Mus. Michigan State Univ., Biol. Series,* 5(3):1–271.

Fleming, T. H.
1971. Population ecology of three species of neotropical rodents. *Misc. Pubs. Mus. Zool., Univ. Michigan,* 143:1–77.

Handley, C. O., Jr.
1966. A checklist of the mammals of Panama. Pages 753–795 in *Ectoparasites of Panama,* edited by R. L. Wenzel and V. J. Tipton. Chicago: Field Museum of Natural History.
1976. Mammals of the Smithsonian Venezuelan Project. *Brigham Young Univ. Sci. Bull., Biol. Series,* 20(5):1–91.

Ojasti, J.
1973. *Estudio Biologico del Chigüuire o Capibara.* Caracas: Fundo Nacional de Investigaciones Agropecuaraias.

Rood, J. P.
1972. Ecological and behavioral comparisons of three genera of Argentine cavies. *Anim. Behav. Monogr.,* 5:1–83.

Smythe, N.
1978. The natural history of the Central American agouti (*Dasyprocta punctata*). *Smithsonian Contribs. Zool.,* 257:1–52.

PETER V. AUGUST
Department of Biology
Boston University
Boston, Massachusetts 02215

Distress Calls in *Artibeus jamaicensis*: Ecology and Evolutionary Implications[1]

ABSTRACT

Artibeus jamaicensis respond to tape-recorded playbacks of the distress calls of conspecifics. The response by *Artibeus* involves close swoops to the source of the sounds but the bats never land on or near the speaker. The response is limited to the vicinity of the source of the distress call and is not a general fleeing response. Visual observations and mist netting show that at least *Phyllostomus hastatus* respond to *Artibeus jamaicensis* distress calls suggesting that there is no species specificity in who responds to these calls. The overall response is similar to avian mobbing behavior and may serve as a mechanism to reduce losses to nocturnal predators of bats.

RESÚMEN

Se discute la respuesta de *Artibeus jamaicensis* al sonido de peligro de conespecies de una grabadora. El *Artibeus* voló cerca de los alto parlantes pero nunca se paró encima o cerca de ellos. Los murciélagos siempre vuelan soló en la vecindad de los altos parlantes y no se observa una reacción de fuga.

[1] This research is dedicated to the memory of Professor Robert L. Packard

Introduction

Research emphasis on bat vocalizations has been strongly oriented towards ultrasonic calls (Simmons et al., 1975; Gould, 1977; Novick, 1977), but there is a growing body of evidence suggesting that bats regularly use low-frequency (below 20 kHz) sounds in social vocalizations (Bradbury, 1977a). There are a number of reports of bats responding to sonic calls given by bats under physical duress (distress calls) (Guthrie, 1933; O'Farrell, and Miller, 1972; Tuttle, 1976). Recently, Fenton et al. (1976) reported the results of four nights of playback experiments to roosting *Myotis lucifugus*. They found an increase in flight activity during playbacks of conspecific distress calls but little or no activity during playbacks of feeding sounds of bats, 45 kHz pure tone, distress calls of *Eptesicus fuscus*, or silence. They did not suggest any possible functions for the behavior.

Artibeus jamaicensis is a frugivorous phyllostomatid bat found from southern Mexico to the Mato Grasso of Brazil (Gardner, 1977; Jones and Carter, 1976). In any one area, *Artibeus* can be especially abundant. A mist net placed under the canopy of a fruiting tree can catch large numbers of *A. jamaicensis* (pers. obs.; Morrison, 1978). When removing these bats from the mist nets or otherwise handling them, they will often produce loud, audible vocalizations sometimes attracting large numbers of other bats into the area (*pers. obs.*; Guthrie, 1933). To my knowledge, there have not been any detailed studies of either this low-frequency signal or the response in any neotropical bat.

The purpose of this research was to determine if tape recorded playbacks of *Artibeus jamaicensis* distress calls, artificial *Artibeus* distress calls, or white noise would elicit an attraction response similar to that observed using hand-held live bats.

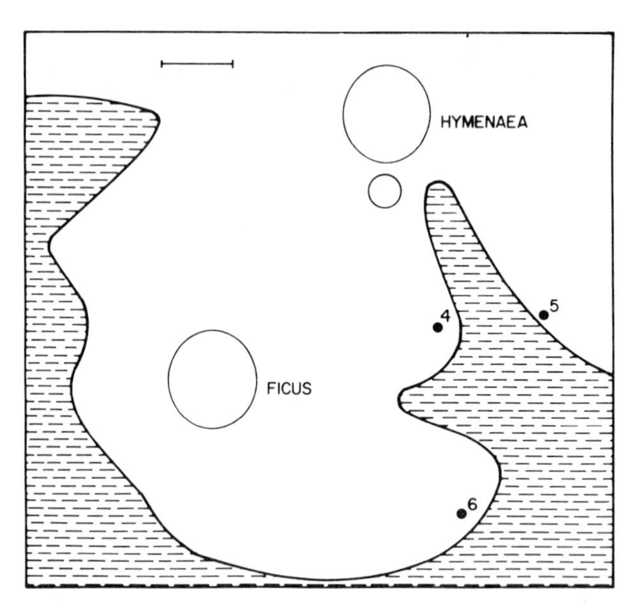

Figure 1. Observer and speaker positions at the *Hymenaea courbaril* experimental area 1. Shaded areas indicate shrub or low tree cover. The bar represents 10 m.

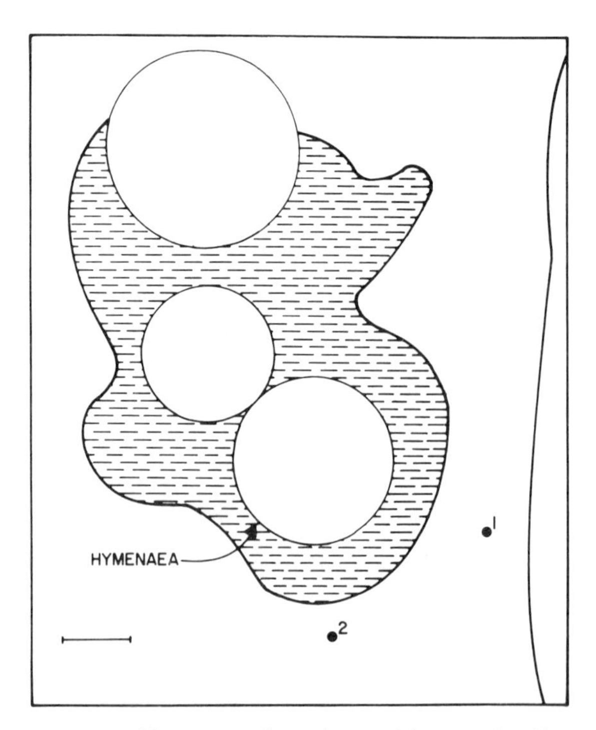

Figure 2. Observer and speaker positions at the *Hymenaea courbaril* experimental area 2. Shaded areas indicate shrubs or low tree cover. The bar represents 10 m.

Materials and Methods

This research was done on Hato Masaguaral, 45 km south of Calabozo, Estado Guarico, Venezuela (8° 33′ N. Lat. × 67° 36′ Long.). Masaguaral lies in the llanos region of Venezuela and is considered dry tropical forest (Ewel and Madriz, 1968). Troth (see p. 17) has described the study area in detail. All experiments were done during the dry season from December 1977 to May 1978.

Playback experiments were conducted at four sites containing dense concentrations of foraging *Artibeus jamaicensus*. Locations one and two were large *Ficus trigonata* trees with numerous ripe fruits. Locations three and four (Figures 1 and 2) were flowering algorobbo trees (*Hymenaea courbaril*). All but four nights of experimentation were done at the *Hymenaea* sites. These were superior experimental areas because they took as long as two to three weeks to complete flowering thus supporting a pool of bats for a prolonged time. A fig tree laden with ripe fruits can be stripped by frugivorous birds and bats in less than a week (*pers. obs.*).

Chiropteran distress calls have been defined as vocalizations produced while a bat is under physical duress (Fenton et al., 1976). All *Artibeus jamaicensis* distress calls were recorded immediately after a captured subject was removed from the mist net. I assume that *Artibeus* do not have any special vocalizations for this circumstance and perceive the bat handler as a potential predator.

All recordings were made at a tape speed of 19 cm per sec on a Uher 4000 Report IC tape recorder using a Uher M514 microphone on Scotch magnetic tape. Playbacks were made through the built-in speaker of the Uher tape recorder. The *Artibeus jamaicensis* distress call used in the experiments was recorded from an adult male on 30 December 1977. Voucher specimens of *Artibeus jamaicensis* were taken and are deposited in The Museum, Texas Tech University, Lubbock, Texas. Other recordings were not used in order to minimize between-recording variance in the results. I made the artificial *Artibeus* distress call (mimic call) by slowing down a recording of an *Artibeus jamaicensis* distress call to 2.4 cm per sec (one-eighth normal speed) and practiced imitating the slowed version. When I was sufficiently able to duplicate the pattern of sounds, I recorded my imitation at a speed of 2.4 cm per sec. This was speeded up to 19 cm per sec for the experimental playbacks. The slowed *Artibeus* call sounded like a nasal "neah-neah-neah" while the call played at normal speed was a rapid, high pitched "eee-eee-eee". The white noise recording was made from interference static at 7.2 MHz on a short wave radio (see Marcellini, 1977 for a discussion of this technique). Sound spectrographs of these calls will be presented in a later paper.

Each experiment consisted of 80 seconds of silence (PRECALL); 80 seconds of *Artibeus jamaicensis* distress call, mimic call, or white noise (CALL); and 80 seconds of silence (POSTCALL). Experiments started between 1930 and 2230 hours. On most nights six replicates of a given paradigm were tested, each separated by a five-minute pause. An increase in bat activity was judged a response to a playback. Bat activity was quantified by holding a sheet of cardboard paper, with a 9.5 cm square cut in the middle, 15 to 20 cm from the observer's face. All bats that passed through this field of view during the precall, call, or postcall portions of an experiment were noted.

Four paradigms were tested: 1) responses of bats to playbacks of the *Artibeus jamaicensis* distress call; 2) responses of bats to playbacks of the mimic of an *Artibeus* distress call; 3) responses of bats at a position approximately 30 m from the speaker during playbacks of the mimic call; and 4) responses of bats to playbacks of white noise. In paradigms one and two, the observer was stationed adjacent to the speaker approximately 30 m from the *Hymenaea* or *Ficus* tree. In paradigm three, one observer was stationed at the sound source (position 4 or 6 in Figure 1, or position 1 in Figure 2) and a second observer at a position either away from the speaker but equidistant to the feeding tree (position 5 in Figure 1 or position 2 in Figure 2) or at a different distance from the feeding tree (position 4 and 6 in Figure 1). In these two observer experiments, a seventh replicate was run each night with both observers at the sound source to determine if there was any bias between observers in counting bat passes. The white noise experiments (paradigm 4) were done over two nights. On each night there were three replicates of the white noise playback and three replicates of the *Artibeus jamaicensis* distress call. The sequence of playbacks was determined with a random numbers table. In these, bat activity was quantified by shining a low intensity lantern beam over the speaker and counting the number of bats passing through the illuminated area during a playback.

Results

Table 1 summarizes the results of the playback experiments using the *Artibeus jamaicensis* distress call. Clearly, more bats passed in the vicinity of the speaker during the call portion of the experiment than before or after. The greater activity during the postcall portion, as compared to the precall portion, was due to bats lingering in the area after the call phase of the tape. There was a marked response to the playback of the mimic call (Table 1). The distribution of bat passes over the three segments of the experiments was the same as observed for the *Artibeus* distress call.

The results of the experiments with two observers at different distances from the feeding tree (Table 2) and equidistant from the feeding tree (Table 3) show that the response by bats was limited to the area near the speaker. Two-way analysis of variance revealed highly significant position, call portion, and interaction effects (Tables 4 and 5). The results of the control experiments determining any bias in counting bat passes between two observers are given in Table 6. Two by three tests of independence using the G statistic for each replicate showed no significant difference between observers.

Table 1. Results of playbacks using tapes of the *Artibeus jamaicensis* distress call and the mimic call[1]

Call	*No. of reps.*	*No. of nights*	*Total bat passes*			*Probability of random dist.*			
			PRE	*CALL*	*POST*	*n.s.*	*<.05*	*<.01*	*<.001*
Mimic	42	7	18	1727	189	0	1	0	41
Artibeus	28	5	21	675	295	7	1	1	19

[1] The probability, that bat passes were randomly distributed over the three segments of each replicate, was determined by a G-test of goodness of fit.

Table 2. Distribution of bat passes in two-observer experiments where observers are at different distances from the feeding tree[1]

Observer	*Total bat passes*		
	PRE	*CALL*	*POST*
Near speaker	19	413	240
Away from speaker	1	5	10

[1] N = 12 replicates.

Table 3. Distribution of bat passes in two-observer experiments where observers are equidistant from the feeding tree[1]

Observer	*Total bat passes*		
	PRE	*CALL*	*POST*
Near speaker	4	592	78
Away from speaker	9	48	13

[1] N = 18 replicates

Table 4. Results of two-way analysis of variance of the effects of playback segment (PRE, CALL, POST) and the position of observers in the experiments with two observers at different distances from the feeding tree—ANOVA Test

Source of variation	*df*	*SS*	*MS*	*F*
Subgroups	5	12480.4	2496.1	—
segment	2	3326.8	1663.4	38.1[1]
observer position	1	5994.9	5994.9	137.2[1]
segment × observer interaction	2	3158.8	1579.4	36.1[1]
Error	426	18599.3	43.7	—
Total	431	31079.8	—	—

[1] p = <0.001

Table 5. Results of two-way analysis of variance of the effects of playback segment (PRE, CALL, POST) and the position of observers when the two observers were equidistant from the feeding-tree experiments—ANOVA Test

Source of variation	*df*	*SS*	*MS*	*F*
Subgroups	5	14825.7	2965.0	—
segment	2	6487.2	3243.6	313.7[1]
observer position	1	3377.9	3377.3	326.7[1]
segment × observer interaction	2	4960.6	2480.3	239.9[1]
Error	642	6641.0	10.3	—
Total	647	21466.7	—	—

[1] p = <0.001

Table 6. Results of playback experiments with two observers at the speaker[1]

Observer	*Total bat passes*		
	PRE	*CALL*	*POST*
A	0	82	24
B	0	67	20

[1] N = 4 replicates

Table 7. Results of control experiments using playbacks of the *Artibeus jamaicensis* distress call and white noise[1]

Sound	*Total bat passes*		
	PRE	*CALL*	*POST*
Artibeus distress call	5	77	16
white noise	4	5	5

[1] N = 6 replicates

The results of the playback experiments using white noise during the call portion of the experiment are given in Table 7. It is evident that the bats were not responding to merely a high noise level.

There was a general pattern of declining intensity in response over the six replicates on a given night of experimentation (Figure 3). This either reflects habituation to the playbacks or possibly a decline in the number of bats in the area that can potentially respond to the tape. In experiments done at *Ficus* locality one, I noticed a sharp decline in the intensity of foraging activity at the fig tree after the playback experiments each night. It was difficult to judge changes in foraging activity at the *Hymenaea* sites because the trees were so large and had so many bats flying through them.

Mist netting near the *Hymenaea* locations showed that *Artibeus jamaicensis* was not the only species of bat using the tree as a foraging site. One 12.8 × 2.1 m mist net was set 30 m from the *Hymenaea* for three hours after sunset for three nights. The following bats were captured: *Artibeus jamaicensis*, 50; *Carollia perspicillata*, 8; *Phyllostomus discolor*, 17; and *P. hastatus*, 12. During playbacks at both *Hymenaea* sites I observed *Phyllostomus hastatus* responding to *Artibeus* distress calls and the mimic call. They were easily distinguished from the other taxa by their large size and red color. On one occasion I played the 80 seconds recording of the *Artibeus jamaicensis* distress call behind a mist net placed midway between the *Ficus* and the *Hymenaea* at locality 3. During one playback: four *Carollia perspicillata*, four *Artibeus jamaicensis*, and four *Phyllostomus discolor* were captured. It was not possible to continue mist netting during the playback experiments as there was insufficient time between replicates to remove captured bats from the net.

On a number of occasions, after completing a night's experiments, I observed bats swooping toward the recorder during a playback and generally, this activity was in the close vicinity of the speaker. Bats would fly toward the speaker and pass within one or two meters but would never land on or near it. During experiments, bats would often pass close enough to me to generate a perceptible draft in the area. In-flight collisions were sometimes heard when there were large numbers of bats responding to a playback.

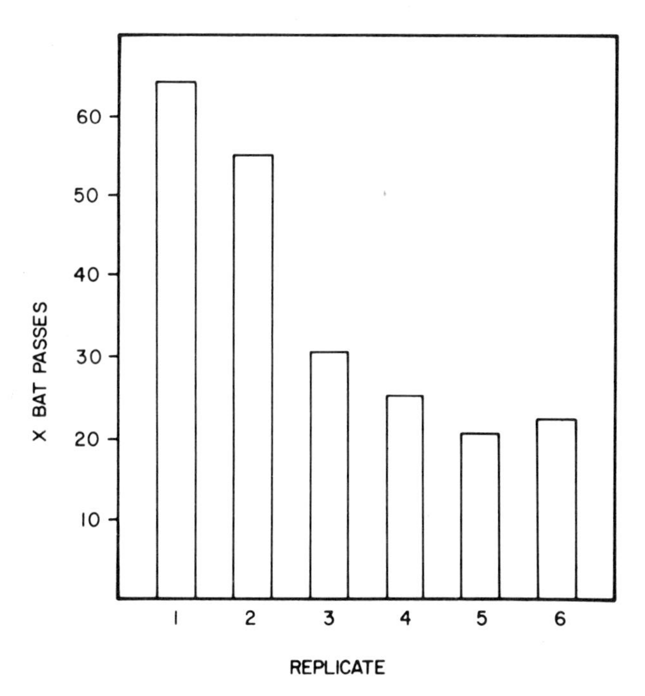

Figure 3. Response intensity over the six nightly replicates for all playback experiments using the *Artibeus jamaicensis* distress call and the *Artibeus* mimic call.

Discussion

The results indicate that tape recorded playbacks of *Artibeus jamaicensis* distress calls and mimic sounds of an *Artibeus* distress call attract bats to the sound source. When the playback of recorded sounds was completed, bats left the vicinity of the speaker. This response is limited to the area around the speaker and is not a general fleeing response. This is quite similar to the response obtained using a live, hand-held *Artibeus* that is giving distress calls (*pers. obs.*). These results are consistent with the Fenton et al. (1976) description of *Myotis lucifugus* responding to distress calls of conspecifics. Observations of bat activity at the sound source show that the response includes swoops at the speaker but bats never landed on or near it. Experiments using playbacks of white noise show that the bats were not just responding to increased noise levels.

It is likely that there is no species specificity in responses to distress calls of some bats. Multispecies responses to distress calls have been noted in bats (Tuttle, 1976) as well as in birds (Marler, 1957). It is interesting to note that all of the bats attracted to the *Artibeus jamaicensis* distress call or the mimic call in my experiments were mainly frugivorous or florivorous (Gardner, 1977) and often forage in the same foot patches.

Manufacture specifications of the tape recorder used in these experiments indicate that this model has an upper response limit of 20 kHz. This implies that the sounds reproduced during experimental playbacks are low-frequency calls and do not contain ultrasonic components. The tremendous response to the mimic of an *Artibeus* distress call confirms that high-frequency components are not necessary to elicit responses in bats. The response to the mimic call also indicates that considerable variability is allowable in the signal.

Fenton et al. (1976), using a night vision scope, observed the flight behavior of *Myotis lucifugus* to playbacks of conspecific distress calls. They found that during these calls "bats altered their flight patterns and began to swoop down towards the speaker, often approaching it within 10 cm." O'Farrell and Miller (1972) reported similar responses by *Pipistrellus hesperus* to the calls of other pipistrelles contained in a cloth collecting sack. They noted the close swooping behavior and occasionally found bats to land on the sacks and then take flight again. Kulzer (1961) found that *Rousettus aegyptiacus* was attracted to young contained in cloth sacks as did Nelson (1965) with *Pteropus poliocephalus*. Tuttle (1976) noted that the distress calls of small bats attract larger species as well as those the size of the caller while the calls of larger bats, such as *Phyllostomus hastatus* bring in other *P. hastatus* but possibly scare the smaller taxa away.

A number of functions have been suggested for the use of low-frequency vocalizations in bats. Mohres (1967) reported that *Rhinolophus ferrumequinum* use vocalizations to locate roost mates, although ultrasonics may also be important in this role. O'Farrell and Miller (1972) suggested that pipistrelle vocalizations may serve to attract members of the opposite sex. On one occasion they observed a male *Pipistrellus hesperus* come to a sack containing only pregnant females. They were unable to present a tenable hypothesis explaining this observation. Bradbury and Emmons (1974) reported that male *Saccopteryx bilineata* responded to low-frequency (under 30 kHz) recordings of females. Sonic "honks," "buzzes," and "clings" are an integral part of the mating display of *Hypsignathus monstrosus* (Bradbury, 1977b).

The only circumstance in which I have regularly heard *Artibeus jamaicensis* produce these low-frequency vocalizations has been while removing them from mist nets or handling them. Not every bat would call but there was no apparent pattern of unusually vocal animals with respect to age or sex. In view of the limited context in which these calls are regularly given, it is unlikely that they are produced to call in potential mates. These vocalizations are most likely given as distress calls. This is not to say that these are the only low-frequency vocalizations produced by *Artibeus jamaicensis*. The complete vocal repertoire of social calls in these bats most likely contains other low-frequency signals but the precise function and nature of these calls will remain obscure until there is considerably more data on all aspects of the social biology of *Artibeus*.

Morrison (1978) has described the foraging behavior of *Artibeus jamaicensis* in Panama. These bats leave their day roosts and fly directly to the fruit tree visited on the previous day. After arriving at the foraging site, they establish a feeding roost, often on the frons of a nearby spiny palm. During the night these bats make several feeding passes to the fruiting tree, pick a fruit, and carry it back to the feeding roost to eat. It has been suggested that the feeding roost serves to minimize the risk of predation by stalking predators (Fenton and Fleming, 1976; Morrison, 1978). There is a notable reduction in activity by *Artibeus jamaicensis* on bright, moonlit nights (Morrison, *in press*). Morrison suggested that this may be a strategy to reduce losses to visually oriented nocturnal predators.

Predation risks must be sufficiently high to justify these behavioral modifications. In my study area I have seen *Didelphis marsupialis* eat *Artibeus jamaicensis* captured in the lower pockets of mist nets. Others have noted *Didelphis*, as well as *Philander*, attacking and consuming bats (Fleming, 1972; Morrison, 1978). Other possible predators of fruit bats found in my study area include hawks, falcons, owls, other bats (eg. *Vampyrum spectrum*), and arboreal snakes (Fenton and Fleming, 1976; Thomas, 1974; Lemke, 1978).

Distress calls in bats may be a strategy to reduce mortality through predation. Loud, low-frequency vocalizations may serve to startle a predator thus affording the captured bat an opportunity to escape. The intense activity by other bats in the vicinity of the source of the distress call may be similar to the mobbing behavior seen in many bird taxa. It is generally agreed that mobbing acts to confuse and drive away a predator (Wilson, 1975). To scare a potential predator away from an intensively used foraging site would be to the benefit of all bats in the area as there would be a notable per capita reduction in the probability of being taken by the predator. The risks to either the calling bat or the bats mobbing the predator are minimal. The calling bat would be in a position to lose all if not startling vocalization were given. By calling, it has at least some chance of being released. My observations of bats responding to distress calls have shown that they do not land at the speaker but merely pass close by. The chances of a predator actually catching a passing bat are quite low.

Close swoops at the source of a distress call may also benefit responding bats in a number of other ways. All of the bats in the area have very accurate information concerning the location of the predator for the duration of the call. The attraction to the call may enhance predator surveillance by responders. It may also be possible that the intense flight activity of bats at the source of the distress call may inform the predator that these prey are aware of its presence. If the element of surprise is an important part of the hunting strategy of the predator, continued stalking of bats may not be profitable (Smythe, 1970, 1977).

The use of low-frequency vocalizations over ultrasonics in distress calls is quite logical. It has been shown that high-frequency sounds are extremely directional (Kinsler and Frey, 1962) and have high attenuation with distance (Morton, 1975). Low-frequency sounds have low attenuation with distance and are relatively omnidirectional. As most predators of bats hear in the low-frequency range, it is not surprising that calling bats use sonic calls to startle predators. Owing to the omnidirectional properties of sonic calls, a greater number of bats in a given location are alerted to the presence of a potential predator.

Artibeus jamaicensis often forage with other frugivorous bats in the same tree. A potential predator of *Artibeus* would, in most cases, be a potential predator of *Carollia* or *Phyllostomus* as well. It would be to the advantage of all bats feeding at the same location to react and drive the predator away. Once the predator was removed from the area, all taxa would be able to continue foraging with a reduced probability of being consumed.

There is a paucity of data concerning what bat taxa produce and respond to low-frequency distress vocalizations. Baker (personal communication) has observed *Brachyphylla cavernarum* responding to distress calls of conspecifics on small predator-free islands in the Lesser Antilles. This is especially interesting in view of the fact that *Brachyphylla*, a genus endemic to the Antilles, has retained this possible predator defense mechanism while undergoing considerable morphological changes throughout its evoluntionary history. This may indicate that a number of low-frequency vocalizations are produced in social contexts and are not limited to antipredator strategies. Fenton (1977) has found that the social calls of *Myotis lucifugus* tend to be of lower frequency than echolocation calls. The presence of distress call-mobbing response in both temperate vespertilionids (Guthrie, 1933; Fenton *et al.*, 1976) as well as tropical phyllostomatids suggests that this behavior was present in bats ancestral to these groups and has been retained by many modern taxa.

In summary, the results reported herein indicate that *Artibeus jamaicensis* distress calls attract large numbers of *Artibeus* and other bat taxa to the source of the sound. This response is quite localized and is similar to avian mobbing behavior. The function of the distress call may be to startle the predator that has captured the calling bat. Other bats making swoops at the source of the distress call may be attempting to drive the predator away from the area that is being intensively used by the bats. Low-frequency vocalizations are adaptively superior to ultrasonic sounds in distress calls because of their potential to alarm predators and alert nearby bats to the presence of a predator. *Artibeus* distress calls do not seem to be species specific in eliciting responses of other bats. This would be expected if, by participating in the mobbing of a predator, the per capita probability of being captured by the predator is reduced in all taxa.

Much remains to be learned about distress calls in bats. Further experiments are needed to determine which bat taxa respond to whose calls. It would be interesting to see if multispecies responses are limited to bats of the same feeding guild as the caller. Experiments are needed to document potential predator behavior during mobbing by bats. Research is needed to determine if the reduction in response intensity over repreated distress calls, as observed in this study, is due to habituation to the call or if bats are leaving the area. Further studies on low-frequency calls of bats will likely show that sonic sounds constitute an important component in the total bat vocal repertoire.

Acknowledgments

It is my pleasure to acknowledge the gracious hospitality and support given to me by Tomás Blohm. E. Mondolfi, R. Ohrta, and R. Rudran provided logistic aid. W. Mader and L. August served as second observers. John Robinson assisted in some of the experiments and provided stimulating discussion. W. Mader, J. Robinson, R. H. Wiley, T. Kunz, R. J. Baker, M. B. Fenton, and F. Wasserman commented on the manuscript. This research was made possible through the support and guidance given to me by John Eisenberg. Financial support was provided by a Predoctoral Fellowship from the Smithsonian Institution.

Literature Cited

Bradbury, J. W.
1977a. Social organization and communication. Pages 1–72 in *Biology of Bats*, edited by W. A. Winsatt. New York: Academic Press.
1977b. Lek mating behavior in the hammer-headed bat. *Zeit. f. Tierpsychol.*, 45:225–255.

Bradbury, J. W. and L. H. Emmons.
1974. Social organization of some Trinidad bats. I. Emballonuridae. *Zeit. f. Tierpsychol.*, 36:137–183.

Ewel, J. J. and A. Madriz.
1968. *Zonas de Vida de Venezuela.* Caracas: Ministario de Agricultura y Cria.

Fenton, M. B.
1977. Variation in the social calls of little brown bats (*Myotis lucifugus*). *Canadian J. Zool.*, 55:1151–1157.

Fenton, M. B. and T. H. Fleming.
1976. Ecological interactions between bats and nocturnal birds. *Biotropica*, 8:104–110.

Fenton, M. B., J. J. Belwood, J. H. Fullard, and T. H. Kunz.
1976. Responses of *Myotis lucifugus* (Chiroptera:Vespertilionidae) to calls of conspecifics and to other sounds. *Canadian J. Zool.*, 54:1443–1448.

Fleming, T. H.
1972. Aspects of the population dynamics of three species of opossums in the Panama Canal Zone. *J. Mammal.*, 53:619–623.

Gardner, A. L.
1977. Feeding habits. Pages 293–350 in *Biology of Bats of the New World Family Phyllostomatidae. Part II*, edited by R. J. Baker, J. Knox Jones, and D. C. Carter. Lubbock, Texas: Special Publication of the Museum, Texas Tech University.

Gould, E.
1977. Echolocation and communication. Pages 247–279 in *Biology of Bats of the New World Family Phyllostomatidae. Part II*, edited by R. J. Baker, J. Knox Jones, and D. C. Carter. Lubbock, Texas: Special Publication of the Museum, Texas Tech University.

Guthrie, M. J.
1933. Notes on the seasonal movements and habits of some cave bats. *J. Mammal.*, 14:1–18.

Jones, J. K. and D. C. Carter.
1976. Annotated checklist, with keys to subfamilies and genera. Pages 7–38 in *Biology of Bats of the New World Family Phyllostomatidae. Part I*, edited by R. J. Baker, J. Knox Jones, and D. C. Carter. Lubbock, Texas: Special Publication of the Museum, Texas Tech University.

Kinsler, L. E. and A. R. Frey.
1962. *Fundamentals of Acoustics.* New York: John Wiley and Sons, Inc.

Kulzer, R.
1961. Über die Biologie der Nil-Flughunde (*Rousettus aegyptiacus*). *Nat. Volk.*, 91:219–228.

Lemke, T. O.
1978. Predation upon bats by *Epicrates cenchris cenchris* in Colombia. *Herp. Rev.*, 9:47.

Marcellini, D. L.
1977. The function of a vocal display of the lizard, *Hemidactylus frenatus* (Sauria: Gekkonidae). *Anim. Behav.*, 25:414–417.

Marler, P. R.
1957. Specific distinctness in the communication signals of birds. *Behav.*, 11:13–39.

Mohres, F. P.
1967. Communicative character of sonar signals in bats. Pages 939–948 in *Animal Sonar Systems*, edited by R. G. Busnel. Joy-en-Jasas, France: Lab. Physiol. Acoust.

Morrison, D. W.
1978. Foraging ecology and energetics of the frugivorous bat, *Artibeus jamaicensis. Ecology*, 59:716–723.
In press Lunar phobia in a neotropical fruit bat, *Artibeus jamaicensis. Anim. Behav.*,

Morton, E.
1975. Ecological sources of selection on avian sounds. *Amer. Nat.*, 109:17–34.

Nelson, J. E.
1965. Behavior of Australian Pteropodidae (Megachiroptera). *Anim. Behav.*, 13:544–557.

Novick, A.
1977. Acoustic orientation. Pages 74–187 in *Biology of Bats*, edited by W. A. Wimsatt. New York: Academic Press.

O'Farrell, M. J. and B. W. Miller.
1972. Pipistrelle bats attracted to vocalizing females and to a blacklight insect trap. *Amer. Midl. Nat.*, 88:462–463.

Simmons, J. A., D. J. Howell, and N. Suga.
1975. Information content of bat sonar echoes. *Amer. Sci.*, 63:204–215.

Smythe, N.
1970. On the existence of "pursuit invitation" signals in mammals. *Amer. Nat.*, 104:491–494.
1977. The function of mammalian alarm advertising: Social signals or pursuit invitation. *Amer. Nat.*, 111:191–194.

Thomas, M. E.
1974. Bats as a food source for *Boa constrictor*. *J. Herp.*, 8: 188.

Troth, R. G.
1979. Vegetational types on a ranch in the central llanos of Venezuela. Pages 17–30 in *Vertebrate Ecology in the Northern Neotropics*, edited by John F. Eisenberg, Washington, D.C.: Smithsonian Institution Press.

Tuttle, M. D.
1976. Collecting techniques. Pages 71–88 in *Biology of Bats of the New World Family Phyllostomatidae. Part I*, edited by R. J. Baker, J. Knox Jones, and D. C. Carter. Lubbock, Texas: Special Publication of the Museum, Texas Tech University.

Wilson, E. O.
1975. *Sociobiology, the New Synthesis*. Cambridge, Massachusetts: Belknap Harvard University Press.

CHARLES A. BRADY
National Zoological Park
Washington, D. C. 20008
and
Department of Zoology
Ohio University
Athens, Ohio 45701

Observations on the Behavior and Ecology of the Crab-Eating Fox (*Cerdocyon thous*)

ABSTRACT

Three adult pairs and six juvenile free-ranging, crab-eating foxes were observed in the llanos of central Venezuela during wet and dry seasons. Pairmates inhabited the same home range, and hunted together but not cooperatively. Some home range overlap between adjacent pairs occurred, and several interpair encounters were observed. Juveniles were most often in the home range of their presumed parents, but also hunted individually. Food items included small vertebrates, insects, fruit, carrion, and crabs. Pairmates urine-marked on the same spot in sequence with either sex initiating these bouts. Close-contact behaviors between pairmates were primarily sniffing and licking toward the partner's head, and parent-juvenile interactions were most often associated with food-begging attempts by the juveniles.

RESÚMEN

Tres parejas de zorros adultas y seis zorros jovenes en morada natural fueron observado en los llanos de Venezuela durante la época mojada y seca. Las parejas, machos y hembras, vivían en al mismo territorio y cazaban juntas pero sin mucho cooperación. Algunos territorios fueron usados por parejas contiguas y se observacion varios encuentros con intercambio de parejas. Los jóvenes estaban la mayoría del tiempo en el territorio de sus padres, pero también cazaban solas. Su alimentación es en base a vertebrados pequeños, insectos, frutas, carroña, y cangrejos. Las parejas orinaban en el misma lugar, empezando esta actividad cualquiera de los dos sexos. Las actividades entre padres y jóvenes generalmente estaban asociades con intentos de pedidos de alimentación por parte de los jóvenes.

Introduction

The crab-eating fox, *Cerdocyon thous,* is a small canid (6–7 kg) with a widespread distribution in South America. It ranges from Colombia and Venezuela south to northern Argentina and Paraguay (Stains, 1975). Specimens have been collected both from the llanos which is open palm-shrub habitat and also from forested regions (Langguth, 1975). It is morphologically unspecialized when compared with the several highly differentiated genera of South American canids, but does have short robust legs in comparison to the *Dusicyon* species (Langguth, 1975). Clutton-Brock et al. (1976) suggests that *Cerdocyon* be incorporated into the *Dusicyon* group based on a taxonomic review of the family; this issue has not yet been resolved.

There are few published accounts of the ecology or field behavior of the crab-eating fox. The behavior of captive foxes is reported by Brady (in preparation), and aspects of reproduction and parental care are briefly discussed in Brady (1978) and Coimbra-Filho (1966). A similar sized South American canid, the pampas fox (*Dusicyon gymnocercus*) has a quite variable diet which consists of insects, fruit, small vertebrates, and carrion, and it is most often found in open terrain (Crespo, 1971).

In this study, observations on the behavior, movements, and diets of free-ranging, crab-eating foxes are reported. The foxes inhabited the llanos of central Venezuela and were observed in both wet and dry seasons.

Methods

Two three-week field surveys, one in the wet season of 1976 (June) and the second in the dry season of 1977 (April), were conducted in central Venezuela. The study site was llanos habitat, and contained both highland and lowland areas. It was primarily open terrain with occasional palm trees and shrub rows. The study site was charted with the aid of a jeep odometer and compass and is illustrated in Figure 1.

The resident foxes were individually identified by a composite of differences which included size, pelage markings and wounds. Photographs were taken of each animal during the wet season to assist in identification of the foxes during the subsequent year's survey. Age and sex were determined by the presence or absence of adult urine-marking postures which are sexually dimorphic. The composition of the crab-eating fox groups on the study site is presented in Table 1.

The observation technique was to drive through the study area from 1700 through 2400 hours and search for foxes with the vehicle head lights and a portable

Table 1. The sex, age class, and group affiliation of the crab-eating foxes on a study site in central Venezuela

Animal number	*Sex*	*Age*	*Group affiliation*
1	Male	Adult	Pair 1
2	Female	Adult	Pair 1
3	Female	Juvenile[1]	Presumed progeny of pair 1
4	Male	Juvenile[2]	Alone
5	Male	Adult	Pair 2
6	Female	Adult	Pair 2
7	Female	Juvenile[2]	Presumed progeny of pair 2
8	?	Juvenile[2]	Presumed progeny of pair 2
9	?	Juvenile[2]	Presumed progeny of pair 2
10	?	Juvenile[2]	Presumed progeny of pair 2
11	Male	Adult	Pair 3
12	Female	Adult	Pair 3

[1] Observed only during the wet season.
[2] Observed only during the dry season.

Table 2. The composition of crab-eating fox sightings for three known pairs in central Venezuela[1]

	Pairs			*Total*
	1	*2*	*3*	
Pair together	171	108	56	335
Female leads the pair	49	27	9	85
Male leads the pair	50	31	14	95
Lead exchanges	35	19	3	57
One pair member alone	49	23	14	86

[1] The numbers represent locational points and do not take into account the duration of a sighting. The sex of the lead fox was determined whenever possible by the sexual dimorphism in urine-marking postures.

Figure 1. The 180 hectare study site in the llanos of central Venezuela. The circles represent location points of crab-eating fox pair no. 1, the triangles those of pair no. 2, and the squares those of pair no. 3. Solid symbols are dry season location points and open symbols are wet season location points.

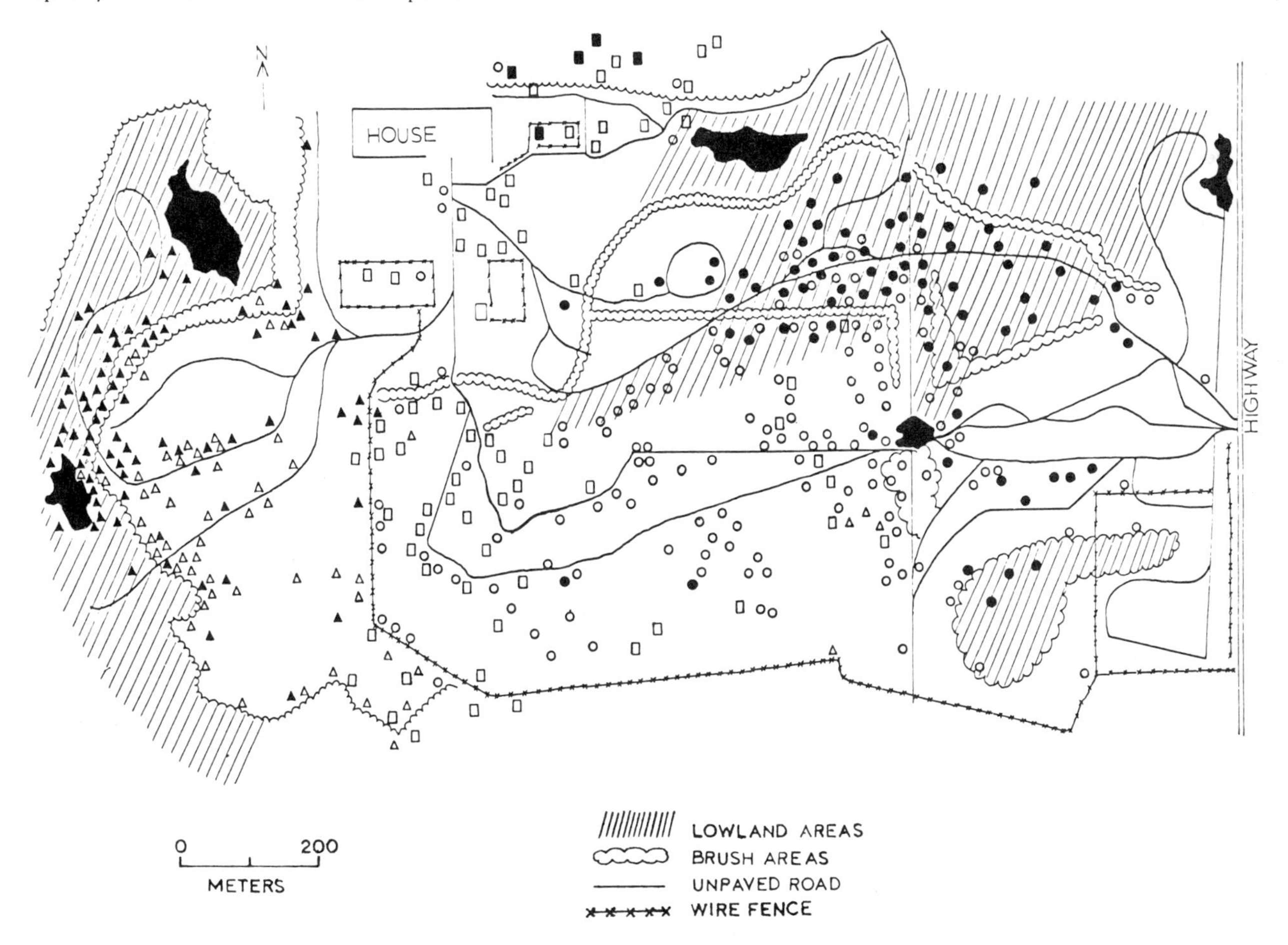

40-watt spotlight. Once a fox group was sighted, it was followed at a distance of 10 m to 30 m for as long as possible. The foxes quickly adapted to this technique and were often followed for up to one-half hour. When the terrain was too difficult for the jeep, the foxes were followed on foot and observed with the aid of a flashlight. In both cases, data were dictated into a portable cassette recorder.

A color-coded marker was dropped each time a fox group was sighted. In addition, the 180 ha study area was divided into 12 roughly equal subsections, and each time a fox group moved into a new subsection another locational marker was dropped. This enabled an estimate of the relative use of different sections of the study area. The markers were retrieved the following morning and the locations marked on the study map. Fox sightings were also made at this time and occasionally at other times of the day.

Results

General Habits

Adult crab-eating foxes most often traveled in pairs, but the distance between partners differed in the wet and dry seasons (see Pair Behavior). The composition of sightings is presented in Table 2, and the data are conservative since the durations of sightings are not taken into account. Table 2 also contains data that indicates that either sex led the pair's foraging movements, and that frequent changes in the lead occurred. Generally when the lead fox caught a prey item and stopped to consume it, the pairmate took over the lead. Other surveys of crab-eating fox populations in the llanos, conducted during the wet and dry seasons of 1974 and 1975, also found that foxes were most often sighted in pairs or as pairs with juveniles (Eisenberg and Kleiman, unpublished data; Montgomery and Lubin, in preparation).

The foxes usually foraged from 1800 through 2400 hours with intermittent rest periods. The majority of sightings outside this time block was of pairs resting together. The activity period appeared to be related to problems with thermoregulation since the few foxes observed foraging during the day showed evident signs of overheating, and quickly returned to cover. Fox pairs began activity periods on regular foraging routes, and utilized the same resting areas between bouts. Most social interactions between pairmates occurred during these rest periods. Generally there were scat latrines near these areas which were utilized either before or after the rest.

During the wet season, foxes spent the day under brush in the highland areas, while in the dry season they crawled into clumps of matted grass in the lowlands. These grass shelters were used repeatedly and some were quite extensive with several entrance holes. Ranchers in the area located pups in such a matted grass structure in January, 1977 (presumably the litter of pair no. 2). The youngsters still had dark pelage which indicated that they were less than 45 days of age and possibly were born at that site. Crab-eating foxes have not been reported to excavate their own underground burrows, but do occasionally utilize abandoned burrows of other animals (Juan Gomez-Nuñez, personal communication).

Home Range

The home ranges of the three study pairs were in close proximity to one another and some overlap occurred (Figure 1). Pair no. 1 had an approximate home range of 96 ha based on 220 location points. This estimate was calculated by joining the peripheral points from both wet and dry season surveys and measuring the enclosed area. Pair no. 2 had an approximate home range of 60 ha which was estimated in the same manner (N = 13 location points). Pair no. 3 was not sighted regularly during the dry season survey, therefore the range estimate of 54 ha was primarily from wet season location points (N = 70). The few dry season sightings of pair no. 3 were tentative, as I was unable to positively identify the pair as the one observed during the previous wet season.

The home ranges of each of the study pairs encompassed different types of terrain including lowland sections which were wet, muddy, and in some places submerged with one to ten centimeters of standing water during the wet season, and highland sections which remained relatively dry. The highland areas were mostly open, short grasslands. The lowland and highland areas on the study site are marked in Figure 1. During the wet season 85 percent of the fox sightings were in the dry highlands (Figure 1). Foxes utilized highland trails when traversing the wet areas and rarely stopped to forage. The situation was reversed during the dry season with 66 percent of the fox sightings occurring in the dry lowlands (Figure 1). The reason for the above shift is related to seasonal changes in prey availability which will be discussed further (see Food Habits).

Home Range Overlap

Several intergroup encounters were observed and most occurred during the wet season when range overlap was greatest. The first encounter occurred at the northeastern end of the study area, and involved pair no. 1 and the male of pair no. 3. The male of pair no. 1 initiated a brief chase when he sighted the male. The two males then faced one another, reared, grappled and muzzle wrestled briefly before the male of pair no. 1 returned to his mate. Pair no. 1 moved off without further incident. The second occasion occurred in the same area and involved pair no. 1 and pair no. 3. As in the previous encounter, the male of pair no. 1 chased the male of pair no. 3; however, no close-range interaction occurred. Soon after the initial chase, both pairs began to forage on palm fruit beneath the same tree. After approximately 10 minutes pair no. 1 departed, ending the encounter. The third wet season intergroup encounter occurred in the center of the home area of pair no. 1, and involved all three pairs as well as the juvenile of pair no. 1. It occurred around a goat carcass, and pair no. 1, the juvenile, and pair no. 3 were already feeding when I arrived. The carcass was quite dispersed, so perhaps some fighting may have occurred during the early division. However, the two pairs and the juvenile foraged singly without incident for 30 minutes at which point pair no. 2 arrived. All the foxes watched the pair approach, and the male of pair no. 1 ran towards them. He did not, however, engage them in a close-range interaction, and returned to the portion of carcass on which he had been feeding. Soon, all the animals resumed feeding. Foxes without food sat and waited for another to leave a piece of the carcass before darting in and taking it. A feeding fox gaped and growled at any other, including its mate, that approached too closely. The animals fed on the carcass for about 4 hours with no further close-range interactions other than several food-begging attempts by the juvenile towards its mother. In addition to the above sightings, pairs often came within 30 m of one another, but did not appear to be aware of each other's presence.

In comparison, intergroup encounters appeared to be more aggressive during the dry season survey. Pair no. 1 encountered juvenile no. 4 on five occasions during the dry season survey at the eastern end of their home area. The male of pair no. 1 chased the juvenile on each occasion but never overtook it. Once he pursued the juvenile in tight circles before giving up the chase. Both males showed piloerection of the back and tail during the encounters. After the chases the male of pair no. 1 circled the area several times with his head lowered and tail arched before returning to the female.

Food Habits

The crab-eating fox is an opportunistic hunter and consumes small vertebrates, insects, and other invertebrates, as well as carrion and fruit when available. Mondolfi (Walker, 1976) reported that the stomachs of 19 foxes contained in order of abundance: small rodents, insects, fruit, lizards, frogs, crabs, and birds. My observations during the wet season survey suggested that the foxes relied primarily on insects which were abundant and easily captured in the dryer areas of their home ranges. Palm fruit and fruit from other indigenous trees that dropped during the wet season was another favored food. During the dry season when insects became scarce, the foxes shifted their hunting activities to the lowlands where crabs and vertebrates were abundant. The crabs were easily captured because their movements were restricted to scattered tussocks of tall, matted grass, apparently because these areas contained the last available moisture.

The percentage of each food type taken by the foxes during both surveys is shown in Figure 2. The increase in the number of vertebrates captured in the dry season derived in part from the large number of lizard nestlings that emerged during the survey period. Foxes spent a long time waiting around emergence holes, and captured the young lizards as soon as they appeared. The types of carrion consumed included domestic hoofstock, road kills, and human garbage. In addition to the above items, ranchers in the area reported that foxes often dug up tortoise eggs and occasionally killed domestic fowl.

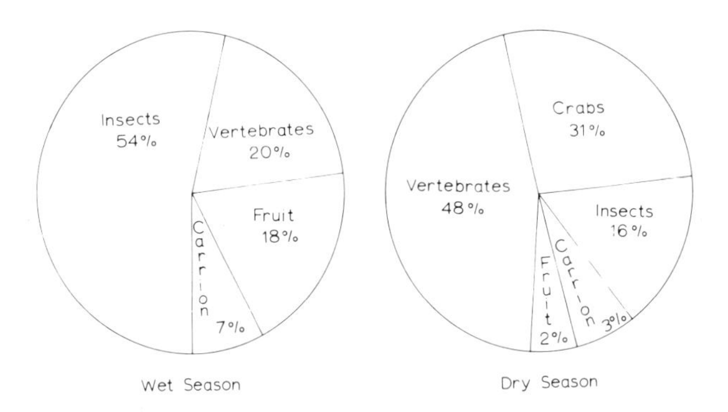

Figure 2. The diets of 12 crab-eating foxes on llanos terrain in central Venezuela based on 165 prey captures. Each instance of feeding on carrion was considered a bout regardless of the length of time spent with the item.

I observed the foxes execute many prey-captures at close range, and in Table 3 the number of attempted and successful captures of each prey type are listed. The foxes utilized distinct capture patterns for each prey type and this was a valuable tool in assessing feeding ecology.

When attempting to catch a vertebrate, a fox stopped and oriented towards the prey, then sprang forward and pounced upon it. The prey was trapped between the forepaws and simultaneously secured with the teeth (Figure 3: c, d, e). If the prey struggled, the fox used head shakes to subdue it, and if they prey eluded the grip, the fox chased it with head low to the ground and mouth open. This sequence along with a prolonged consumption time and shearing with the carnassial teeth were good indicators that the prey was a vertebrate. The types of vertebrates captured included lizards, snakes, and rodents.

The pattern for catching crabs was quite different. Foxes did not appear to search the tussocks of grass for crabs; rather they suddenly deflected from their course and approached a clump of grass. They then reached in with the muzzle and seized the crab (Figure 3: f, g). Once the crab was captured the fox quickly maneuvered it to the back of the mouth and crushed the carapace with its molars. This pattern suggests that the foxes were utilizing an auditory cue (i.e., rustle of the substrate) to locate the crabs.

The third major prey category, insects, was captured with a variety of patterns which were fairly distinct from the patterns previously discussed. The fox generally stood over the insect with its chin tucked then stabbed at it with a slightly opened mouth sometimes four or five times in quick succession (Figure 3: a, b). Presumably this immobilized the prey which was then picked up and quickly consumed. If the insect moved off, the fox bowed and trapped it between the forelegs, and if it moved into a crevice, the fox attempted to excavate it. Occasionally a fox used a pounce to capture an insect, but in these cases the consumption time was brief (<60 seconds).

The foxes had a characteristic head posture when masticating which was used to gauge consumption times. The head was lowered and the ears turned backwards while chewing. The ears were returned

Table 3. The numbers of attempted and successful prey captures by crab-eating foxes in central Venezuela

Animal number	Prey types					
	Crabs		Insects		Vertebrates	
	Attempts	*Captures*	*Attempts*	*Captures*	*Attempts*	*Captures*
1	11	11	16	5	16	9
2	17	15	26	18	13	7
3[1]	—	—	3	2	5	3
4[2]	8	6	1	1	1	0
5	—	—	28	15	27	18
6	2	2	30	9	22	15
7, 8, 9, 10[2]	2	—	5	3	16	15
11	—	—	14	10	11	3
12	—	—	2	2	3	1
Total	40	34 (72)	125	65 (52)	114	71 (62)

Numbers in parentheses indicate the percent captured.
[1] Observed only during the wet season.
[2] Observed only during the dry season.

foreward when ingestion was complete. Nel (1978) used the same method to measure the actual feeding time of bat-eared foxes (*Otocyon megalotis*). He found that the mouths of foraging animals were often obscured by vegetation, but he was always able to obtain the time interval between changes in ear position.

Although foxes most often hunted individually, cooperative hunting by pairs has been observed. R. Rudran (pers. com.) observed a pair of foxes capture a Tegu lizard (*Tupinambis sp.*). One fox feinted a frontal attack while the pairmate rushed in from the rear and delivered bites to the body. This pattern was repeated over ten times until the lizard was killed. One fox then carried it off, and the pairmate followed closely behind. This incident occurred in February which is the birth season (Brady, 1978), and was probably the incentive for cooperative hunting. At other times of the year, a fox generally did not interfere with a pairmate's capture; rather it sat and waited for the partner to finish its meal or continued hunting. A fox that was consuming a prey-item gaped and growled at its pairmate if the latter approached. This type of interaction was only observed when foxes were foraging on carrion.

Crab-eating foxes have also been observed caching food items in June and July (Eisenberg & Kleiman, unpublished data). The pattern used was similar to that of other canids. The fox first dug a shallow hole with the forepaws and placed the food into it, then it swept soil and debris on top of the hole with the muzzle. In captivity, foxes cached surplus food items throughout the year, and females showed a sharp increase in the activity around parturition (Brady, 1978). The caches were not regularly marked with urine as are those of red foxes (*Vulpes vulpes*) (Henry, 1977).

Reproduction

The reproductive season of crab-eating foxes in the wild has not yet been thoroughly analyzed. In captivity, foxes can produce two litters of three-to-six pups a year with a seven-to-eight month interbirth interval (Brady, 1978). In both field surveys, juveniles of approximately five months of age were present, which would set parturition sometime in January or February. However, the breeding season is extended since a dark juvenile with a parent was sighted in April and the carcass of a lactating female was found in June. The meager information available suggests a peak in births around January and February, but it is not known whether another peak occurs in July and August. One possible factor limiting wet season births could be the scarcity of high, dry denning areas.

Figure 3. The techniques crab-eating foxes used to capture insects (a, b), vertebrates (c, d, e), and crabs (f, g).

Pair Behavior

Pairmates of the three study pairs generally only interacted with one another at the end of rest periods, or when greeting one another after a period of separation. The pairs also had a high rate of completed sequence urinations (see below).

During the wet season pairs were most often within five m of one another as they zig-zagged through the highland areas hunting insects. The distance between pairmates was greater during the dry season and they often became separated by 50 to 75 m. In cases where the pairmates lost visual contact with one another they would siren-howl back and forth until reunited. The siren-howl is the long-distance, contact-promoting vocalization of crab-eating foxes (Brady, in prep.), and bouts were heard on three occasions during the wet season and on 11 occasions during the dry season. When reunited, pairmates approached one another with tails raised vertically and sniffed and licked one another's faces.

During both surveys, 51 close-range contacts were observed. The body target distribution for bouts of sniffing and licking was as follows: head region 59; scruff region 4; inguinal region 8; flank region 5; and anal region 10. This corresponds to what has been reported in captive fox pairs. The head is the primary target of licking and sniffing, especially the muzzle and cheeks (Brady, in prep.). The duration of contacts was generally brief, although on seven occasions the greetings transformed into prolonged interactions when one of the pair crouched low to the ground, flattened its ears laterally, and crawled under its mate while wagging its tail. From this position the animal would roll onto its back and continue to wag its tail and bite and lick the partner's inguinal region and hind legs. This type of interaction has been called active submission (Schenkel, 1967) and is the same pattern used by juveniles when food begging (see *Parent-Young Interactions*). Other close-range interactions included several bouts of muzzle wrestling when pairs were reclining together and two instances when one pair-member stood with its forepaws on the other's back.

Another important pair behavior was sequence urinations. Adults of both sexes used a raised-leg posture when urine marking. Males adopted the usual male canid stance with the hindleg back and slightly outward and the females raised their hindleg in a foreward direction (Figure 4). Often the leading member of a pair urinated on a clump of grass and the partner covered it with a urine mark as it passed. This was considered a completed sequence urination bout. The number of urinations observed in this and other contexts is presented in Table 4. The estimate of 73 percent completion was conservative since a fox was scored as initiating a sequence urination bout whenever it urinated within an estimated 10 m in front of its partner. On several occasions the partner did not pass near the mark, but it was difficult to assess whether or not the mark was detected.

In the dry season lone foxes often appeared to be following their partner's scent trail. The fox moved along with its nose near the ground and frequently stopped and sniffed an area, then urinated upon it. Unfortunately, I was not able to verify the use of urine trails in this context. However, on eight occasions, a lone fox that appeared to be following a urine trail came upon its mate.

The temporal pattern of urine marking was highly variable, and foxes marked whenever active. However, the majority of sequence marks occurred in bouts of three to five when a pair first became active after resting, when moving from one foraging area to another, and when separated from each other. The intermark distance for 21 sequence urine marks was 18.2 m (Range 9.5–21.3 m). Also, males appeared to mark more frequently than females (Table 4), but it was not determined whether both sexes were observed for equal time periods. Males and females initiated and completed sequence urinations at about equal rates.

Figure 4. The urine marking postures most often utilized by adult crab-eating foxes.

Parent-Young Interactions

The juveniles present on the study site during the two surveys are listed in Table 1. The presumed filial associations were based upon the following criteria: (1) the youngster(s) shared the home area of their presumed parents; (2) they were occasionally sighted traveling with the pair, and (3) the youngsters interacted with the presumed parents in an amicable manner. Juvenile male no. 4 did not fit the above criteria, although he was observed in the eastern end of pair one's range on 19 occasions. He was therefore considered a transient rather than progeny of pair one. All

Table 4. The frequency of initiated and completed sequence urinations by three pairs of crab-eating foxes and the frequency of urinations in other contexts

Animal number and sex	*Number of initiated sequence urinations*	*Number of completed sequence urinations*	*Percent of initiated sequence urinations completed*	*Number of urinations while animal trailed its pair mate*	*Number of urinations while animal was alone*	*Total number of urinations*
1 Male	57	36	63	30	44	167
2 Female	49	41	84	9	55	154
5 Male	28	25	89	21	61	135
6 Female	26	15	58	16	21	78
11 Male	11	9	82	4	25	49
12 Female	12	8	67	2	8	30
Total Male	96	70	73	55	130	351
Total Female	87	64	74	27	84	262
Total	183	134	73	82	214	613

the remaining juveniles listed in Table 1 fit the criteria. The grouping patterns of the juveniles appeared to reflect the changes in prey distribution from wet to dry season. During the wet season, juvenile female 3 was observed trailing 10 meters or so behind pair 1 on 27 occasions. She was sighted resting with the pair on 5 occasions and foraging alone on 9 occasions. During the dry season the four juveniles associated with pair 2 were sighted with the parents on only three occasions. These occurred around lizard emergence holes and were for extended time periods (ca. 1 hour). The juveniles were sighted foraging alone on 19 occasions.

Both juvenile female 3 and the juveniles associated with pair 2 made attempts to take food items from their presumed parents. A food-begging juvenile chased and overtook the parent and nudged the corner of its mouth or grabbed at the food item if it was exposed. The juveniles crouched low to the ground with ears flattened back and tails wagging. The parent punished the food-begging juvenile on all observed occasions by clasping the juvenile's muzzle with its mouth. The juvenile's reaction was to roll on the back, continue tail wagging, and whine. Juvenile female 3 made five food-begging attempts and was successful twice. One of the juveniles associated with pair two was observed attempting to food beg from an adult on 11 occasions and was successful only once.

Other parent-young interactions observed were six bouts of facial grooming by juvenile three towards one of her parents. Also an adult chased a juvenile on 24 occasions. The chases were brief and neither participant showed piloerection of the fur. The chasing bouts were terminated on ten occasions when the youngster rolled on its back in submission, and on six occasions with brief bouts of muzzle wrestling and grappling. The remaining chases terminated without any close-range interaction.

Inter-Specific Relationships

In central Venezuela crab-eating foxes were not extensively hunted or trapped by man but were often killed by automobiles. To what extent adult foxes were taken by indigenous carnivores was difficult to access. The foxes on the study site generally ignored other larger or equal size mammals and reptiles. They did, however, tend to avoid areas where there were large domestic hoofstock.

Small carnivorous mammals such as the raccoon (*Procyon cancrivorus*), skunk (*Conepatus semistriatus*) and opossum (*Didelphis marsupialis*) were regularly sighted on the study site and other small carnivores were occasionally sighted (see Chapter 14). These predators could probably take unguarded fox whelps especially if they were above ground. Domestic dogs were also potential predators and chased fox pairs twice during the surveys.

Discussion

The locational data indicated that the male and female of a pair inhabited the same home area, and most often foraged close to one another. These observations and the actual home range estimates of one to two km^2 were probably dictated by prey availability, distribution and abundance. Although no quantitative measures of these parameters were collected, insects, the major prey item of the wet season, had much higher densities and were more evenly distributed than the crab and vertebrate prey of the dry season. This difference allowed pairmates to hunt side by side in the wet season but not in the dry season. Although further apart, the pairmates still appeared to be aware of one another's movements in the dry season.

The home range estimates are rough approximations due to the short time periods sampled. In addition, it is unknown whether the densities observed on the study site were typical for crab-eating foxes on llanos terrain. Pair no. 3 could have been yearling progeny of either pair no. 1 or no. 2 since they were young animals, and had range overlap with both pairs.

The seasonal shift in home area utilization also appeared to be correlated with changes in prey availability and distribution. Insects were scarce during the dry season, and the crab population became vulnerable to predation because their movements were restricted. This seasonal shift from high ground in the wet season to low ground during the dry season is probably widespread among fox populations in the llanos. Large areas of the llanos become flooded during the wet season and foxes as well as many other vertebrate species are restricted to high ground with trees, termed "matas". A fox would have difficulty surviving in the wet lowlands since the prey capture techniques described would be ineffective on wet terrain.

Langguth (1975) reported that in many areas of its range the crab-eating fox is restricted to forested areas and savanna edges especially where it occurs sympatrically with the pampas fox (*Dusicyon gymnocercus*). The two species utilize similar food types (Crespo, 1971), and perhaps competitive exclusion occurs in some areas. The habitat of the crab-eating fox probably varies in different parts of its range, and the results presented here apply only to foxes inhabiting llanos terrain with marked wet and dry seasons. It would be interesting to study fox diets and examine the relationship between pairmates in different habitats, especially forested areas.

The home range overlap and the tolerance of adjacent pairs towards one another during the wet season sharply contrasted with the aggressive interactions between unfamiliar animals during the dry season. This seasonal difference was difficult to evaluate since it only involved male no. 1's attitude towards intruders. One reason for the lack of intergroup contacts and range overlap during the dry season was that the pairs used areas of their ranges remote from the home range interfaces. However, the differences in tolerance of intruders could be related to changes in diet or reproductive condition. Also, it could result from different levels of familiarity between the interactants.

The three study pairs appeared to have well established pair bonds based on the criteria presented in Kleiman & Brady (1978). They found that unfamiliar captive crab-eating fox pairmates (less than 45 days together) had a high rate of both aggressive and nonaggressive contacts, and a low rate of completed sequence urinations. Well established pairmates (over one year together) had a low rate of interaction, almost all of which were amicable, and a high rate of completed sequence urinations.

The formation of pair bonds is common in the family Canidae, and the adult male is thought necessary to help protect and provision the young litters (Kleiman & Eisenberg, 1973). The exact role of the crab-eating fox male was not determined since no newborn youngsters were present during the surveys. In captivity, both the male and female bring food to the youngsters, retrieve them when they become displaced and guard the nest area (Brady, 1978).

An interesting aspect of paired crab-eating foxes is the high rate of completed sequence urinations. Urine marking with a raised-leg is uncommon among females of the family Canidae, and many do not urine mark regularly outside of their estrous period (Anisko, 1978; Kleiman, 1966). A possible function of sequence urine marking is that it helps maintain the pair bond. A close look at the marking patterns of newly formed captive pairs indicated that both male and female marked at high frequencies and both attempted to exclude the partner from their mark sites. As the pair became familiar with one another they no longer excluded one another from the mark sites; therefore, more sequence marks occurred (Brady; in prep.). Perhaps the merging of the pairmates' urine marks sets up a common, familiar olfactory field. Golden jackals, (*Canis aureus*) have also been reported to sequence urine mark regularly (van Lawick and van Lawick-Goodall, 1971). During pair formation female golden jackals initiate most of the sequence urine bouts, and afterwards the male initiates most (Golani & Keller, 1975). Sequence urine marking probably also helps the pair orient in their home area and relocate one another when separated. All these proposed functions are compatible with what has been established for mammals in general (Eisenberg & Kleiman, 1972; Ralls, 1971).

The urine marks did not appear to prevent neighboring foxes or transients from entering occupied home areas, and residents did not increase their rate of marking during intergroup encounters. The foxes did not concentrate their urine marks around the periphery of their home ranges as has been reported for the grey wolf (*Canis lupus*) by Peters & Mech (1975). These observations do not, however, exclude the possibility that the urine marks identify the resident's presence to intruders.

The behavior and movements of the juveniles on the study site was in line with what has been proposed for many young carnivores (Ewer, 1973). The youngster remained in the home area of the parents for some time after weaning presumably to perfect their hunting skills. The age of dispersal for young crab-eating foxes was not determined but is quite possibly around six months based on their physical maturity at this time (Brady, 1978). If so, young foxes that disperse in June would find abundant insect food, but perhaps high dry hunting areas would be limited.

Acknowledgments

I am grateful to D. Kleiman, J. Eisenberg and G. Svendsen for their advice throughout the project and to M. Hartman for typing the manuscript and S. Brady for the illustrations. This research is a section of a Ph.D. thesis from the Dept. of Zoology, Ohio University and was supported by a Smithsonian Research Foundation Grant to D. Kleiman and J. Eisenberg and by NIMH grant 27241-03 to D. Kleiman. I would also like to thank Señor Tomás Blohm for allowing the surveys to be conducted on his ranch and S. Harding and M. O'Connell for their assistance with the field work.

Literature Cited

Anisko, J. J.
1976. Communication by chemical signals in Canidae. Pages 283–292 in *Mammalian Olfaction, Reproductive Processes and Behavior,* edited by R. Doty. New York: Academic Press.

Brady, C. A.
1978. Reproduction, growth and parental care in crab-eating foxes (*Cerdocyon thous*) at the National Zoological Park, Washington. *Int. Zoo Yb.*, 18:130–134.
In prep. Mechanisms of communication in the crab-eating fox (*Cerdocyon thous*), the maned wolf (*Chrysocyon brachyurus*) and bush dog (*Speothos venaticus*). Ph.D. thesis, Ohio University, Athens.

Clutton-Brock, J.; Corbet, G. B.; and Hills, M.
1976. A review of the family Canidae, with a classification by numerical methods. *Bull. Br. Mus. Nat. Hist. (Zool.)*, 29:119–199.

Coimbra Filho, A. F.
1966. Notes on the reproduction and diet of Azara's fox *Cerdocyon thous azarae* and the hoary fox *Dusicyon vetulus* at the Rio de Janeiro Zoo. *Int. Zoo Yb.*, 6:168–169.

Crespo, J. A.
1971. Ecologia del zorro gris en La Pampa. *Rev. Mus. Arg. Cs. Nat. Ecol.*, 1:147–205.

Eisenberg, J. F., and Kleiman, D. G.
1972. Olfactory communication in mammals. *Ann. Rev. Ecol. Syst.*, 3:1–32.

Ewer, R. F.
1973. *The Carnivores.* Ithaca: Cornell University Press.

Golani, I., and Keller, A.
1975. A longitudinal field study of the behavior of a pair of golden jackals. Pages 303–335 in *The Wild Canids*, edited by M. W. Fox. New York: Van Nostrand Reinhold Co.

Henry, J. D.
1977. The use of urine marking in the scavenging behavior of the red fox (*Vulpes vulpes*). *Behaviour*, 61:82–105.

Kleiman, D. G.
1966. Scent marking the Canidae. *Symp. Zool. Soc. London*, 18:167–177.

Kleiman, D. G., and Brady, C. A.
1978. Coyote behavior in the context of recent canid research: problems and perspectives. Pages 163–188 in *Coyote Biology*, edited by M. Bekoff. New York: Academic Press.

Kleiman, D. G., and Eisenberg, J. F.
1973. Comparisons of Canid and Felid social systems from an evolutionary perspective. *Anim. Behav.*, 21:637–659.

Langguth, A.
1975. Ecology and evolution in the South American Canids. Pages 192–206 in *The Wild Canids*, edited by M. W. Fox. New York: Van Nostrand Reinhold Co.

Montgomery, G. G., and Lubin, Y.
In prep. Social structure in crab-eating fox (*Cerdocyon thous*).

Nel, J. A. J.
1978. Notes on the food and foraging behaviour of the bat-eared fox, *Otocyon megalotis*. Pages 132–137 in Ecology and Taxonomy of African Small Mammals, Edited by D. Schlittler. *Bull. Carnegie Mus. Nat. Hist.* No. 6, Pittsburgh.

Peters, R. P., and Mech, L. D.
1975. Scent marking in wolves. *Am. Sci.* 63:628–637.

Ralls, K.
1971. Mammalian scent marking. *Science*, 171:443–449.

Schenkel, R.
1967. Submission: its features and functions in the wolf and dog. *Am. Zool.*, 7:319–329.

Stains, H. J.
1975. Distribution and taxonomy of the Canidae. Pages 3–26 in *The Wild Canids*, edited by M. W. Fox. New York: Van Nostrand Reinhold Co.

van Lawick, H., and van Lawick-Goodall, J.
1971. Pages 49–148 in *Innocent Killers.* Boston: Houghton Mifflin.

Walker, E. P.
1968. Page 1162 in *Mammals of the World*, vol. II. Baltimore: Johns Hopkins Press.

D. G. KLEIMAN
J. F. EISENBERG
E. MALINIAK
National Zoological Park
Washington, D.C. 20008

Reproductive Parameters and Productivity of Caviomorph Rodents

ABSTRACT

Data on estrous cycle length, seasonality of reproduction, gestation, litter size, and growth rates are presented for species of caviomorph rodents studied in the field and the laboratory. When species of mammals are compared, litter size tends to decrease with increasing mean body size and gestation tends to increase. Some families of caviomorph rodents, such as the Caviidae and Dasyproctidae, show rather uniform trends in reproductive rate and growth of the young which would be predicted from a knowledge of the mean body size of the adults. On the other hand, several families of caviomorph rodents exhibit atypical reproductive trends when gestation length and litter size are regressed against mean body size of the adult. Perhaps basic differences in metabolic rate among these species account for part of the variation shown. In any event, the degree of divergence in reproductive rates within some families is quite great. An analysis of productivity for caviomorph species suggests that some forms have undergone extensive "r" selection. High productivity typifies the spiny rat, *Proechimys semispinosus*, and the capybara, *Hydrochaeris hydrochaeris*.

RESÚMEN

Se presentan datos de la longitud del ciclo de estro, época de reproducción, gestación, número de crías, y crecimiento de éstas para las especies de roedoros cávidos, en el campo y el laboratorio. Cuando se comparan especies en los mamíferos, el número de crías tiende a disminuir aumentando su tamaño medio y la gestación tiende a aumentar. Algunas familias de roedoros cávidos, como los Caviidae y Dasyproctidae, tienen una velocidad de reproducción y crecimiento, que se puede pronosticar al conocer el tamano mediano del cuerpo de los adultos. Por otra parte, varias familias de roedoros cavidos tienen un proceso de reproducción muy raro cuando la gestación y el tamaño de las crías son comparadas con el tamaño media del cuerpo del adulto. Es posible que diferencias basicas en el metabolismo entre estas especies sea responsable de algunas de las diferencias mostradas. En todo caso, el grado de diferencia de reproducción, en algunas familias, es bastante grande. Un análisis de productividad de roedoros cavidos indica que algunos tienen un alta selección "r." Alta productividad es tipica del roedor macanques, *Proechimys semispinosus* y el capibara, *Hydrochoerus hydrochaeris*.

Introduction

The neotropical caviomorph rodents are an interesting mammalian group, having radiated into a variety of niches in South and Central America. This group also contributes significantly to the vertebrate biomass of South and Central America, and many species are regularly hunted for food (Smythe, 1978). Indeed, the guinea pig (*Cavia porcellus*) was originally domesticated as a food resource, and several South American governments are exploring the potential of domesticating other caviomorph species as a protein source.

The caviomorphs as a group possess a set of unique reproductive characteristics, including long and variable estrous cycles, long gestations, relatively precocial offspring, and a moderate to small litter size. Most species are also long-lived, relative to the myomorph rodents. This complex of characters persists in the suborder despite considerable differences in the behavioral ecology of the members of different families.

The biology of the New World Caviomorpha and Old World Hystricomorpha has recently been reviewed in Rowlands and Weir (1974), with Weir (1974) concentrating on reproductive characteristics, and Kleiman (1974) on aspects of reproductive behavior. The purpose of this review is to compare the reproductive parameters of caviomorph species from the northern neotropics with other close relatives. We will also concentrate on aspects of reproductive biology and behavioral ecology that have relevance for long-term captive breeding or cropping programs since the larger caviomorphs, at least, are potentially a major food source for rural populations.

Methods

The data presented in this account derive from unpublished information gathered during nearly a decade of captive breeding of numerous caviomorph species at the National Zoological Park (NZP) as well as published accounts by other authors. The data from published accounts have had to be extrapolated or estimated in some cases. Adult weights are usually based upon averages from both sexes. If a genus exhibits considerable sexual dimorphism, as in *Proechimys,* only average female weights have been used. An exception was for *Hydrochaeris* where average female weights were unavailable.

The weight data are from both captive and field observations. Growth rates are always from mother-reared young, except for *Coendou.* The data and references may be found in Appendix I.

Behavioral Ecology

Caviomorphs live in nearly all potential South American habitats and include aquatic, terrestrial, scansorial, fossorial, and rock-dwelling forms (Kleiman, 1974). Species may be diurnal, crepuscular, or nocturnal in habit. Within each family there is a tendency for species to exhibit similar adaptations in locomotion, feeding habits, and habitat preferences, thus related species often occupy a series of similar niches. Some exceptions to this generalization are of interest because they suggest that some families have undergone a more extensive radiation than others. For example, in the Chinchillidae, *Chinchilla* and *Lagidium* are both high-altitude montane forms, while *Lagostomus* is an open grassland fossorial type. The capromyid species also exhibit considerable variation in habitat preferences and social organization. Of course, the capromyids are a family whose radiation has been restricted mainly to islands in the West Indies. By contrast, the Dasyproctidae and Caviidae are each relatively tightly knit families in terms of feeding habits, social organization, and behavioral ecology.

Figure 1 presents examples of the social systems of caviomorphs, plotted against the increasing mean size of genera, based on body weight. It is apparent that, within the suborder, every major form of social organization is represented, but that there is no relationship between size alone and the social system. There is, however, a tendency for genera within each family to be similar in their basic social adaptations. Thus, caviids and the related *Hydrochaeris* usually exist in polygamous or polygynous social groups, while the dasyproctids are relatively asocial, existing as solitary individuals or pairs. The Chinchillidae and Ctenomyidae tend to be colonial, regardless of whether individual genera are communal and polygamous or solitary and polygamous.

Kleiman (1974) suggested that a majority of caviomorphs may exhibit a tendency towards sociality or coloniality. This may in part be an anti-predator adaptation, deriving from the tendency for young to be born in a precocial state and to be mobile long before they reach adult size. Moreover, the caviomorphs are major prey species for both large and small carnivores in the neotropics.

Reproductive Parameters

The Estrous Cycle and Seasonality

Weir (1974) has reviewed the known reproductive cycles of caviomorph species. Figure 2 details the percentage of births or of pregnant females during the annual cycle in selected species in the wild and captivity. In the wild, most caviomorphs exhibit seasonality in their reproduction. However, species vary with respect to how synchronized the birth season is. For example, *Lagostomus maximus* has a highly synchronized single birth season each year (Weir, 1971), while *Dasyprocta punctata* and *Hydrochaeris* exhibit reproduction throughout the annual cycle, but with a birth peak (Figure 2). Since the caviomorphs are all thought to be potentially polyestrus, species which exhibit a single synchronized birth season in the wild often show a greater spread of births after adapting to captivity, e.g., *Lagostomus* (Weir, 1971) (Figure 2). There do not appear to be any caviomorphs which are monoestrus in an obligate sense, i.e., can have only a single estrus each year.

Most forms from the northern neotropics (*Dasyprocta, Myoprocta, Hydrochaeris, Proechimys*) do exhibit an extended birth season in the wild. In *Hydrochaeris* and *Dasyprocta*, more births occur and more young apparently survive during the wet season (Ojasti, 1973; Smythe, 1978). Thus, reproduction is inhibited and juvenile survivorship lower during the period of reduced food availability.

One would predict for most species and even populations of the same species that the degree and timing of seasonality in reproduction would be closely attuned to the degree of local change in essential resources, such as food availability. For example, *Hydrochaeris* may exhibit stronger seasonality in the llanos where the availability of water may decrease dramatically in the dry season, compared with regions with more permanent water sources. Also, *Dasyprocta* may be more seasonal in regions of highly deciduous forest relative to tropical evergreen rain forest.

Breeding data available for *Octodon degus* suggest a major difference in the seasonality of two populations in captivity which may derive from differences in the original capture location and thus reflect an intrinsic control of ovarian activity. Weir (1970) found her captive degus to be highly seasonal, with births occurring first only in June, but after adaptation to the shift across the equator, only in December. In North America, colonies at several institutions report births during every month of the year, but with birth peaks in both December and July to August at the University of Vermont (Woods and Boraker, 1975), and December/January, March, and July at the U.S. National Zoological Park. The origins of the two colonies are unknown but field observations suggest that Central Chilean degus breed only once a year in September while in northern Chile, there is an extended breeding season from November to April (Fulk, 1976). Thus, some differences in the timing of reproduction in the wild may continue to be apparent under captive con-

Figure 1. The social organization of some caviomorph rodents presented in order of increasing weight of genera. Cte = *Ctenomys*; Spa = *Spalacopus*; Oct = *Octodon*; Mic = *Microcavia*; Dip = *Diplomys*; Gal = *Galea*; Pro = *Proechimys*; Cav = *Cavia*; Myo = *Myoprocta*; Pla = *Plagiodontia*; Lag = *Lagidium*; Dol = *Dolichotis*; Das = *Dasyprocta*; Lgs = *Lagostomus*; Coe = *Coendou*; Cap = *Capromys*; My.C. = *Myocastor*; Ago = *Agouti*; Din = *Dinomys*; Hyd = *Hydrochaeris*.

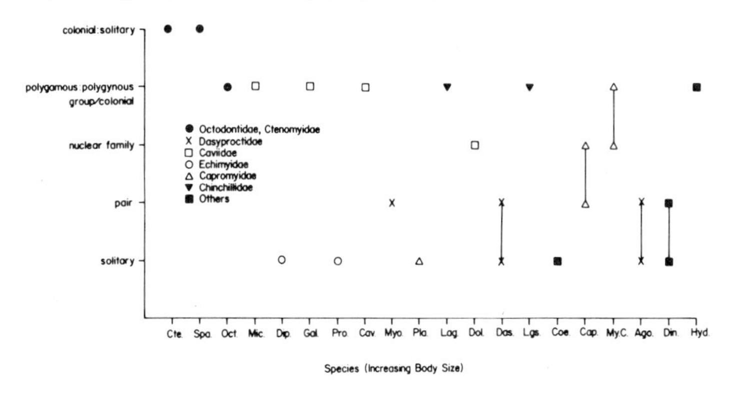

Figure 2. The annual reproductive cycle and degree of seasonality in selected caviomorph rodents in the wild and in captivity. The annual cycle is divided into quarters. All examples except for *Hydrochoerus* give the percent of births per quarter.

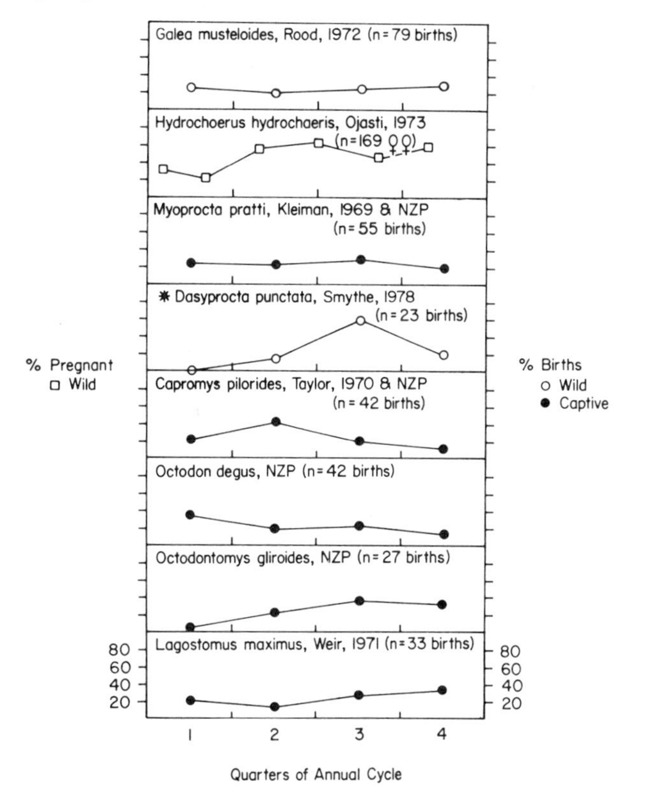

ditions, despite the generally flexible nature of reproduction in many caviomorphs in captivity.

All caviomorphs so far examined, except *Myocastor*, have a vaginal closure membrane which typically seals the vagina except at estrus or parturition. Estrous cycles of caviomorphs range from an average of 16 days in *Cavia porcellus* to 22 days in *Proechimys guariae* (Weir, 1973) to 40 days in *Myoprocta pratti* (Kleiman, 1970), based on intervals between perforation of the vaginal closure membrane. Caviids tend toward shorter cycles while chinchillids and dasyproctids have longer cycles. The distribution and range of cycle lengths is highly variable within most species, when compared with some polyestrous myomorph rodents. Weir (1974) and Lusty and Seaton (1978) have suggested that some species, e.g., *Octodon degus, Proechimys guariae* and *Ctenomys talarum*, do not have a regular estrous cycle at all, but are dependent on the presence of the male to stimulate ovarian activity. In determining the degree of lability in the occurrence of estrus and ovulation, several separate issues must be resolved.

First, it must be determined whether the relationship between vaginal opening and ovarian activity is the same in all caviomorphs. If vaginal closure membrane perforation reflects ovarian activity, it must be determined whether the presence of the male is necessary to stimulate vaginal opening, i.e., an estrous condition. Finally, it must be determined whether ovulation and therefore the estrous cycle are spontaneous, induced by the presence of the male or specifically induced by copulation.

It is probably inappropriate to try to categorize most caviomorphs with respect to the spontaneity of estrous cycling and/or ovulation, given our present knowledge. Weir (1974) suggests the existence of a continuum with respect to ovulation type, but fails to differentiate between species in which copulation induces ovulation rather than male presence inducing or maintaining the ovarian cycle. If there is a continuum with respect to the dependence of the estrous cycle and ovulation on male presence and/or copulation when species are compared, there may also be a similar continuum within a species. Thus not all individuals within a species will respond similarly when in apparently identical conditions. For example, at the National Zoological Park, we have had isolated female degus exhibit an apparently normal cycle in the frequency of vaginal opening, with intervals ranging from 20 to 30 days, while other females exhibit no apparent cyclicity at all. Weir (1974) had reported that *Octodon* does not exhibit an estrous cycle in the absence of the male, and thus felt that male presence was necessary to maintain the estrous cycle.

Support for the hypothesis that male presence is

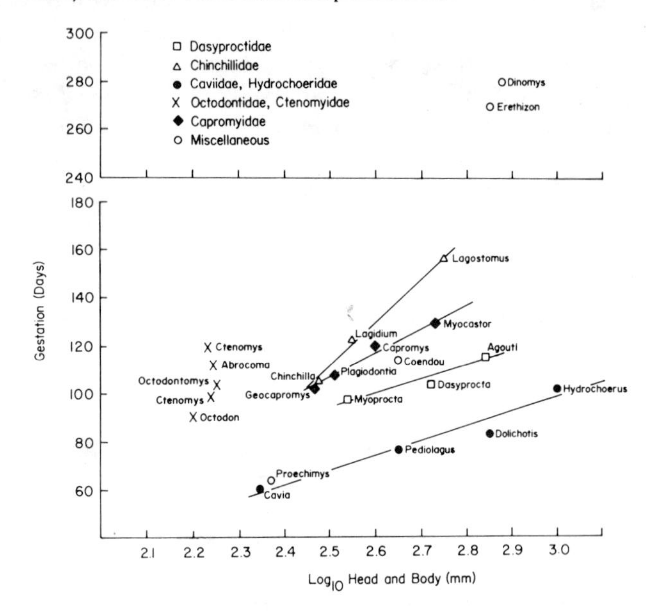

Figure 3. The relationship between gestation length and body size in selected caviomorph rodents.

important to stimulate ovarian activity and estrus in some caviomorph species comes from behavior studies which have shown that courtship behavior in caviomorphs is exceedingly complex compared with most mammal groups (Kleiman, 1971, 1974). It would be useful to correlate the type of estrous cycle with the form and frequency of male courtship in a series of caviomorph species, when there is additional information available on estrous cycles in caviomorphs.

Gestation

Weir (1974) has summarized the known gestation lengths of caviomorphs. They are long relative to other rodents and range from a minimum of 53 days in *Galea musteloides* (Rood, 1972) to 100 days in *Myoprocta pratti* (Kleiman, 1970), to well over 225 days in *Dinomys branickii* (Collins and Eisenberg, 1972). Gestation length appears to be positively correlated with body size within each family (Figure 3). Thus, the caviids and *Hydrochaeris* fall along the same line, as do the dasyproctids and the chinchillids. This positive correlation between body size and gestation length is generally true for most mammals (Kihlström, 1972).

As Weir (1974) has pointed out for the suborder as a whole, gestation length does not appear to be correlated with (1) habitat type, (2) litter size, or (3) neonatal weight relative to adult weight. However, if one considers each family separately, some generalizations can be established. Interestingly, some families are more cohesive than others which suggests that the

Figure 4. The relationship between gestation length and (a) litter size, (b) the ratio of neonatal to maternal weight, and (c) the ratio of total litter weight to maternal weight.

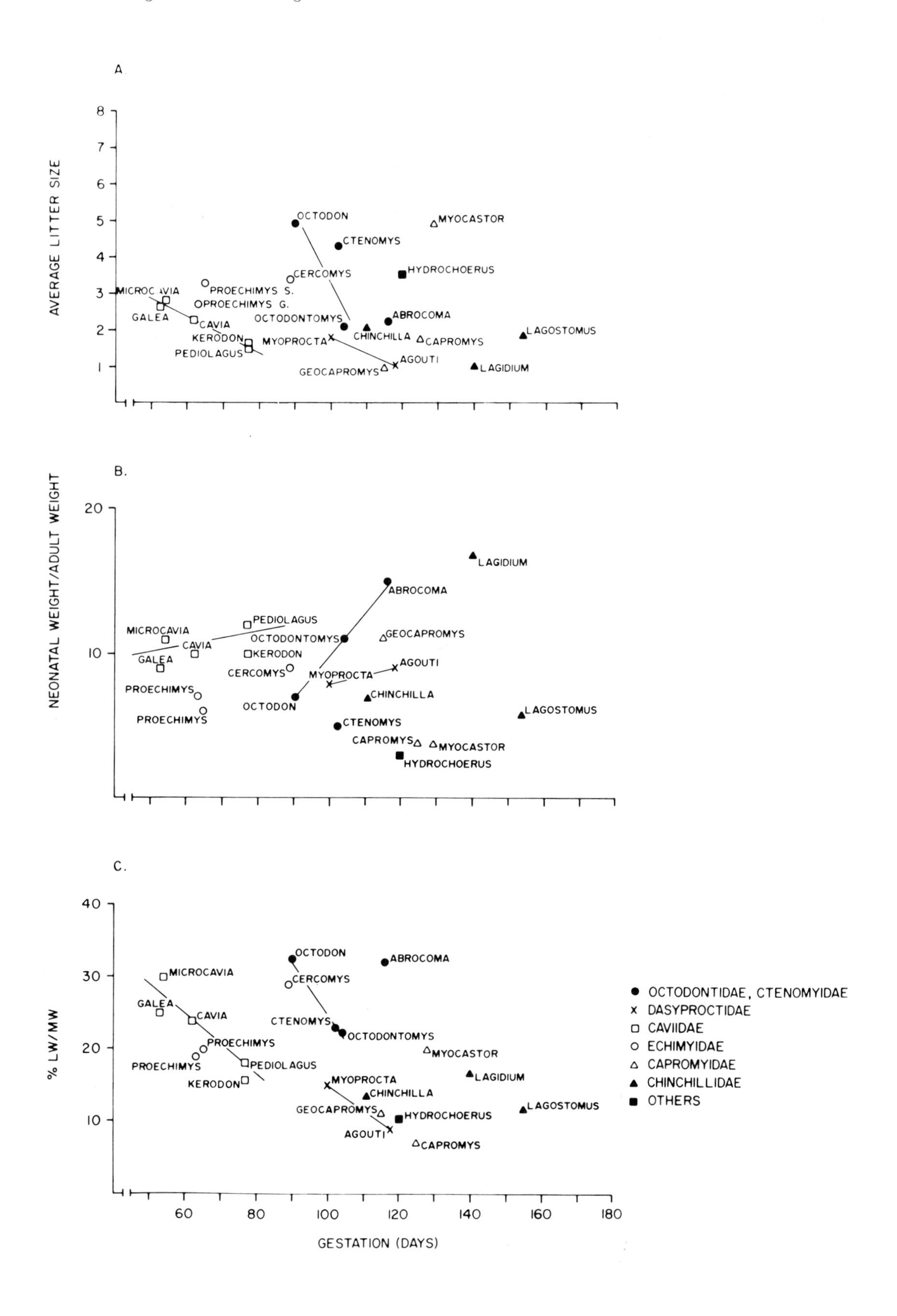

Figure 5. The mean and range of litter sizes in selected caviomorph rodents in order of increasing size. C.t. = *Ctenomys talarum*; A.c. = *Abrocoma cinerea*; O.g. = *Octodontomys gliroides*; O.d. = *Octodon degus*; M.a. = *Microcavia australis*; Pr.g. = *Proechimys guariae*; D.d. = *Diplomys darlingii*; G.m. = *Galea musteloides*; H.g. = *Hoplomys gymnurus*; Pr.s. = *Proechimys semispinosus*; C.a. = *Cavia aperea*; G.i. = *Geocapromys ingrahami*; M.p. = *Myoprocta pratti*; P.a. = *Plagiodontia aedium*; L.p. = *Lagidium peruanum*; P.s. = *Pediolagus salinicola*; L.m. = *Lagostomus maximus*; C.pr. = *Coendou prehensilis*; C.p. = *Capromys pilorides*; A.p. = *Agouti paca*; H.h. = *Hydrochaeris hydrochaeris*.

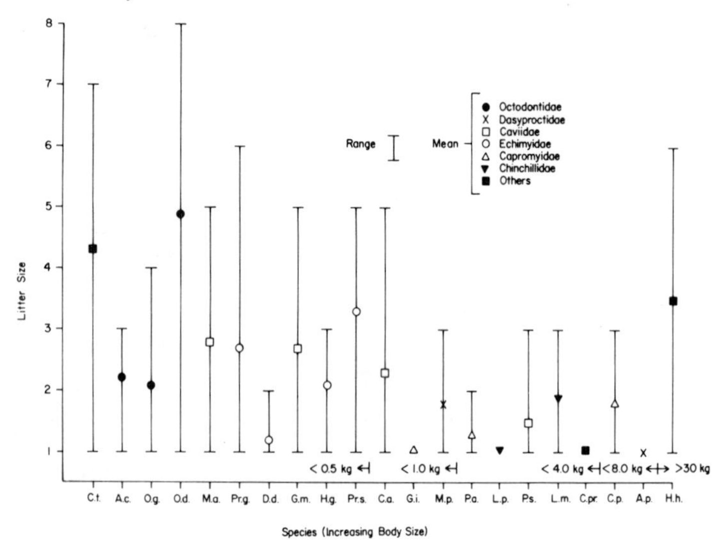

Figure 6. The ratio of total litter weight to maternal weight in selected caviomorph species in order of increasing body size. Species abbreviations are as in Figure 5.

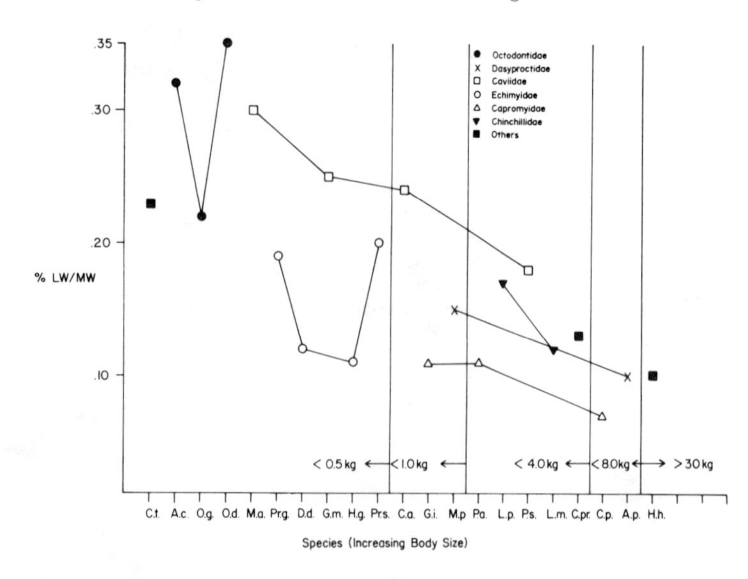

degree of genetic divergence has differed within each family. For example, Figure 4 presents the relationship between (a) gestation length and litter size, (b) gestation length and the ratio of neonatal weight to adult body weight, and (c) gestation length and the ratio of total litter mass to adult body weight.

In the suborder as a whole, there is a negative correlation between the ratio of total litter mass to adult body weight and the gestation length. This is seen most clearly in the individual families, Caviidae and Dasyproctidae. Thus, the longer the gestation, the lighter the litter is relative to adult weight. Since the gestation length is positively correlated with body size within each family and since the ratio of litter weight to adult weight is negatively correlated with body size (see below), the above relationship is not surprising. This also relates to a negative correlation between average litter size and gestation length (Figure 4a). These relationships cannot be clearly seen among the Capromyidae and Chinchillidae which suggests that there is a greater difference among the genera in these families than among the dasyproctids and caviids. Also, among the Octodontidae, *Abrocoma* stands out as an exception. Too few data are available to indicate whether the Echimyidae are a cohesive family group with respect to the relationship between body size, gestation length, litter size and relative litter mass.

Litter Size

Figure 5 presents the average litter size relative to body weight in selected caviomorph species. For the suborder, there is a negative correlation among these characteristics, i.e., the larger caviomorphs tend to have smaller litters. There are, however, two major exceptions, *Hydrochaeris* and *Myocastor* (litter size averages 5). Moreover, within several families, this relationship does not hold or is the reverse, i.e., the Octodontidae, Echimyidae, Capromyidae, and Chinchillidae. The Caviidae and Dasyproctidae are cohesive families in this respect.

The weight of the neonatal mass relative to adult body size is also negatively correlated with body size. The most anomalous families are the Octodontidae and Echimyidae (Figure 6).

Growth Rates

Figures 7 and 8 present the relative growth rates (percent of adult weight relative to age) of caviomorphs within each family. The Octodontidae, Caviidae, and Dasyproctidae exhibit the fastest relative growth rates and the species within these families show the most similarity, even though adult size in the

Figure 7. Relative growth rates (percent of adult body weight) in the Echimyidae, Octodontidae, and Caviidae. Absolute rates were extrapolated from Rood, 1972 (Caviidae), Weir, 1973 (*Proechimys guariae*), Tesh, 1970b (*Hoplomys gymnurus*), and the National Zoological Park (NZP) (Octodontidae and *Proechimys semispinosus*).

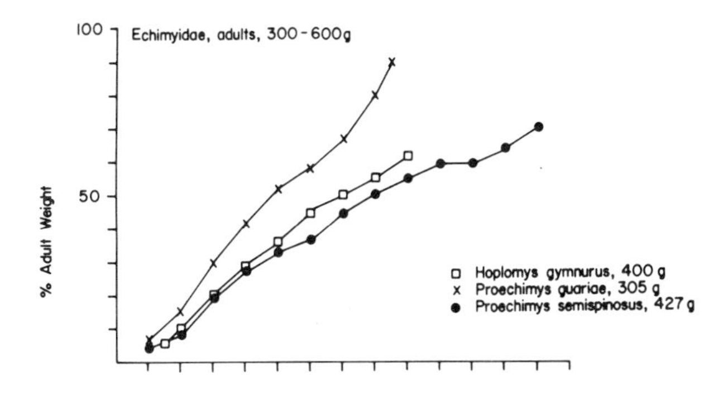

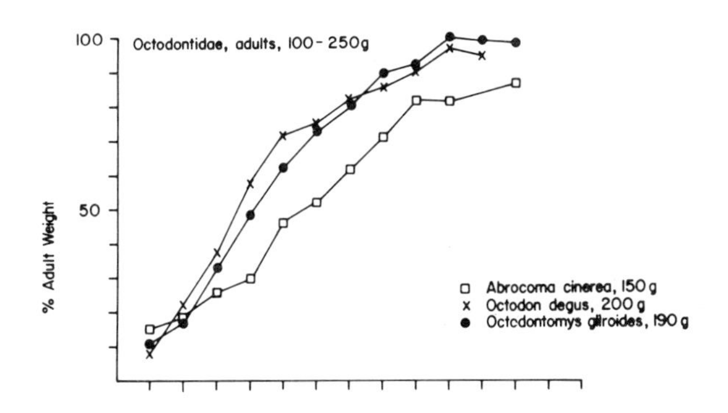

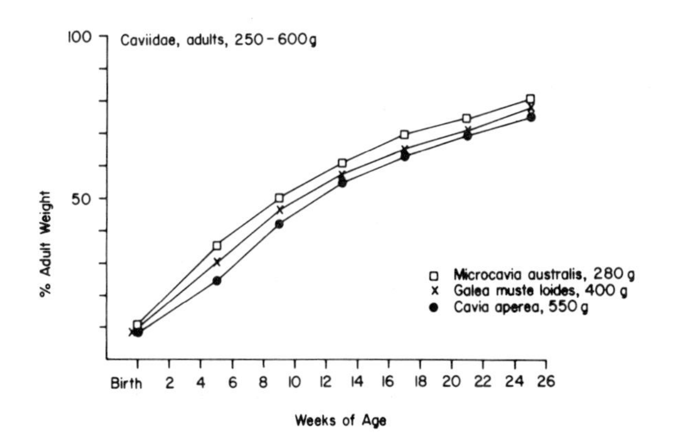

Figure 8. Relative growth rates in the Dasyproctidae, Capromyidae and miscellaneous caviomorph rodents. Absolute growth rates were extrapolated from Smythe, 1978 (*Dasyprocta punctata*); Kleiman, 1969, 1970 (*Myoprocta pratti*); Weir, 1971 (*Lagostomus maximus*); Ojasti, 1973 (*Hydrochaeris hydrochaeris*); Howe and Clough, 1971 (*Geocapromys ingrahami*); and NZP (*Agouti paca, Coendou prehensilis, Lagidium peruanum, Capromys pilorides,* and *Plagiodontia aedium*).

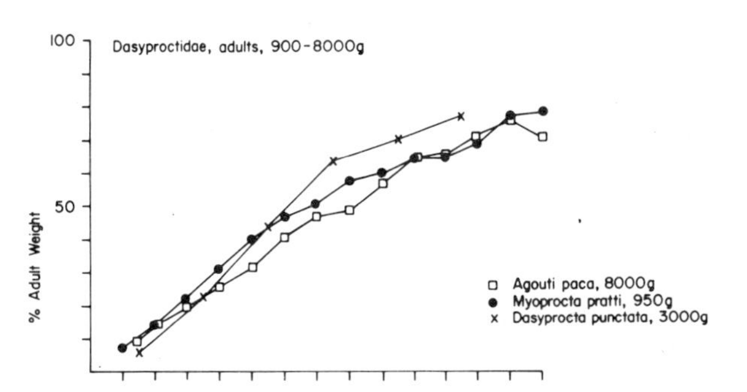

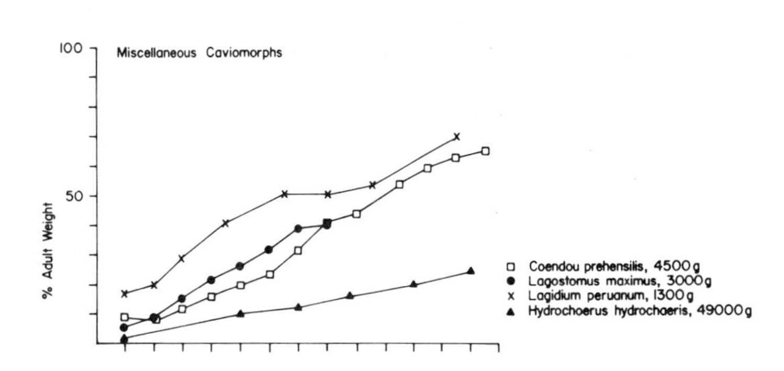

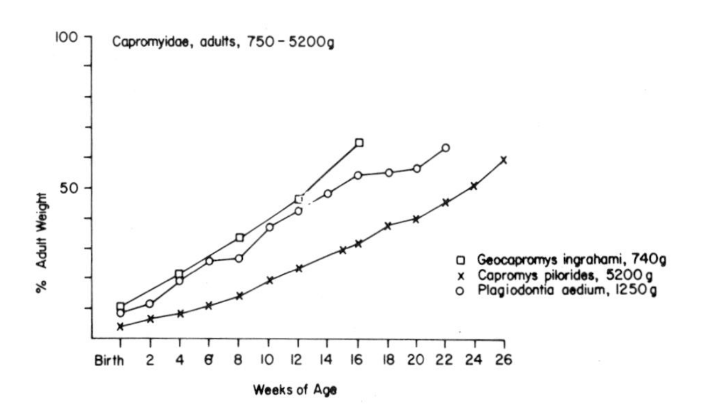

Dasyproctidae differs by a factor of 5 to 10 from the smallest to the largest genus. The slowest relative growth rates may be found among the capromyids and in *Hydrochaeris*. To some degree growth rates are related to body size with the larger species growing at a relatively slower rate. This can be seen in Figure 9 which details the age in weeks when species reach 50 percent of adult size, relative to adult body weight. Genera which grow slowly relative to their body weight include *Abrocoma, Capromys, Lagostomus* and *Hydrochaeris. Hydrochaeris* is especially slow, with young taking well over one year to reach even half of adult body weight.

Figure 9. The age in weeks when juvenile caviomorph rodents reach 50% of adult weight, in order of increasing body size. Species abbreviations as in Figure 5.

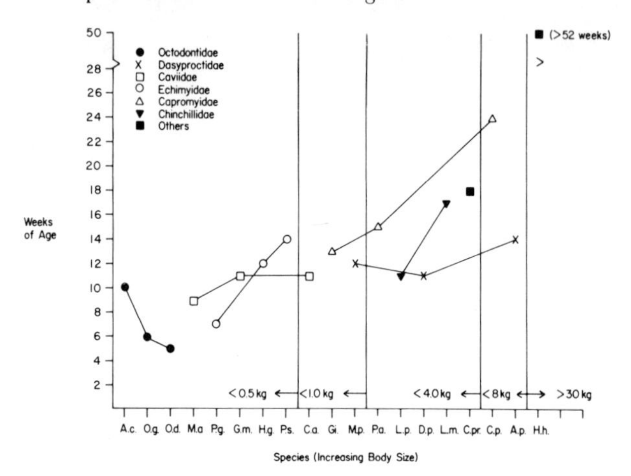

Productivity

Productivity by wild caviomorph populations has been estimated for *Proechimys semispinosus* in Panama (Gliwicz, 1973) and *Hydrochaeris hydrochaeris* (Ojasti, 1973). Ojasti demonstrated that net production by capybaras ranged from 740 to 1,220 kg/km^2 at Hato el Frio in the State of Apure, Venezuela. Gliwicz calculated productivity for *Proechimys semispinosus* in Panama at 4.1 kg/ha with a turnover of 1.4 kg/ha. Net productivity would then total some 2.7 kg/ha or 270 kg/km^2. Given the data recently published by Smythe (1978) for *Dasyprocta punctata* in Panama together with data from our captive studies on *Agouti paca* and our live-trapping data from Guatopo National Park, Venezuela, it is possible to estimate theoretical production values for *Agouti paca, Dasyprocta aguti,* and *Proechimys semispinosus* and compare these values with theoretical productivity for *H. hydrochaeris.*[1]

Based on captive studies, the gestation period for *Proechimys semispinosus* is approximately 64.8 days; for *Dasyprocta* sp., 112 days; and for *Agouti paca*, 115 days. The average number of young per litter for the three species is 3.0, 1.5 and 1.5, respectively. The interbirth interval for a given female may exceed the gestation significantly if she undergoes a lactation anestrus. For example, the paca shows an interbirth interval of 215 days and the agouti around 180 days. By contrast, the spiny rat, *Proechimys,* shows no apparent lactation anestrus in captivity and its interbirth interval approximates the gestation length or 65 days.

The number of young produced per adult female per year may be calculated if we make several assumptions. We must assume that reproduction is not seriously interrupted by drastic seasonal differences in the abundance of food. This latter assumption certainly does not hold for all areas where the three species occur sympatrically (see above). If we assume that a resident adult female lives for at least one year in the wild and if we divide 365 days by the interbirth interval and then multiply this by the average number of young per birth per female, we obtain the following figures for annual production of young: 16.8/yr for *Proechimys,* 3 young/yr for *Dasyprocta*, and 2.5 young/yr for *Agouti paca.* In prime habitat, the male and female agouti occupy approximately 2 hectares; this means that for 100 hectares or 1 square kilometer, there would be on the average 50 adult reproducing female agoutis. Given the annual productivity, one square kilometer could give a maximum yield of 150 agoutis per year. If we assume that one-third of the 50 resident adult females die each year on the average, then 16.7 of the females born each year will replace a resident female. Out of the 150 agoutis produced per year, 75 should be females. For a steady state population, approximately 59 females per year will either emigrate, serve as food for predators, or succomb to disease and other "accidents." If we assume a slightly higher mortality for resident males and apply similar reasoning, we could conclude that only 38 surplus males could be produced per year/km^2. This would mean that, given steady state population, a total of approximately 97 young agoutis will either die from accidents, disease or human and/or animal predators per year/km^2. In the absence of external predators and assuming an arbitrary 20 percent cropping rate, a rural human population could take approximately 20 agoutis per km^2 per year in prime habitat. If we assume that the mean weight of agouti at the time of catching is approximately 1.9 kg, then the harvestable biomass per year from a square kilometer at a 20

[1] Preliminary values were presented by Eisenberg and O'Connell, 1976.

percent cropping rate would be roughly 38 kg per year.

In primary habitat an adult male and female paca share approximately 4 to 5 hectares. If we assume a 50:50 sex ratio for adults, then there will be 25 adult female pacas/km^2. Given the annual productivity for the paca at 1.5 young per year, then approximately 37 young could be produced per km^2/yr. Assuming an arbitrary low cropping rate of 20 percent and further assuming that approximately 11 young are necessary to replace adult losses, then approximately 5.3 young could be taken for human use, or 30.6 kg/km^2 have been produced for human consumption.

The spiny rat, *Proechimys,* no doubt bears the major predation pressure in the lowland tropical ecosystem. *Proechimys* can reach high densities and in an optimal habitat it is not unreasonable to predict 150 females/km^2. Given the annual productivity of approximately 16.8 young per female, then 2,526 young spiny rats could be produced on one square kilometer per year. If we assume as low as a 20 percent predation rate on the original hypothetical production, then 505 spiny rats could be supplied to the larger predators per annum. This converts to 151 kg of surplus biomass for predator utilization per km^2/yr, if we assign 0.3 kg as an average weight value for a spiny rat at the time of cropping.

Even given the slow growth of the young capybara, the large litter size of this species permits a high annual productivity. Ojasti (1973) estimates net productivity at 740 kg/km^2 for the year March 1967 to April 1968 in the llanos. Even harvesting at 20 percent would yield 148 kg of harvestable biomass. Clearly, the large *Hydrochaeris hydrochaeris* has been under intensive r selection with increased fecundity far beyond what may have been predicted from its absolute size. From March 1969 to April 1970, the same area yielded a net productivity of 1,220 kg/km^2 which if cropped at 20 percent would have yielded 244 kg/km^2. In spite of its low natural density, the capybara has a remarkable capacity to contribute to the net productivity in the llanos. The productivity of the capybara in prime habitat may exceed the combined productivity of the three common forest-dwelling caviomorph rodents for a comparable area.

Reproduction and growth studies for other caviomorph rodents can serve as a basis for further calculations if certain essential pieces of field data were available. These would of course include mean litter size in the field, mean number of litters per year, and mortality data without which it is difficult to calculate net productivity. In some parts of Central and South America, the spiny rats are eaten by rural populations. In most parts of the neotropics, agoutis and pacas serve as major items of food in the diets of rural people. These species are often the most important small game mammal, even in areas with rather intensive cultivation. In unexploited habitats, the spiny rat, agouti and paca are basic elements in the food chain of tropical forests. These species contribute to the diets of the bush dog, *Speothos venaticus*; the ocelot, *Felis pardalis*; the jaguaroundi, *Felis yagaoroundi*; and in Panama, even the coati, *Nasua narica* (Smythe, 1978). The extent to which avian predators utilize the terrestrial forest caviomorphs is as yet unassessed. We trust that this chapter can serve us the basis for future calculations in estimating carrying capacities not only for rural people, but for predators in intact ecosystems.

Summary

The reproductive characteristics of caviomorph species are compared, concentrating on the relationships between gestation length, body size, litter size, and litter weight. Growth rates of species are also presented. It is proposed that some individual families within the suborder (e.g., Caviidae and Dasyproctidae) are more cohesive with respect to their reproductive characteristics, which suggests that some caviomorph families have undergone a less extensive radiation than others.

The annual productivity of four caviomorph species from the northern neotropics is compared, with the finding that *Hydrochaeris'* contribution to net productivity in its prime habitat probably exceeds that of the other three species combined (*Agouti, Dasyprocta,* and *Proechimys*).

Acknowledgments

We are grateful to Lucinda Taft for summarizing a part of the caviomorph reproductive data. Larry Newman, Betty Howser, Todd Davis, Susan Wilson and Josh Lipsman also provided aid in data collection and analysis. We especially thank Michael Deal, John Hough, and Larry Newman for their careful management of the caviomorph collection over the years.

Literature Cited

Collins, L. R. and J. F. Eisenberg
1972. Notes on the behavior and breeding of pacaranas, *Dinomys branickii,* in captivity. *Internat. Zoo Yb.*, 12: 108–114.

Eisenberg, J. F. and M. A. O'Connell
1976. The reproductive characteristics of some caviomorph rodents and their implications for management. In *II Seminario Sobre Chigüires Y Babas.* Maracay: Consejo Nacional de Investigaciones Cientificas y Tecnologicas.

Fulk, G. W.
1976. Notes on the activity, reproduction, and social behavior of *Octodon degus*. *J. Mammal.*, 57:495–506.

Gliwicz, J.
1973. A short characteristics of a population of *Proechimys semispinosus* (Tomes 1860)–a rodent species of the tropical rain forest. *Bull. Acad. Polonaise Sciences, Ser. Sci. Biol.*, 21:413–418.

Howe, R. and G. C. Clough
1971. The Bahaman hutia, *Geocapromys ingrahami*, in captivity. *Internat. Zoo Yb.*, 11:89–93.

Kihlström, J. E.
1972. Period of gestation and body weight in some placental mammals. *Comp. Biochem. Physiol.*, 43-A:674–679.

Kleiman, D. G.
1969. The reproductive behaviour of the green acouchi, *Myoprocta pratti*. Ph.D. Thesis, University of London.
1970. Reproduction in the female green acouchi, *Myoprocta pratti* Pocock. J. Reprod. Fertil., 23:55–65.
1971. The courtship and copulatory behaviour of the green acouchi, *Myoprocta pratti*. *Zeit. f. Tierpsychol.*, 29:259–278.
1974. Patterns of behaviour in hystricomorph rodents. *Symp. Zool. Soc. London*, 34:171–209.

Lander, D. E.
1974. *Observaciones preliminaires sobre: Lapas, Agouti paca (Linne 1766) (Rodentia, Agoutidae) en Venezuela.* Maracay: Univ. Central Venezuela, Fac. Agronomia.

Lusty, J. A. and B. Seaton
1978. Oestrus and ovulation in the casiragua *Proechimys guariae* (Rodentia, Hystricomorpha). *J. Zool., London*, 184:255–265.

Maliniak, E. and J. F. Eisenberg
1971. Breeding spiny rats, *Proechimys semispinosus*, in captivity. *Internat. Zoo Yb.*, 11:93–98.

Ojasti, J.
1973. *Estudio Biologico del Chigüire o Capibara.* Caracas: Fondo Nacional de Investigaciones Agropecuiarias.

Rood, J. P.
1972. Ecological and behavioural comparisons of three genera of Argentine cavies. *Anim. Behav. Monogr.*, 5: 1–83.

Rowlands, I. W. and B. J. Weir (editors)
1974. *The Biology of Hystricomorph Rodents.* New York: Academic Press.

Smythe, N.
1978. The natural history of the Central American agouti (*Dasyprocta punctata*). *Smithsonian Contribs. Zool.*, 257:1–52.

Taylor, R. H.
1970. Reproduction, development, and behavior of the Cuban hutia conga, *Capromys p. pilorides*, in captivity. M. S. Thesis, University of Puget Sound.

Tesh, R. B.
1970a. Observations on the natural history of *Diplomys darlingi*. *J. Mammal.*, 51:197–198.
1970b. Notes on the repoduction, growth, and development of echimyid rodents in Panama. *J. Mammal.*, 51:199–202.

Weir, B. J.
1970. The management and breeding of some more hystricomorph rodents. *Lab. Anim.*, 4:83–97.
1971. The reproductive physiology of the plains viscacha, *Lagostomus maximus*. *J. Reprod. Fertil.*, 25:355–363.
1973. Another hystricomorph rodent: Keeping casiragua (*Proechimys guariae*) in captivity. *Lab. Anim.*, 7:125–134.
1974. Reproductive characteristics of hystricomorph rodents. *Symp. Zool. Soc. Lond.*, 34:265–301.

Woods, C. A. and D. K. Boraker
1975. *Octodon degus. Mammalian Species*, 67:1–5.

Appendix 1. Reproductive data for caviomorph rodents (in order of increasing weight)

Species	*Adult weight (g)*	*Birth weight (g)*	*Neonatal wt/ Adult wt*	*Mean litter size (range)*	*Litter wt/ Adult wt*	*Age when 50% adult wt achieved (wks)*	*Gestation (days)*	*References*
Ctenomys talarum	150	8.0	.05	4.3 (1–7)	.23		102	Weir 1974
Abrocoma cinerea	150	22.0	.15	2.2 (1–3)	.32	10	116	NZP[1]
Octodontomys gliroides	190	20.0	.11	2.1 (1–4)	.22	6	104	NZP
Octodon degus	200	14.4	.07	4.9 (1–8)	.35	5	90	NZP
Microcavia australis	280	30.0	.11	2.8 (1–5)	.30	9	54	Rood 1972
Proechimys guariae	305	20.9	.07	2.7 (1–6)	.19	7	63	Weir 1973
Cercomys cunicularis	335	29.0	.09	3.4 (1–5)	.29		89	NZP
Diplomys darlingei	350	35.0	.10	1.2 (1–2)	.12			Tesh 1970a
Galea musteloides	400	37.0	.09	2.7 (1–5)	.25	11	53	Rood 1972
Hoplomys gymnurus	450	24.3	.05	2.1 (1–3)	.11	12		Tesh 1970b
Proechimys semispinosus	450	26.5	.06	3.3 (1–5)	.20	14	65	Tesh 1970b; Maliniak and Eisenberg 1971; NZP
Chinchilla lanigera	500	35.0	.07	2.0	.14		111	Weir 1974
Cavia aperea	550	57.6	.10	2.3 (1–5)	.24	11	62	Rood 1972; Weir 1970
Geocapromys ingrahami	740	85.0	.11	1.0	.11	13	115	Howe and Clough 1971
Kerodon r. rupestris	800	80.0	.10	1.6 (1–2)	.16		77	NZP
Myoprocta pratti	950	77.0	.08	1.8 (1–3)	.15	12	99	Kleiman 1969, 1970
Plagiodontia aedium	1267	110.0	.09	1.3 (1–2)	.11	15		NZP
Lagidium peruanum	1300	226.0	.17	1.0	.17	11	140	NZP; Weir 1974
Pediolagus salinicola	1700	199.0	.12	1.5 (1–3)	.18		77	NZP
Dasyprocta punctata	3000			(1–2)		11		Smythe 1978
Lagostomus maximus	3000	193	.06	1.9 (1–3)	.12	17	154	Weir 1971
Coendou prehensilis	3100	390	.13	1.0	.13	18		NZP
Capromys pilorides	5200	200	.04	1.8 (1–3)	.07	24	125	Taylor 1970; NZP
Myocastor coypu	6000	225	.04	5.0	.20		129	Weir 1974
Agouti paca	8000	710	.09	1.0	.09	14	118	Lander 1974; NZP
Dinomys branickii	13000	900	.07	(1–2)	.14		223–280	Collins and Eisenberg 1972
Hydrochaeris hydrochaeris	49000	1400	.03	3.5 (1–6)	.10	52	120	Ojasti 1973

[1] NZP = Records of the U.S. National Zoological Park.

SECTION 5:

A Comparison of Llanos and Rain Forest Mammal Faunas

Odocoileus virginianus gymnotis, the white-tailed deer. In suitable habitats, this species may be a dominant component of the savanna ecosystem although the capybara forms the dominant mammalian component of the biomass in the low, wet llanos.

JOHN F. EISENBERG
National Zoological Park
Washington, D. C. 20008

M. A. O'CONNELL
Department of Biological Sciences
Texas Tech University
P. O. Box 4149
Lubbock, Texas 79409

PETER V. AUGUST
Department of Biology
Boston University
2 Cummington Street
Boston, Massachusetts 02215

Density, Productivity, and Distribution of Mammals in Two Venezuelan Habitats

RESÚMEN

Se compara y se hace un contraste entre la fauna de mamíferos en Parqúe Nacional de Guatopo y el Fundo Pecuario Masaguaral. Los llanos tienen una diversidad menor de mamíferos que no vuelan. Se presenta la densidad estimada de mamíferos que no vuelan para los dos lugares y se cacula la biomasa. Los cálculos de biomasa para los dos lugares fueron luego comparados con datos previos de la isla de Barro Colorado en Panamá. Se puede decir que bosques secundarios y lugares que tienen extrema sequía muestran una biomasa de mamíferos arboreos menor que en lugares con bosques antiguos que tienen una producción de plantas más o menos igual por año. La biomasa de mamíferos terrestres puede alcanzar niveles muy altos en los llanos de Venezuela. El capibara es un contribuyente de alto valor en sitios donde no es cazado. La mayoría de los mamíferos dentro de cualquier communidad son herbívoros.

Un análisis preliminaro de productividad demuestra que los monos aluatinos pueden tener una biomasa muy alta en sitios óptimos de morada pero la producción anual es baja en comparación al venado y al capibara. El resultodo es que el capibara es un animal que tiene un potencial de productividad alta asi como el mentenimiento de una biomasa muy alta.

ABSTRACT

The mammalian fauna for Guatopo National Park and Fundo Pecuario Masaguaral is compared and contrasted. The llanos habitat at Masaguaral exhibits a lower diversity of non-volant mammalian species. Density estimates for non-volant mammals were developed for the two habitats. From these density estimates, biomass values were calculated. The biomass values for the two Venezuelan study sites were then compared with previous biomass estimates for Barro Colorado Island, Panama. It is concluded that second growth forests and areas subjected to extreme seasonal drought may show lower arboreal mammalian biomass values than areas of mature forest growth with a less seasonal schedule of plant productivity. Standing crop biomass of terrestrial mammals may reach rather high levels in the llanos of Venezuela. The capybara is a significant contributor to standing crop mammalian biomass in areas where it has not been extensively hunted. Standing crop biomasses of individual species, when analyzed from the standpoint of trophic strategy, indicate a pyramidal structure. Herbivores comprise the bulk of mammalian biomass within any community.

A preliminary analysis of productivity demonstrates that, although the howler monkey may have a high standing crop biomass in areas of optimal habitats, its annual productivity is rather low when compared to that of white-tailed deer and capybara. The capybara again shows a potential for high productivity as well as the maintenance of high standing crop biomasses.

Introduction

Research on the ecology and behavior of vertebrates in the neotropics has had a long and distinguished history. Long-range research on populations of vertebrates carried out in areas where relevant environment data were being collected in parallel have, however, been a rarity. Part of the reason for this has resulted from the fact that there are few research stations in the neotropics that have a continuity in scientific activity. The Smithsonian Tropical Research Institute in Panama has been one location where vertebrate studies have been carried out for decades. Most of the research efforts were conducted on Barro Colorado Island, a biological preserve, since the formation of the Panama Canal. In the early 1970s a number of us realized that it would be useful to develop study sites for investigations of neotropical vertebrate populations which were not confined to an insular environment. Vegetational succession, local extinctions, and the loss of large predators on BCI had produced a somewhat artifical faunal assemblage (Eisenberg and Thorington, 1973). In our attempts to establish research areas in northern South America, a number of national parks were investigated. Since vertebrate ecology in the extensive savannas of northern South America had barely been studied, it soon became apparent that it would be useful to establish study areas in at least two different habitat types. We hoped to establish a forested study area and a savanna study area close enough to permit interchange of personnel. Through the efforts of Sr. Edgardo Mondolfi and Sr. Tomas Blohm, it was decided to establish a study area in Guatopo National Park and on the ranch of Sr. Blohm, Fundo Pecuario Masaguaral.

Figure 1. Range and mean precipitation per month for Guatopo National Park (south end). Note the tendency for February through April to show the least variation and the minimum monthly averages. Precipitation at the south end of Guatopo is similar to the pattern shown at Masaguaral, but the dry season is shorter and less severe than is the average case in the llanos.

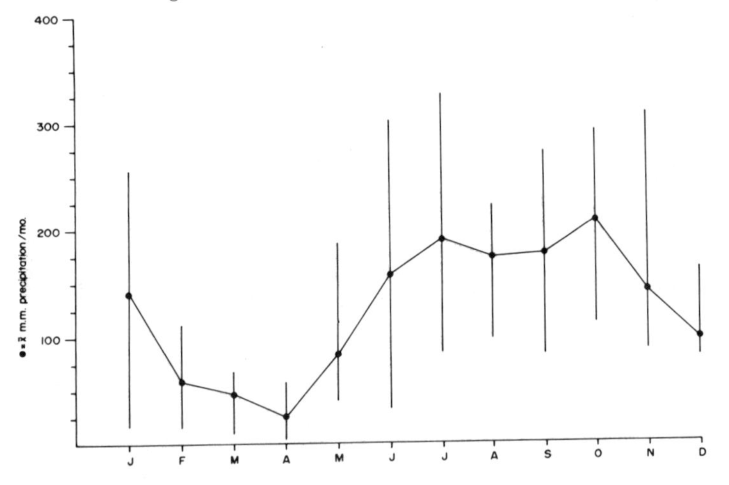

Figure 2. Geographical location and park boundaries (1976) for Parque Nacional Guatopo. Dotted line indicates road from Santa Teresa to Alta Gracia which bisects the park.

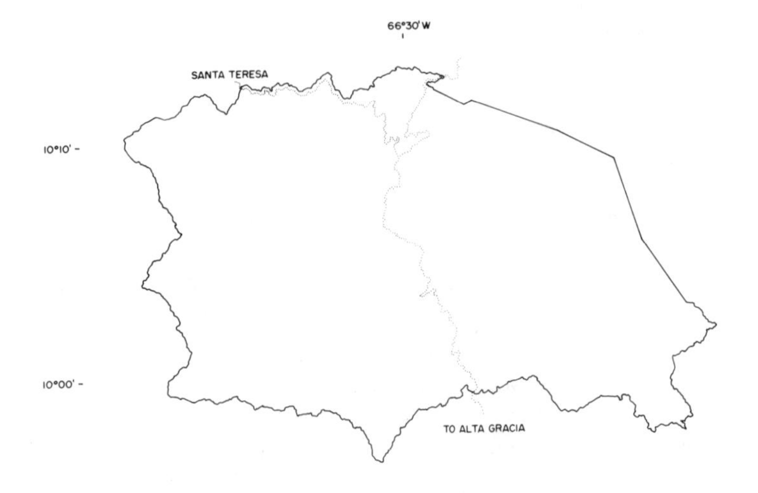

Guatopo National Park is situated in the southern range of the north coastal mountains of Venezuela and is approximately 40 km south-southeast of Caracas. The topography is submontane to montane with elevations ranging from 250 to 1500 meters. The southern end of Guatopo was the focus of our studies (Figure 1). This area of the park receives about 1500 mm of precipitation per year (Figure 2). The rainfall in any given month is quite variable but a definite dry season of variable duration usually occurs between January and May. The south slopes of the north coastal range tend to be in a rain shadow and support a semi-deciduous to deciduous forest. The mountain tops and the windward side tend to be heavily forested to a lower elevation. The faunal diversity in the north coast ranges appears to be markedly higher than that found in a comparable area of the llanos.

The primary llanos study area was at Masaguaral, a working ranch, and a nature preserve. Located some 45 km south of Calabozo in the State of Guarico, the ranch lies at about 100 m elevation (see Figure 3). Rainfall averages about 1500 mm per year but is strongly seasonal with a pronounced drought beginning in November and often extending well into May (Figure 4). During the rainy season, part of the habitat is subjected to flooding and this profoundly affects the form of the vegetation (see Troth, p. 17) and the habitat utilization patterns of terrestrial mammals (see Brady, p. 161, and O'Connell, p. 73).

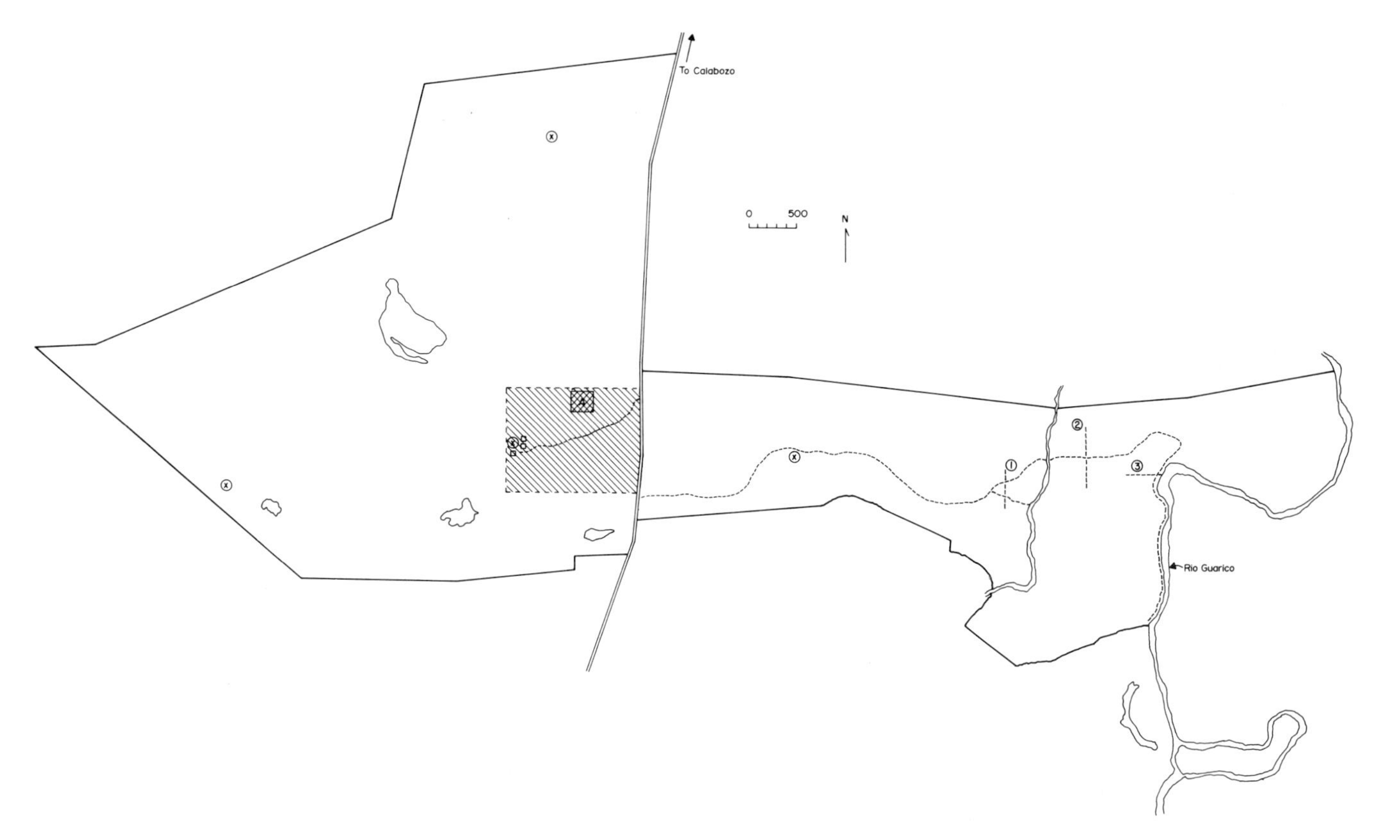

Figure 3. Outline map of the boundaries of Fundo Pecuaria Masaguaral some 45 km south of Calobozo, Estado Guarico, Venezuela. Boundaries as of 1975. x = location of major windmills. Cross-hatched area = approximate location of intensive study sites on the west side of the ranch. Small square within study area indicates location of 4-hectare grid established by O'Connell. Major study areas on the east side of the ranch were carried out on the west side of the Caño Caricol.

Hato el Frio in the State of Apure served as a secondary research site. This ranch run by the Maldonado family is situated in an area of lower elevation and, during the summer rainy season, is subject to more extensive flooding than is the case at Hato Masaguaral. Studies on *Podocnemus* and *Caiman* were conducted intensively during two seasons at El Frio. The habitat is described by Ojasti (1973).

Density

Of special interest to our group was to establish estimates of density for various vertebrates and thereby calculate their biomass. These biomass calculations could then be compared with similar values derived from studies on Barro Colorado Island (see Eisenberg and Thorington, 1973). Biomass values for vertebrate communities can give some indication of the carrying capacity for different habitats. A three-way comparison among study sites on Barro Colorado, at Guatopo and at Hato Masaguaral also permitted us to understand the differential utilization of niches as a function of drastic alterations in habitat structure. For example, in Barro Colorado, the forest is rather mature and continuous in its coverage; in the areas of Guatopo National Park where we undertook our studies, the

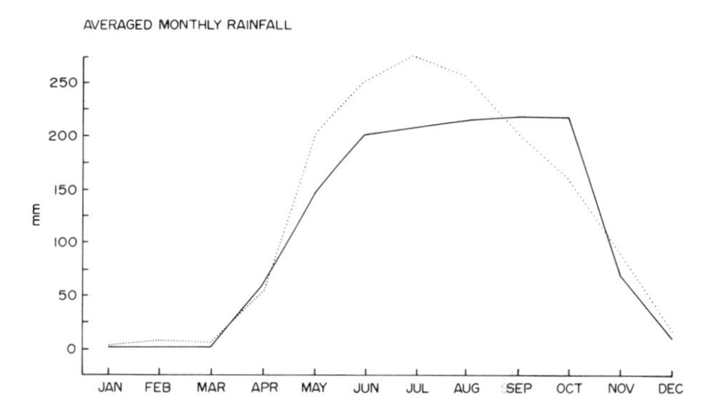

Figure 4. Average monthly rainfall for Corozo Pondo. Solid line indicates years 1972 to 1977; dotted line indicates years 1952 to 1971.

forest is much younger second growth and subject to a more prolonged period of seasonal aridity; the llanos has discontinuous forest cover, forest islands and extensive areas of seasonally inundated grassland. With such a diversity in habitat types, we felt that an understanding of shifts in substrate utilization might be detected, giving us some estimation of what impact land clearing would have on vertebrate faunas. In other words, by cross-comparing these various vegetational types, we were in a way sampling a natural experimental situation.

The Diversity of Terrestrial Mammals

Tables 1 and 2 present the lists of mammalian species recorded for Guatopo National Park and Masaguaral. Our bat species lists were tabulated through the efforts of Drs. T. Fleming and R. J. Baker. In some cases the species identifications are tentative. The species of bats as determined for Masaguaral are probably nearly complete. We estimate that only 60 percent of the bats existing in Guatopo have been netted. If we concentrate on the non-volant mammals, then 40 species are known to exist in Guatopo and only 28 in the llanos habitat. The structural complexity of the Guatopo habitats probably permits a higher species diversity. The influence of habitat diversity on small mammal diversity can be demonstrated even within the Masaguaral habitat types. Table 3 compares the abundance and diversity of small, trapable mammals for three habitat types. The vegetational simplicity of the palm-grass savanna correlates with the single semi-arboreal rodent species, *Oryzomys*. The structurally complex formations of matas, palms and strangler figs contain up to seven species of small mammals.

The Densities of Non-volant Mammals

For the purpose of this discussion, Masaguaral will be considered in two parts: the west side and the east side. The road from Calabozo to San Fernando cuts through the ranch and serves as a convenient reference for this subdivision. The east side of the ranch extends to the Rio Guarico and supports a typical semi-deciduous tropical forest on the high ground (see Troth, p. 17). Table 4 presents the density and biomass data for Masaguaral and Table 5 portrays the same values for Guatopo. Two types of density and biomass values are offered: Crude and Ecological. Crude density does not correct for microhabitat preferences and presents the total population in terms of the total base area surveyed. Ecological densities are densities calculated from a base area corrected to include only suitable habitat for the species in question. The methods em-

Figure 5. Plot of density versus absolute body size. Species are arranged according to rank order of abundance. Species list for Masaguaral (see Table 4).

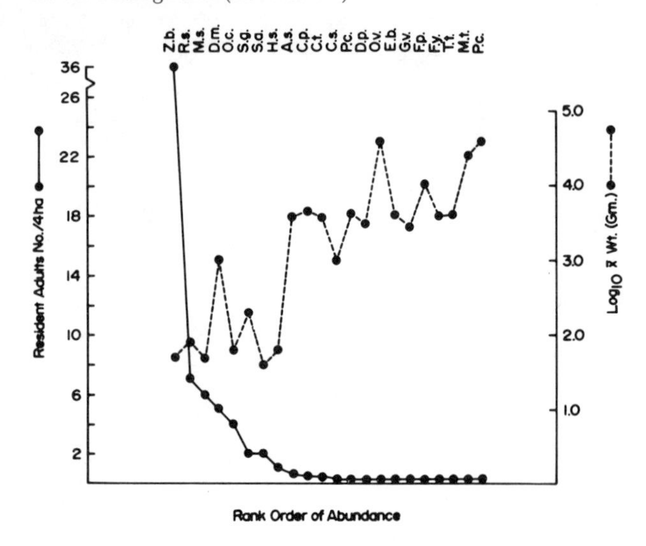

ployed in developing these population estimates are included in Appendix I.

Population density is related to absolute body size with large mammals generally exhibiting a low numerical density. Figure 5 illustrates this trend with a portrayal of the mammalian densities for Masaguaral. The same trends are demonstrable for Guatopo. Yet the inverse relationship between body size and density is not a linear one and the trophic strategy of a species profoundly affects the density it can reach. Within the same size class, herbivores generally exist at higher densities than do carnivores. A better estimate of the impact a species may have within the ecosystem is to convert the density estimate to a biomass estimate by multiplying density by a "unit weight" for each species. The unit weight reflects the average weight of an individual whether it is an adult or juvenile and is an abstract value derived from an estimate of the proportion of the various age classes. These data are also presented in Tables 4 and 5.

Comparison of Censusing Results between Panama and Venezuela

Transect censusing of the mammalian fauna of Barro Colorado Island, Panama, was undertaken by Eisenberg in 1964. This transect walking technique was repeated in 1965, 1970 and 1974. The combined results from Eisenberg and Thorington's efforts through 1972 have been published (Eisenberg and Thorington, 1973). The number of mammal sightings (either individuals or troops per km walked for each successive

Table 1. List of mammals for Guatopo National Park, 1975–1978

Marsupialia

1. *Didelphis marsupialis*
2. *Chironectes minimus*
3. *Marmosa murina*
4. *Marmosa fuscata*
5. *Marmosa cinerea*
6. *Monodelphis brevicauda*
7. *Caluromys philander*

Edentata

8. *Bradypus infuscatus*
9. *Tamandua tetradactyla*
10. *Dasypus novemcinctus*
11. *Cabassous unicinctus*
12. *Priodontes gigantea*

Chiroptera (South End)

Emballonuridae

13. *Saccopteryx bilineata*
14. *Saccopteryx leptura*

Mormoopidae

15. *Pteronotus daveyi*
16. *Pteronotus parnellii*
17. *Phyllostomus discolor*
18. *Phyllostomus hastatus*
19. *Carollia perspicillata*
20. *Glossophaga soricina*
21. *Anoura caudifer*
22. *Sturnira lilium*
23. *Sturnira ludovici*
24. *Sturnira tildae*
25. *Uroderma bilobatum*
26. *Uroderma magnirostrum*
27. *Artibeus jamaicensis*
28. *Artibeus lituratus*
29. *Artibeus cinereus*
30. *Artibeus phaotis*
31. *Sphaeronycteris toxophyllum*
32. *Ametrida centurio*
33. *Vampyrodes caraccioloi*
34. *Vampyressa pusilla*
35. *Chiroderma trinitatum*
36. *Vampyrops helleri*

Vespertilionidae

37. *Myotis simus*
38. *Rhogeessa tumida*
39. *Eptesicus furinalus*
40. *Lasiurus borealis*

Molossidae

41. *Eumops bonairiensis*

Primates

42. *Alouatta seniculus*
43. *Cebus nigrivittatus*
44. *Ateles belzebuth*

Carnivora

45. *Eira barbara*
46. *Conepatus semistriatus*
47. *Procyon cancrivorus*
48. *Potos flavus*
49. *Felis pardalis*
50. *Felis yaguaroundi*
51. *Puma concolor*
52. *Panthera onca*

Lagomorpha

53. *Sylvilagus brasiliensis*

Rodentia

54. *Sciurus granatensis*
55. *Heteromys anomalus*
56. *Oryzomys albigularis*
57. *Oryzomys capito*
58. *Oryzomys concolor*
59. *Oryzomys bicolor*
60. *Akodon urichi*
61. *Neacomys tenuipes*
62. *Rhipidomys venezuelae*
63. *Proechimys semispinosus*
64. *Echimys semivillosus*
65. *Dasyprocta aguti*
66. *Agouti paca*

Perissodactyla

67. *Tapirus terrestris*

Artiodactyla

68. *Mazama americana*
69. *Tayassu tajacu*

Table 2. List of mammals for Fundo Pecuario Masaguaral, Guarico, Venezuela 1975–1978

Marsupialia

1. *Didelphis marsupialis*
2. *Marmosa robinsoni*

Edentata

3. *Dasypus novemcinctus*
4. *Tamandua tetradactyla*
5. *Myrmecophaga tridactyla*

Chiroptera

Emballonuridae

6. *Rhynchonycteris naso*
7. *Saccopteryx bilineata*
8. *Saccopteryx canescens*
9. *Saccopteryx leptura*

Noctilionidae

10. *Noctilio leporinus*
11. *Noctilio albiventris*

Phyllostomatidae

12. *Micronycteris minuta*
13. *Micronycteris nicefori*
14. *Micronycteris megalotis*

15. *Tonatia minuta*
16. *Tonatia venezuelae*
17. *Mimon crenulatum*
18. *Phyllostomus discolor*
19. *Phyllostomus hastatus*
20. *Phyllostomus elongatus*
21. *Trachops cirrhosus*
22. *Vampyrum spectrum*
23. *Carollia perspicillata*
24. *Glossophaga soricina*
25. *Glossophaga longirostris*
26. *Sturnira lilium*
27. *Sturnira ludovici*
28. *Artibeus jamaicensis*
29. *Sphaeronycteris toxophyllum*
30. *Ametrida centurio*
31. *Uroderma magnirostrum*

Desmodontinae

32. *Desmodus rotundus*

Vespertilionidae

33. *Myotis nigriscans*
34. *Myotis albescens*
35. *Eptesicus furinalus*
36. *Rhogeesa tumida*
37. *Lasiurus ega*
38. *Lasiurus borealis*

Molossidae

39. *Molossops terminckii*
40. *Molossops greenhalli*
41. *Molossus molossus*
42. *Molossus pretiosus*
43. *Eumops bonariensis*
44. *Eumops dabbeni*
45. *Eumops glaucinus*
46. *Eumops hansae*
47. *Eumops auripendulus*

Primates

48. *Cebus nigrivittatus*[1]
49. *Alouatta seniculus*

Carnivora

50. *Conepatus semistriatus*
51. *Grison vittatus*
52. *Procyon cancrivorus*
53. *Cerdocyon thous*
54. *Felis yaguaroundi*
55. *Felis paradalis*
56. *Puma concolor*[1]
57. *Eira barbara*

Lagomorpha

58. *Sylvilagus floridanus*

Rodentia

59. *Sciurus granatensis*
60. *Heteromys anomalus*
61. *Rhipidomys* sp.
62. *Oryzomys bicolor*
63. *Zygodontomys brevicauda*
64. *Sigmomys alstoni*
65. *Coendou prehensilis*
66. *Echimys semivillosus*[1]
67. *Dasyprocta agouti*[1]
68. *Agouti (Cuniculus) paca*[1]
69. *Hydrochaeris hydrochaeris*

Artiodactyla

70. *Odocoileus virginianus*
71. *Tayassu tajacu*[1]

[1] Predominately in the Gallery Forest on the Rio Guarico.

Table 3. Densities of small mammals on three habitats in the llanos[1]

Palm-grass savanna		*Palms and matas*		*Deciduous forest*	
Species	*No./km²*	*Species*	*No./km²*	*Species*	*No./km²*
Oryzomys bicolor	80	*Marmosa robinsoni*	150	*Marmosa robinsoni*	280
		Sciurus granatensis	50	*Sciurus granatensis*	40
		Heteromys anomalus	25	*Heteromys anomalus*	110
		Rhipidomys sp.	187	*Zygodontomys brevicauda*	60
		Oryzomys bicolor	60	*Echimys semivillosus*	40
		Zygodontomys brevicauda[2]	375		
		Sigmomys alstoni	50		

[1] Mean number known alive per month, small mammals defined as <500 grams.

[2] *Zygodontomys* fluctuates dramatically in this habitat; the value may drop to less than 50 or exceed 500.

Table 4. Masaguaral non-volant mammalian densities and biomasses

Species	*Unit wt. (kg)*	*West side*				*East side*			
		Density #/km²		*Biomass kg/km²*		*Density #/km²*		*Biomass kg/km²*	
		Ecol.	*Crude*	*Ecol.*	*Crude*	*Ecol.*	*Crude*	*Ecol.*	*Crude*
Didelphis marsupialis	1.00	122.00	52.50	122.0	53.0	122.00	52.50	122.0	53.0
Marmosa robinsoni	.05	150.00	64.50	7.5	5.3	280.00	120.50	14.0	6.0
Dasypus novemcinctus	3.80	12.00	5.00	45.6	19.0	13.00	10.00	49.4	38.0
Tamandua tetradactyla	4.00	3.00	2.00	12.0	8.0	3.00	2.00	12.0	8.0
Myrmecophaga tridactyla	27.00	.18	.12	4.9	3.2	.18	.12	4.9	3.2
Cebus nigrivittatus	2.60	.00	.00	0.0	0.0	44.00	19.00	114.0	50.0
Alouatta seniculus	4.30	150.00[1]	41.00	645.0	176.3	50.00	<20.00	215.0	86.0
Conepatus semistriatus	1.20	12.50	6.00	15.0	7.2	12.50	1.20	15.0	1.4
Grison vittatus	3.20	2.40	1.10	7.6	3.5	2.40	1.10	7.7	3.5
Procyon cancrivorus	4.70	10.00	6.24	47.0	29.4	10.00	6.20	47.0	29.0
Cerdocyon thous	4.00	4.00	2.50	16.0	10.0	4.00	2.50	16.0	10.0
Felis yagouaroundi	4.00	.25	.25	1.0	1.0	.25	.25	1.0	1.0
Felis pardalis	12.00	.25	.25	3.0	3.0	.25	.25	3.0	3.0
Puma concolor	40.00	?	?	?	?	.09	.02	3.6	<1.0
Eira barbara	4.00	?	?	?	?	2.00	1.00	8.0	2.0
Sylvilagus floridanus	.80	10.00	5.00	8.0	4.0	35.00	11.00	20.0	9.0
Sciurus granatensis	.25	50.00	26.50	12.5	6.7	40.00	17.20	10.0	4.3
Heteromys anomalus	.07	25.00	10.70	1.8	.7	110.00	47.30	7.7	3.3
Rhipidomys sp.	.09	187.00	80.00	16.3	7.2	?	?	?	?
Oryzomys bicolor	.06	60.00	30.00	3.6	1.8	?	?	?	?
Zygodontomys brevicauda	.05	375.00	187.00	18.8	9.3	60.00	26.00	3.0	1.3
Sigmomys alstoni	.04	50.00	25.00	2.0	1.0	50.00	25.00	2.0	1.0
Coendou prehensilis	5.00	7.00	3.50	35.0	17.5	7.00	3.50	35.0	17.5
Echimys semivillosus	.20	?	?	?	?	40.00	17.20	8.0	3.4
Dasyprocta aguti	2.00	>3.00	>1.50	6.0	3.0	80.00	40.00	160.0	80.0
Agouti paca	8.00	?	?	?	?	25.00[3]	12.00	200.0	96.0
Hydrochaeris hydrochaeris	30.00	10.00[2]	.29	300.0	8.7	23.00	10.00	690.0	300.0
Odocoileus virginianus	40.00	3.00	2.50	120.0	100.0	4.00	2.00	160.0	80.0
Tayassu tajacu	23.00	?	?	?	?	12.00	8.50	276.0	195.0
				1,450.6 (1,170.6)[1]	478.8			2,203.9	1,085.9

[1] Calculated at the smallest base area @ 44 ha; if assume base of 100 ha, then 85/km² and 365.5 kg/km.²

[2] Numbers have increased steadily; this is the 1978 estimate.

[3] Number may be an overestimate as it is based on track frequency. The figure is very tentative.

Table 5. Guatopo non-volant mammalian densities and biomasses

Species	*Unit wt. (kg)*	*Density #/km²*		*Biomass kg/km²*	
		Ecol.	*Crude*	*Ecol.*	*Crude*
Didelphis marsupialis	1.00	65.0	41.00	65.0	51.00
Marmosa fuscata	.06	97.0	61.00	5.8	3.70
Marmosa cinerea	.15	40.0	22.00	6.0	3.30
Monodelphis brevicaudata	.08	37.0	23.00	2.9	1.80
Caluromys philander	.17	<20.0	<13.00	3.4	2.20
Bradypus infuscatus	3.20	3.7	2.33	11.8	7.50
Tamandua tetradactyla	4.00	9.0	5.67	36.0	22.80
Dasypus novemcinctus	3.80	6.0	3.79	22.8	14.40
Cabassous unicinctus	4.80	1.2	.75	5.8	3.60
Priodontes giganteus	25.00	?	<.20	?	<5.00
Alouatta seniculus	4.30	20.7	16.00	89.0	68.80
Cebus nigrivittatus	2.60	35.0	27.00	91.0	70.20
Ateles belzebuth	5.00	?	5.60	75.0	28.00
Eira barbara	4.00	2.0	1.60	8.0	6.40
Conepatus semistriatus	1.20	9.0	5.67	10.8	6.80
Procyon cancrivorus	4.70	4.2	2.52	19.7	11.80
Potos flavus	2.50	3.7	2.33	9.3	5.80
Felis pardalis	12.00	?	<.50	?	<.60
Felis yagouaroundi	5.00	1.2	.80	4.8	4.00
Panthera onca	55.00	?	< .02	?	1.10
Sylvilagus brasiliensis	.70	5.0	3.40	3.5	2.28
Sciurus granatensis	.25	<40.0	25.00	10.0	6.25
Heteromys anomalus	.07	200.0	126.00	14.0	8.80
Oryzomys capito	.05	220.0	139.00	11.0	6.95
Oryzomys concolor	.06	150.0	95.00	9.0	5.70
Akodon urichi	.04	200.0	126.00	8.0	5.00
Neacomys tenuipes	.02	400.0	252.00	8.0	5.00
Rhipidomys venezuelae	.09	144.0	91.00	12.9	8.20
Proechimys semispinosus	.30	200.0	126.00	60.0	37.80
Dasyprocta aguti	2.00	100.0	63.00	200.0	126.00
Agouti paca	8.00	<30.0	18.00	240.0	144.00
Coendou prehensilis	2.60	6.0	3.78	15.6	9.80
Tapirus terrestris	262.00	.8	.60	209.6	157.20
Mazama americana	15.00	10.0	5.25	150.0	78.80
Tayassu tajacu	18.00	4.0	1.90	72.0	34.20
				1,489.1	946.40

series of censusing in Panama remain remarkably uniform. For example, in 1974 after having walked 48 km, approximately 69 groups or individual mammals were encountered. In 1970, after having traversed 48 km, a total of 72 mammalian groups or individuals were tallied. The uniformity of sightings in Panama encouraged us to employ a similar technique during our censusing operations in Venezuela. Rate of sighting in Venezuela after traversing 48 km was approximately 19 individuals or groups. The sighting rates are approximately 1.4 animal or group per km walked in Panama and .34 per km walked in Guatopo. If we assume an equal detection distance, this implies that Guatopo shows a density approximately one-fourth that of Panama.

The reasons for this discrepancy are a little difficult to pinpoint at this time. It may have to do with the fact that Guatopo is at an earlier stage of vegetational succession than is now the case on Barro Colorado Island. Further, it may be a result of a slightly different faunal balance in Guatopo National Park. For example, small predators are abundant in Guatopo and are obviously having an impact on the smaller diurnal rodents, such as agoutis and squirrels. Since agoutis and squirrels account for the larger number of diurnal sightings during transect walking, it may be no coincidence that sightings of squirrels and agoutis are rarer in Guatopo than is the case on Barro Colorado Island. Finally howler monkeys are sighted with an extremely high frequency on Barro Colorado, but a very low frequency in Guatopo. This is undoubtedly related to the fact that howlers are favored under conditions of more mature forests, and it is believed that the howler density in Guatopo now more nearly approximates the density of howlers during the early 1930s when Carpenter first studied them on Barro Colorado Island. At that time forest succession on Barro Colorado had proceeded to approximately the level of forest succession over much of the second growth areas currently to be found in Guatopo.

At extreme ecological density, Eisenberg and Thorington (1973) estimated that the mammalian biomass of Barro Colorado Island could reach a level of greater than 4,431 kg/km². A more realistic estimate approximating crude density would be a value of about 3,000 kg/km². Extrapolating from sighting frequencies, one could estimate that ecological densities in Guatopo Park would approximate 1,700 kg/km² and that crude density would be 900 kg/km². These data are in reasonable agreement with estimates derived by trap, mark and release and extrapolations based on road kills and sighting frequency (see Table 5).

In 1973, Eisenberg and Thorington published a preliminary analysis of a neotropical mammalian fauna. Combining the density data from Barro Colorado Island (BCI) concerning the larger species of mammals and extrapolating from studies of small mammal density from regions adjacent to the island, Eisenberg and Thorington attempted to construct the probable biomass relationships among the various trophic categories of mammals found on BCI. They relied on the small mammal work by Fleming (1971, 1973), and the work by Oppenheimer (1968), Chivers (1969), Montgomery and Sunquist (1975), and Smythe (1970) with the larger mammals found on BCI. The purpose of the paper was to demonstrate what was known about some key species on BCI and to highlight the species that were poorly studied. Estimates for BCI were compared with the data derived from a rescue operation in Surinam (Walsh and Gannon, 1967). There was a remarkable correspondence between the proportions of various species rescued in Surinam and the presumptive proportion of various dominant species present in the fauna of BCI. It is now possible through the published work of Montgomery and Sunquist (1978), Montgomery and Lubin (1977, 1978), Smythe (1978), Heaney and Thorington (1978), and Glanz (1978), together with further censusing work by Thorington and Eisenberg to present better estimates of the numerical abundance for the dominant terrestrial mammals comprising the community on BCI.

Figure 6 compares ecological and crude biomass estimates for neotropical mammals censused in the tropical evergreen forest on BCI. As one can see, the original relationships established in the paper of 1973 seem to hold up. Herbivores comprise the dominant percentage of mammalian biomass, followed by mixed browsers and frugivores, omnivores, myrmecophages, frugivores, granivores, and carnivores, in descending order. The importance of the arboreal herbivore component in tropical rain forests is clearly emphasized.

Figure 7 presents the biomass data for Guatopo National Park. Although browsers and frugivore browsers contribute heavily to the total biomass, the rank order of the species is quite different from BCI. Terrestrial forms contribute a great deal to the total biomass and the arboreal folivores, such as *Bradypus* and *Alouatta*, are not dominant contributors to the mammalian biomass. This may result from the fact that our major survey areas were covered with second growth forest less than 25 years old. This stage of succession may not favor high densities of *Alouatta* and *Bradypus*.

Figure 6. Biomass estimates for 19 species of mammals found on Barro Colorado Island. The greater biomass is contributed by the folivores (Br) and frugivore-folivore (Br/F) species.

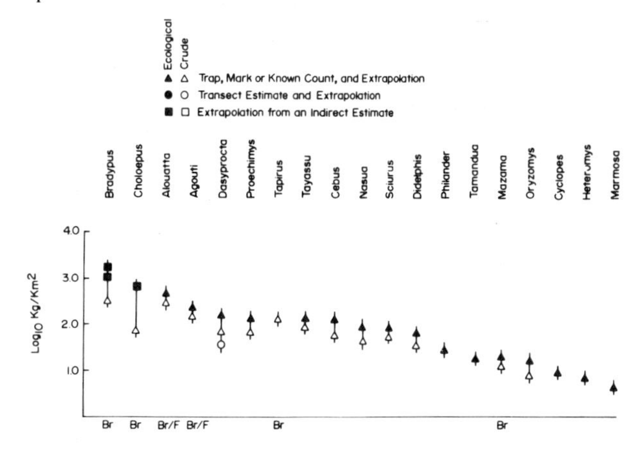

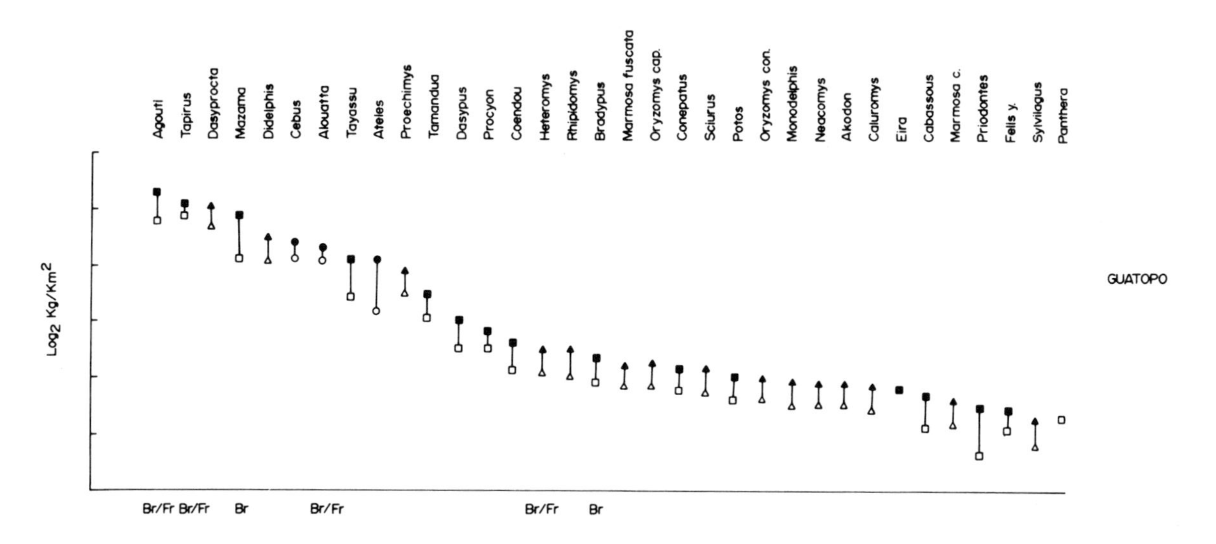

Figure 7. Logarithmic plot of biomass according to rank order of ecological biomass values for the non-volant mammals found in Guatopo National Park, Venezuela. Method of plotting and symbols as outlined in Figure 6.

The data from the llanos are also instructive. The total crude biomass for the west side of Masaguaral is rather low, but in the Bajio-Mata habitat *Alouatta* density can be very high. Figure 8, based on ecological biomass values, demonstrates that *Alouatta* is the dominant contributor to mammalian biomass. Figure 9 offers a comparable plot for the east side of Masaguaral and reflects mainly the non-inundated deciduous forest and the drainage system of the Caño Caracol. The semi-aquatic capybara, *Hydrochaeris*, and the peccary, *Tayassu*, together with the paca, *Agouti*, probably comprise the dominant contributors to biomass.

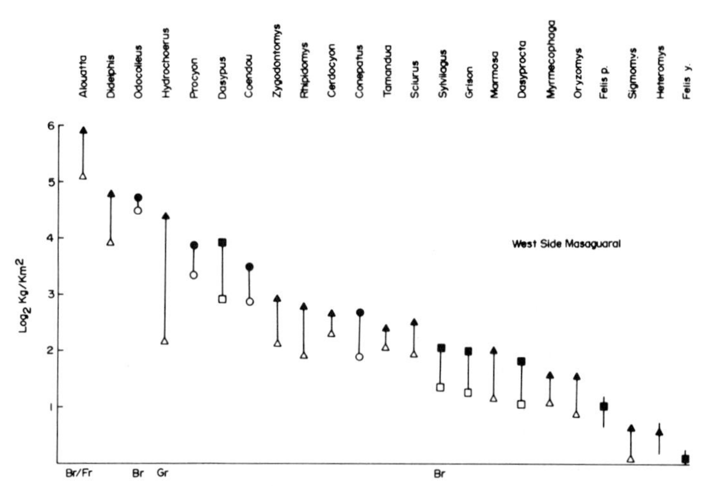

Figure 8. Logarithmic plot of biomass according to rank order of the ecological biomass values for the non-volant mammals found at the west side of Fundo Pecuario Masaguaral, Venezuela. Method of plotting and symbols as outlined in Figure 6. Browsers and grazers contribute heavily to the total biomass value.

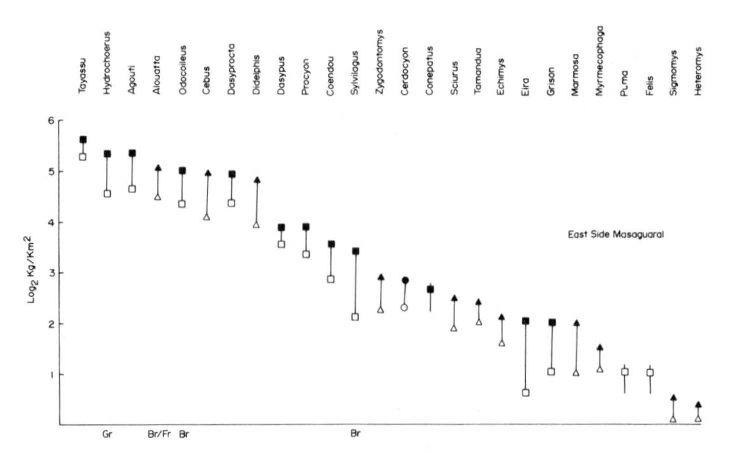

Figure 9. Logarithmic plot of biomass according to the rank order of the ecological biomass values for the non-volant mammals found at the east side of Fundo Pecuario Masaguaral, Venezuela. Method of plotting and symbols as outlined in Figure 6. Browsers and grazers contribute heavily to the total biomass value.

Tables 6 and 7 offer a percentage breakdown of the total mammalian biomass with respect to the trophic and substrate utilization categories as outlined in Eisenberg and Thorington (1973). Arboreal biomass dominates on BCI in Panama. This is a mature, tropical forest with negligible predation by large raptorial birds or human hunters. In the second growth forests of Guatopo, the arboreal biomass is reduced as is the case in all savanna or savanna-forest mosaics. In the latter habitats, scansorial and terrestrial forms predominate (see Table 6). A consideration of Table 7 indicates the pyramidal nature of the trophic categories. In both the llanos and the forest, the herbivores (browsers and grazers) contribute more than 50 percent to the total mammalian biomass.

Clearly the structure and extent of the forest influence the relative density of arboreal forms. Seasonal flooding can also depress the carrying capacity for terrestrial forms. The range of carrying capacities, as inferred from biomass levels, is highly variable when tropical habitats are compared. Such initial efforts as ours are important, however, in elucidating the differences in the structure and impact of mammalian faunas drawn from geographically distinct areas of the tropics. Hendrichs (1978) offers a comparison between a mammalian community in Guatemala and one in Kivu, Zaire. His biomass values for Guatemala as well as his trophic and substrate analyses are remarkably similar to our values for Guatopo. This encourages us to believe that a generalized profile for mammalian communities as a function of (1) vegetational succession and (2) different climax formations may be obtainable for the neotropics.

Livestock (mainly cattle) on Masaguaral averages around 7,600 kg/km^2 crude biomass for both the west side and the east side of the ranch. Since the stock is often given supplementary feeding during the drought, these values do not reflect a true carrying capacity for introduced grazing mammals. Crude biomass values for Hato El Frio in the State of Apure can reach 2,564 kg/km^2 for the grazing capybara (*Hydrochaeris*) and 18,504 kg/km^2 for grazing livestock (horses, mules, and cattle). The high value for livestock is again maintained in part through supplementary feeding during prolonged drought periods. The potential productivity of the low llanos for terrestrial mammals appears to be quite high, but soil fertility is highly variable in the llanos and generalizations for such a mosaic habitat could be premature.

Productivity

Although biomass estimates are extremely useful in assessing the relative dominance of a species within a

Table 6. Percentage of mammalian biomass divided according to substrate preferences

Location	*Arboreal*	*Scansorial*	*Terrestrial*	*Remarks*	*References*
Panama					
Barro Colorado	70	5	25	Mature forest (island)	Eisenberg and Thorington, 1973
Surinam					
Recovery operation	40	1[1]	59	Riverine forest	
Venezuela					
Masaguaral (west)	42	22	35	Developed land and palm forest	
Masaguaral (east)	23	12	65	Riverine llanos	
Guatopo	17	13	70	Second growth forest	
Hato El Frio	6	3	91	Riverine llanos (wet savanna)	
Bolivar	10	3	87	Forest savanna mosaic[2] (recovery operation)	

[1] May be artificially low.

[2] Data courtesy of E. Mondolfi and F. Bruzual.

community, they offer only part of the picture for understanding community dynamics. While it is true that larger species consume larger amounts of food within a 24-hour period, we also know that the correlation between absolute body size and metabolic rate is a negative one. This means that, although larger species consume more food than a smaller species, they consume less than would have been predicted if one assumed that the relationship between the metabolic rate and body size was a direct linear correlation. Eisenberg and Thorington (1973) attempted to calculate the annual caloric consumption for each species of mammal found on Barro Colorado Island and demonstrated that where large species were extremely abundant, indeed, they consumed more energy on an annual basis than did smaller species. However, in some cases, small species because of their numerical abundance had a significant impact on the community in terms of caloric consumption.

Another way to approach the problem rather than calculating average annual caloric consumption for a species within a community is to estimate relative productivity. For the purposes of this discussion, we will consider the mammals on the west side of Hato

Table 7. Percentage contribution to total mammalian biomass, arranged according to trophic categories[1]

Trophic category	*Localities*			
	Panama	*Guatopo*	*Masa-guaral (W)*	*Masa-guaral (E)*
Carnivores/Insectivores				
Top carnivores	1.0%	1.0%	1.0%	1.0%
Myrmecophage	1.0%	3.0%	2.0%	1.0%
Insectivore/Omnivore	.3%	4.0%	6.0%	4.0%
Omnivores				
Frugivore/Granivore	8.4%	22.0%	6.0%	9.0%
Frugivore/Omnivore	9.5%	18.0%	19.0%	31.0%
Herbivores				
Frugivore/Browser	29.0%	25.0%	41.0%	18.0%
Browser/Frugivore	53.0%	27.0%	20.0%	6.0%
Grazer	0	0	4.0%	30.0%

[1] Tabulation based on Tables 4 and 5.

Masaguaral and use only those species whose life history parameters we understand reasonably well. If we take the density of a species in question and, based upon known sex ratios, calculate the mean number of adult females present during a year within a square kilometer, it is then possible to estimate the mean number of young per adult female per year and, for those species that breed within the year of their birth, we can additionally calculate the number of young produced in an average year by the descendents of a founding female. It is then possible to calculate the mean cumulative weight of the young which survive, making suitable allowances for deaths (Table 8). The manner in which these calculations were carried out is included in Appendix I, but the net result is a calculation of productivity in terms of kilograms of biomass produced per square kilometer per year by the species in question. This may be expressed as either crude production, using crude density as the basis for the calculation, or as ecological productivity, utilizing the ecological densities.

If we take the species in question and arrange them according to the rank order of the density of reproductive females, it will be noticed that productivity does not necessarily parallel mean density. The productivity of the capybara *Hydrochaeris hydrochaeris*), the white-tailed deer (*Odocoileus virginianus*), the armadillo (*Dasypus novemcinctus*) and the opossum (*Didelphis marsupialius*) is much higher than would have been predicted from density considerations alone (Figure 10).

Another way of presenting the data is to rank the species according to their mean size and plot productivity in terms of kilograms per adult female per year. When this is done, it is seen that the larger species, such as the capybara and white-tailed deer, indeed produce the highest amount of protoplasm per female per year. It may be noted, however, that the howler monkey is very low; a factor that is also demonstrable in the previous figure (Figure 11). Productivity can also be compared with the standing crop of the species in question. If we take the previous set and arrange the species in the rank order of standing crop biomass and then plot the productivity in terms of kilograms per square kilometer per year, we see many deviations (Figure 12). As might be predicted, although *Alouatta* has the highest standing crop biomass, it has an extremely low productivity relative to its standing crop. The opossum has a high productivity as do *Zygodontomys brevicauda*, *Hydrochaeris hydrochaeris*, and *Marmosa robinsoni*. This reflects the fact that, although *Alouatta* exists at a high density with a high standing crop of biomass, its productivity is very low and very little surplus biomass is generated in any given year. On the other hand, a small species, such as *Zygodontomys brevi-*

Figure 10. Double plot of density and productivity. Species are ordered according to mean density. Z.b. = *Zygodontomys brevicauda*; M.r. = *Marmosa robinsoni*; R.sp. = *Rhipidomys* sp.; D.m. = *Didelphis marsupialis*; A.s. = *Alouatta seniculus*; D.n. = *Dasypus novemcinctus*; O.v. = *Odocoileus virginianus*; C.t. = *Cerdocyon thous*; T.t. = *Tamandua tetradactyla*; H.h. = *Hydrochaeris hydrochaeris*.

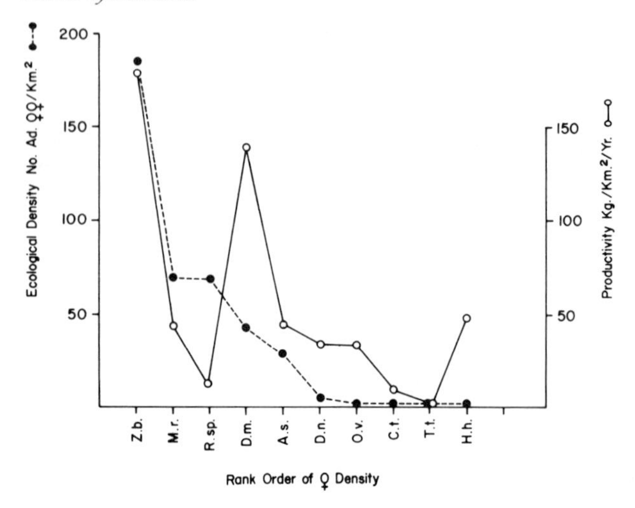

Figure 11. Annual productivity as a function of mean body size. Species are ranked in order from the smallest to the largest. Abbreviations as in Figure 10.

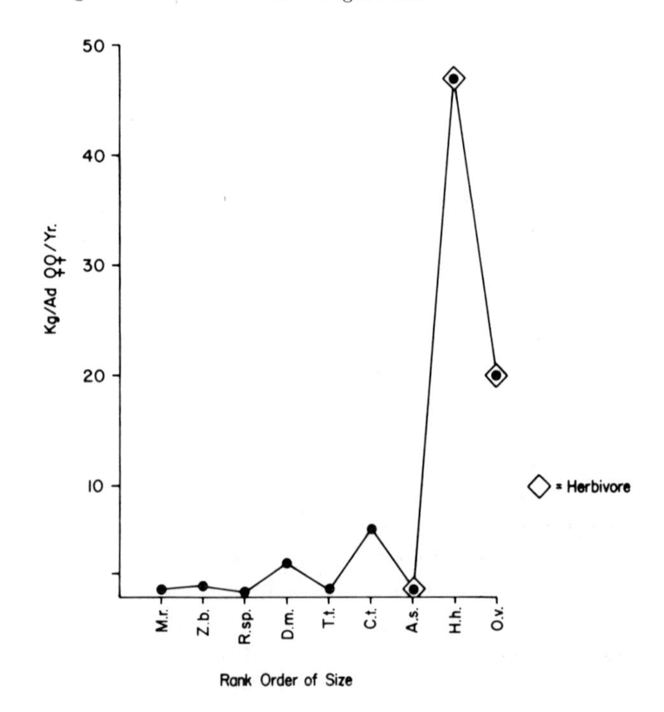

Table 8. Productivity values for some mammals of the llanos[1]

Species	*Biomass kg/km² of population (♂ & ♀)*	*Number of adult ♀/km²*	*Kg/Ad ♀/yr*	*Kg/km²/yr*	*Kg/km²/yr (+ growth of descendents)*
Didelphis marsupialis	122.0	46.5	3.0	140.0	140.0
Marmosa robinsoni	7.5	70.0	0.7[2]	49.0	49.0
Dasypus novemcinctus	45.6	6.0	6.4	38.0	38.0
Tamandua tetradactyla	12.0	1.5	1.0	1.5	2.5
Alouatta seniculus	365.5	31.0	1.5	46.6	66.6
Cerdocyon thous	16.0	1.6	6.0	9.6	11.6
Rhipidomys sp.	16.3	70.1	0.2[2]	14.7	14.7
Zygodontomys brevicauda	18.8	181.0	1.0[2]	181.0	181.0
Hydrochaeris hydrochaeris	90.0	1.0[3]	50.0	50.0	65.0
Odocoileus virginianus	120.0	1.8	20.0	36.0	46.0

[1] Assume at ecological density for Hato Masaguaral-West Side, *Alouatta* taken at 1 km² base area.

[2] Includes assumption that some daughters may breed in the year of their birth.

[3] Number has been increasing steadily since 1976; this assumes the 1976 value.

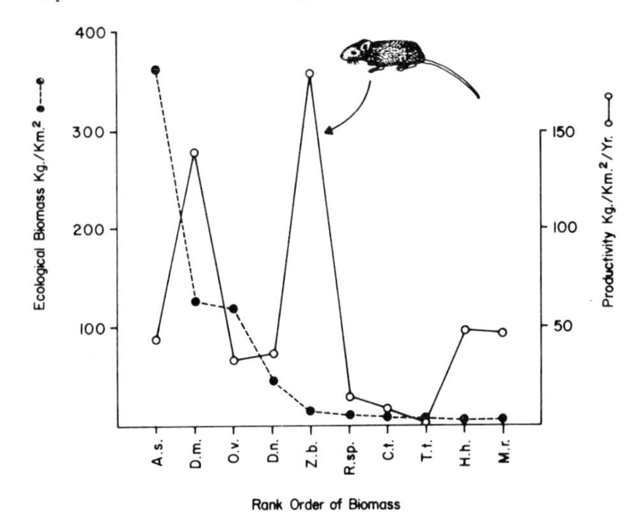

Figure 12. Double plot of annual productivity and average biomass. Species are ranked in order of decreasing standing crop. Abbreviations as in Figure 10.

cauda, while existing at a high density, preoccupies at any point in time a relatively low biomass, but its powers of productivity exceed those of any other mammal on the study area. In effect, it means that, if the population of *Zygodontomys* were to maintain itself at a zero growth value, a tremendous surplus of productivity must either emigrate or be utilized by predators. This is exactly what one would anticipate if, indeed, *Zygodontomys* lies, as we believe, at the bottom of the food chain for many of the small omnivores and small carnivores found at Hato Masaguaral.

The figures would be more dramatic for *Hydrochaeris*, if it existed at a higher density at Masaguaral. As Ojasti (1973) has shown, *Hydrochaeris* has a remarkable ability to convert vegetation into new biomass; indeed, the high reproductive and growth rates of the capybara allows it to be cropped as a meat animal on an annual basis. Even assuming extremely low densities for the capybara in the lowland llanos, a capybara can easily produce 320 kilograms per km² of harvestable young per annum. The capybara at low densities has a productivity in the llanos which exceeds the combined theoretical productivity of the three caviomorph species (agouti, paca, and spiny rat) in typical lowland neotropical forests (see Kleiman, Eisenberg and Maliniak, p. 173).

Summary

In summary, then, although general correlations between the absolute size and productivity can be discerned, each species should be studied in some detail to estimate its role in the ecosystem as a consumer and a producer. Species with a high standing crop of biomass, such as the howler monkey, may exhibit extremely low productivity and low population turnover. A small species, such as the cane rat (*Zygodontomys*) may exhibit a high density in numbers but because of its small size, will show a low standing crop of biomass. On the other hand, its productivity may be exceedingly high and, as such, contributes in no small measure to the maintenance of predators at higher trophic levels. The capybara (*Hydrochaeris*), although quite large, has a remarkably high productivity for its size and as such is probably an example of an "r" selected form capable of rapidly expanding under optimum conditions. The capybara retains a reproductive system somewhat uncharacteristic of an animal in its size class, having far larger litters than the two other large neotropical caviomorphs, the agouti and paca. As such, the capybara probably represents by its reproductive adaptations a species that can sustain high predation levels and expand rapidly into new habitats.

These preliminary studies of productivity in the neotropical habitats are only a beginning but suggest further studies which could assist in not only understanding the dynamics of an ecosystem but also in allowing predictions concerning the management of natural areas in the neotropics, and the development of rational utilization programs when such species are to be harvested for food by rural human populations.

Appendix

Introduction

The initiation of the Venezuela Mammalian Ecology Project began with a trip in April of 1973 by Drs. Kleiman and Eisenberg to assess the feasibility of establishing long-term study areas in Venezuela for the analysis of the resident mammalian faunas.

M. A. O'Connell began field work in Guatopo and at Masaguaral in June of 1975. She completed her efforts in November of 1977. During this interval she trapped, marked and released small mammals on four hectare grids at the two locations. Additionally, she gathered data on sighting frequency, road kill frequency, and animal signs. P. V. August commenced field work in Guatopo in July of 1976. He prepared a trap line designed to catch, mark and release *Agouti*-sized mammals in the drainage basin behind the living quarters (see Figure 13). At Masaguaral, he established a series of trapping grids for small mammals which included the major subdivisions of the mosaic habitat described by Troth (p. 17). Notes concerning mammal sightings were also maintained by August and he concluded his work in June of 1978.

As of the end of January 1978, Eisenberg had spent 197 days in Venezuela, 100 of the 197 days or approximately 50 percent of the time had been spent engaged in field work. The remainder of the time included travel time and administrative activities attendant with setting up the project and maintaining project continuity. Eisenberg concentrated on walking transect surveys in Guatopo and at Masaguaral.

During overall surveys in Guatopo National Park, the most frequently sighted mammalian species was *Cebus nigrivittatus*, followed by *Dasyprocta aguti* and the squirrel, *Sciurus granatensis*. This corresponds well with the frequency of sightings recorded by O'Connell over a much longer period. In her case, the agouti was the most frequently sighted, followed by *Cebus nigrivittatus* and followed again by the squirrel, *Sciurus granatensis* (Table 9). These three species are the most likely to be encountered while walking during daylight hours in the second-growth forests of Guatopo.

During explorations of Masaguaral, the most frequent sightings on the west side of the highway during daylight hours are of groups of the howler monkey, *Alouatta seniculus*. On the east side of the highway, *Dasyprocta* and *Cebus* are the most likely to be sighted, followed by *Alouatta*. *Sciurus* sightings are much rarer at Masaguaral than at Guatopo. During evening drives or walks, the most frequently encountered mammals include the crab-eating fox, *Cerdocyon thous*, white-tailed deer, *Odocoileus*, closely followed by the opossum, *Didelphis* (Table 10).

The average group size of an *Alouatta* troop at first sighting is approximately four individuals. In reality, however, the troop's actual size averages 8.5. Upon first sighting *Cebus nigrivittatus*, one generally sees approximately 4.8 individuals, although the mean troop size appears to be 17.5. Out of 62 sightings of *Cerdocyon thous*, the mean group size was 1.4. In reality the basic social unit is a pair and their juvenile offspring. One can see the interpretation of group size requires a prolonged study because the first encounter usually gives one only a partial impression of the actual biologically stable group. Out of 32 contacts with white-tailed deer at Masaguaral, the range in group size was 1–4 with an average of 1.52. Almost invariably units of greater than one included a female and an attendant subadult or fawn. Comparable data for white-tailed deer encounters from O'Connell's report gives a range

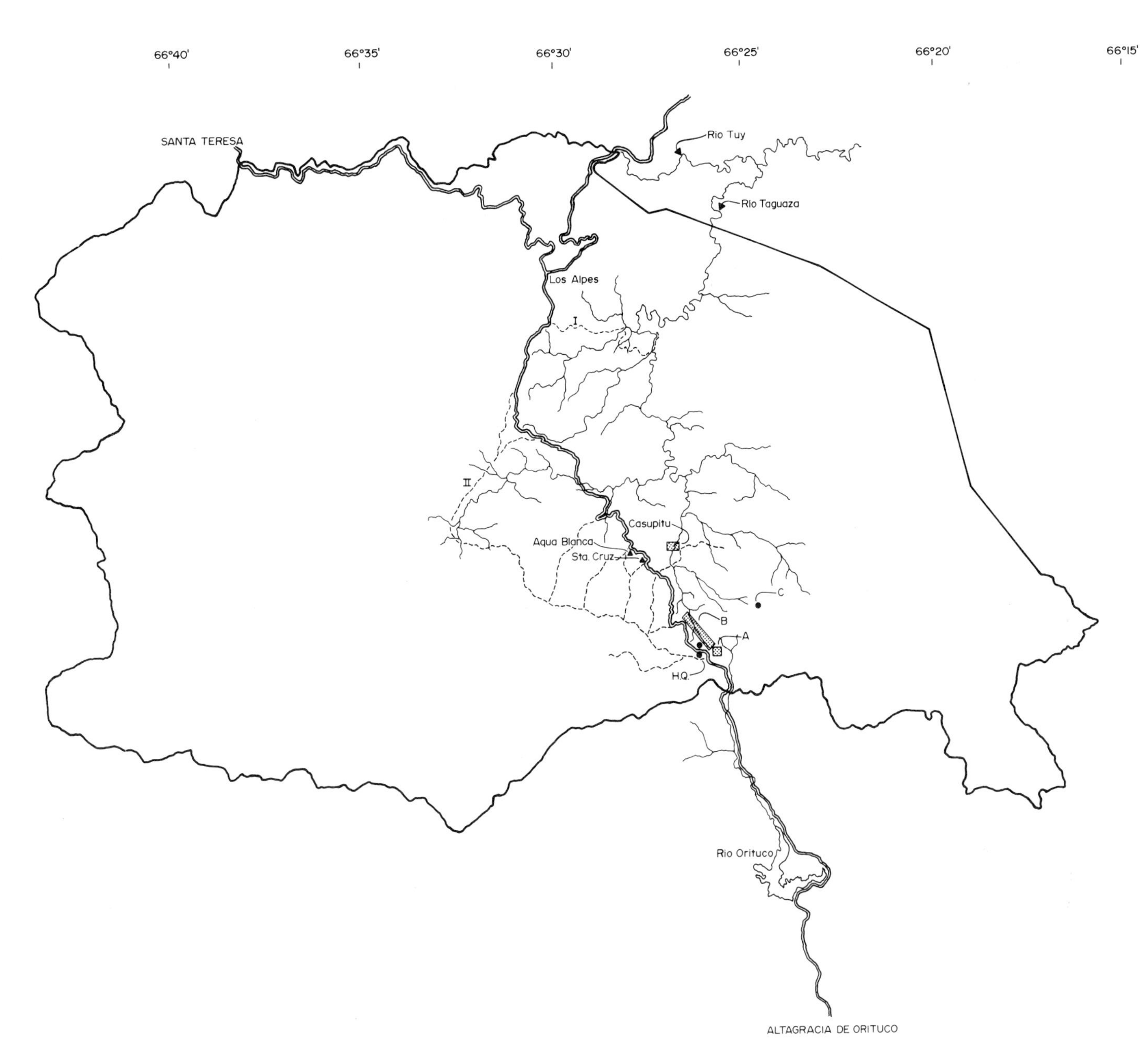

Figure 13. Location of study areas in Guatopo National Park. I = Pico Inos Trail or north survey area. II = Trail 2, central survey area. HQ = park headquarters. A = approximate location of O'Connell study grid. B = approximate location of August live-trapping transect. C = high altitude survey area. Casupitu = original study area for O'Connell and Morton. Park boundaries as of 1976.

of 1–3 with an average of 1.1 for a group size from a total of 48 deer encountered.

Animal Census Techniques and Methods for Fundo Pecuario Masaguaral

Density estimates and population trends for the non-volant mammals found on Masaguaral were made between January 1975 and June 1978 by the combined efforts of Kenneth Green, Margaret O'Connell, Peter August, Dr. R. Rudran, Dr. J. Robinson, Dr. J. F. Eisenberg, and Dr. D. G. Kleiman. In addition, a special study of the crab-eating fox, *Cerdocyon thous*, was conducted by Charles Brady (p. 161), and a study of anteaters by Drs. Montgomery and Lubin (1977).

Table 9. Rank order of sighting frequency for Guatopo National Park

Species	*Number of encounters*	*Total counted*	*$\bar{X}$ Group size*
Cebus nigrivittatus	7	30	4.8
Dasyprocta aguti	6	6	1.0
Sciurus granatensis	4	4	1.0
Alouatta seniculus	4	20	5.0
Ateles belzebuth	2	7	3.5
Agouti paca	1	1	1.0
Tayassu tajacu	1	3	3.0
Panthera onca[1]	1	1	1.0

[1] While driving.

Table 10. Rank order of sighting frequency for Masaguaral

Species	*No. of encounters*	*Total counted*	*$\bar{X}$ Group size*
Nocturnal-Auto Surveys:			
Cerdocyon thous	44	62	1.4
Odocoileus virginianus[1]	21	32	1.5
Didelphis marsupialis	10	10	1.0
Conepatus semistriatus	5	5	1.0
Procyon cancrivorus	3	3	1.0
Coendou prehensilis	3	3	1.0
Tamandua tetradactylus	2	2	1.0
Diurnal or Dusk Foot Surveys:			
Alouatta seniculus[2]	8	40	5.0
Cebus nigrivittatus	3	14	4.8
Dasyprocta aguti	4	4	1.0
Sylvilagus floridanus	3	3	1.0
Tayassu tajacu	1	3	3.0
Felis jaguaroundi	2	2	1.0
Tamandua tetradactyla	1	2	1.5[3]
Dasypus novemcinctus	1	1	1.0

[1] Includes some diurnal encounters.

[2] During formal surveys only; dozens of troops may be encountered in prime habitat on the west side.

[3] Mother and young.

Masaguaral includes approximately 34 km^2 of land. The land is divided by a north-south highway from Calabozo to San Fernando. It is convenient to refer to the ranch in terms of the "east side" and the "west side" of the highway. The east side includes approximately 11 km^2. Of this area, approximately 6 km^2 is tropical, deciduous forest and scrub. Approximately 5 km^2 is land which at certain times of the year is partially covered with water for periods varying from three to ten weeks. During the dry season this area is savanna-like grassland with occasional standing trees and low shrubs. True evergreen gallery forest appears in patches at the extreme eastern boundary of the property which ends at the stream bed of the Rio Guarico. Patches of permanent evergreen forest are associated with the Caño Carricol which meanders in a slightly north-westerly direction from the Rio Guarico.

The west side of the highway includes approximately 23 km^2 of which 13 km^2 are a mixed palm forest, *Copernicia tectorum*, with a few large trees, and a low scrub understory (see Troth, p. 17). Approximately 10 km^2 are open grassland. The mixed palm and tree association on the west side is considered an optimum habitat for many species. As a rule of the thumb, the open grassland or grassland with low scrub does not support high densities of many of the species which are found in association with the tree-palm complexes or dry forests. Density estimates for this area of palm and trees are considered as ecological densities for those species confined to this habitat. Crude densities for the west area are estimated from the ecological density value multiplied by .58, a correction factor based on the proportion of prime habitat on the west side. Similarly, on the eastern side of the ranch, ecological densities are presented as the actual densities for species found in the dry forest, and corrected for the crude density of the entire area by taking .45 of the total, which is the proportional area determined for the dry forest.

All small mammal density estimates, based on trap, mark and release in optimum habitat, are corrected to reflect the mean number of adults known to be alive during the greater part of the annual cycle. This is an attempt to average out the fluctuations of seasonally reproducing species that may show elevated densities resulting from the presence of resident adults and juveniles at certain times of the year.

Table 11 offers ecological densities determined by actual count on the various study grids. The data for *Alouatta* on the west side are from the study by Rudran. John Robinson provided the primate densities on the east side of the ranch. The density value for *Hydro-*

Table 11. Mammalian ecological densities (No./km²) derived from total counts[1]

Species	*Location*	
	Masa-guaral (west)	*Masa-guaral (east)*
Alouatta seniculus	150	50
Cebus nigrivittatus	—	44
Coendou prehensilis	7	—
Odocoileus virginianus	3	—
Hydrochaeris hydrochaeris	10	—

[1] Base area 44–154 ha.

Table 12. Mammalian mean ecological densities (No./km²) derived from trap, mark, and release

Species	*Location*		
	Gua-topo	*Masa-guaral (west)*	*Masa-guaral (east)*
Didelphis marsupialis	65	122	—
Marmosa fuscata	97	—	—
Marmosa robinsoni	—	150	280
Marmosa cinerea	40	—	—
Monodelphis brevicauda	37	—	—
Caluromys philander	< 20	—	—
Sylvilagus brasiliensis	5	—	—
Sciurus granatensis	40	50	40
Heteromys anomalus	200	25	110
Oryzomys capito	220	—	—
Oryzomys concolor	150	—	—
Oryzomys bicolor	—	60	—
Zygodontomys brevicauda	—	375[1]	60
Akodon urichi	200	—	—
Neacomys tenuipes	400	—	—
Sigmomys alstoni	—	50[2]	—
Rhipidomys sp.	144	187	—
Proechimys semispinosus	200	—	—
Echimys semivillosus	—	—	40
Dasyprocta aguti	100	—	—
Agouti paca	< 30	—	—

[1] Two-year average.

[2] Not present on the grid the entire year.

Table 13. Mammalian ecological densities (No./km²) extrapolated from transect sighting frequencies compared to the sighting frequency for a species with a known density

Species	*Locality*	
	Guatopo	*Masaguaral*
Dasyprocta aguti	—	80.0[2]
Conepatus semistriatus	—	12.5
Eira barbara	1.6	2.0
Mazama americana	10.0	—
Tapirus terrestris	.8	—
Tayassu tajacu	4.0	—
Cebus nigrivittatus[1]	35.0	—
Alouatta seniculus[1]	20.7	—
Ateles belzebuth[1]	15.0	—

[1] Transect computation from detection distance and extrapolation to ecological density.

[2] East.

chaeris is from a 1977 count of Lake Guacimo, the preferred habitat on the west side.

Table 12 presents the ecological densities deriving from trap, mark and release studies. The data for the west end are from O'Connell's study and the east side gallery forest data are from the efforts of August.

Table 13 gives the density estimates for those species which were censused by transect walking. For *Conepatus* we took the number of nocturnal sightings and compared them to the sighting rate for *Didelphis*, a species with a known density. If we assume equal probabilities for sighting *Didelphis* and *Conepatus*, then the crude density for *Conepatus* is about one-tenth that of *Didelphis*.

The agouti, *Dasyprocta*, is occasionally seen in the palm forests of the west side. Their numbers are quite low and we have estimated low densities based on a low contact frequency. Contact frequency during transect walking on the east side leads us to believe that the density may be .8 of that value determined for Guatopo.

The density estimate for *Eira* is based on the relative frequency of sighting when compared with the known density for *Sciurus*. This value should be taken as quite tentative.

Table 14 contains the density estimates derived from road kill frequency compared with the frequency of road kills of species with a known density (e.g., *Didelphis*). The data from road kill frequencies are

Table 14. Mammalian ecological densities (No./km^2) extrapolated from road-kill frequency[1]

Species	*Location*	
	Guatopo	*Masaguaral*
Bradypus infuscatus	3.7	—
Dasypus novemcinctus	6.0	—
Tamandua tetradactyla	9.0	—
Cabassous unicinctus	1.2	—
Conepatus semistriatus	9.0	—
Procyon cancrivorus	4.2	10.0
Potos flavus	3.7	—
Felis yagouaroundi	1.2	—
Grison vittatus	—	2.4
Coendou prehensilis	6.0	—

[1] See text for methods.

discussed in the next section. These estimates should be treated with some caution.

The data from Montgomery and Lubin (1977) allowed us to make an estimate of abundance for the anteaters. *Tamandua* has a home range of approximately 200 hectares. *Myrmecophaga* has a home range of approximately 2,500 hectares. For both species, the home ranges of individuals overlap significantly. There may be as many as 5 to 7 *Myrmecophaga* using parts of Masaguaral; this includes adults and young since Masaguaral is 3,403 hectares in extent, it would probably support around 2 *Myrmecophaga* and 25 to 30 *Tamandua*, however, habitat utilization is not uniform over the whole ranch. Probably less than 50 percent of the area is sutiable for *Tamandua*, therefore, the density must be corrected down. The density for *Tamandua* may be as high as 2/km^2 in good habitat in the dry zone and, by the same logic, *Myrmecophaga* could exist at one/8 km^2 or .12 animals/km^2.

On the 4 ha grids established on the west side, a single *Dasypus* was occasionally caught. Sightings of burrow activity would indicate that the home range exceeded 4 ha. On this assumption, we estimated that 12/km^2 at ecological density would not be unreasonable. In the gallery forest, ecological density for *Dasypus* was estimated by counting the number of fresh burrows seen during transect walking during the maximum flooding when the armadillos were assumed to be concentrated on high ground.

The paca, *Agouti paca*, appears to be absent from most of the west side but on the east side its tracks are frequently sighted. Its density has been set at a value of about .30 that of the agouti density which is consistent with estimates for Panama.

Hydrochaeris hydrochaeris is reasonably abundant on the east side. The density estimate was based on a guess[1] based on track frequency for the total number of capybara associated with the Caño Carricol. We believe almost the entire resident capybara population on the east side is associated with the Caño during the dry season.

Montgomery and Lubin (1978) describe the home range characteristics for three *Coendou prehensilis* at Masaguaral. Rudran recorded sighting frequencies on the west side during the course of his *Alouatta* study. Combining these sets of data, we can extrapolate a density value of 7/km^2 in optimum habitat.

The density of *Sylvilagus floridanus* on the west side is a guess based on habitat preferences and sighting frequency. The east side density is set 2.5 times as high as that of the west side based on a higher sighting frequency.

The density estimates for *Odocoileus virginianus* on the west side are based on a known count of individuals occupying the 2 km^2 survey area. This survey area on the west side contains sufficient habitat diversity to permit an extrapolation to the rest of the west side. The density estimate for the east side is set at a slightly higher figure for the ecological density based on a higher sighting frequency during transect surveys.

Density of *Cerdocyon thous* for the west side is based on a known count and study of three families by Brady. On the east side, density estimates are set equivalent to the west side as a result of equivalent sighting frequency.

Density estimates for *Tayassu tajacu* on the east side are based on the habitat use patterns of one large troop that partially uses Robinson's grid.

The density estimates for the carnivores, *Felis yagouaroundi*, *F. paradalis* and *Puma concolor*, are based on the frequency of sighting or frequency of seeing tracks. These are to be treated as very tentative and may be under-estimates.

Animal Census Techniques and Methods for Guatopo National Park

Through a trap, mark and release program carried on by O'Connell and August near the park headquarters

[1] Guesses—When an individual is "thoroughly familiar with an area and its fauna, he may be able to make a realistic guess at the number of animals living there. If an accurate estimate is not required, a guess may serve as an estimate of density." There are many caveats however (see Caughley, 1977).

in Guatopo, we have established relative abundance and density estimates for most of the non-volant small mammals (Table 12). The larger mammals, including primate and carnivore densities, are based on mixed methods and extrapolations. To provide some habitat diversity for the calculation of crude density in the park, a second growth, seasonally inundated area at Casu Pitu (originally studied by O'Connell) is used as a correction factor yielding a 40 percent reduction in abundance for the crude densities of the entire park.

Consistent recaptures of *Dasyprocta* were difficult on both O'Connell's grid and on August's trap line. The density value is an estimate based on O'Connell's grid. *Agouti paca* was difficult to trap but based on its presence and meagre trapping results, we place the density at 25 percent the value for *Dasyprocta*.

The list of densities for *Ateles belzebuth* and *Tayassu tajacu* are based on transect surveys on Trail 1 (Pico Inos) and thus represents densities at the north end of the park. Nevertheless, we have computed the crude density for the entire park as if the faunal complement near the south end of the park includes populations of *Ateles* and *Tayassu*. This latter assumption could be unwarranted. The biomass estimates for Guatopo are a composite and should be treated as such.

Transect walking and estimates of density were conducted for *Alouatta*, *Cebus* and *Ateles*. *Cebus* densities were further augmented by frequency of sighting, duly recorded by O'Connell at both Casu Pitu and her current grid near the park headquarters (Table 13).

The densities of *Eira barbara* and *Mazama americana* were computed assuming equal probabilities for sighting compared with the sightings for *Sciurus granatensis* and a known density for *Sciurus*.

Density estimates for *Bradypus* were derived by assuming equal probabilities for road kills between *Bradypus* and *Didelphis* and computing the density of *Bradypus* at .03 that of *Didelphis*. In a similar manner by adjusting road kills to the standard based on *Didelphis* road kills and the known density of *Didelphis*, we computed the relative density for *Tamandua tetradactyla*, *Dasypus novemcinctus*, *Cabassous unicinctus*, *Conepatus semistraitus*, *Procyon cancrivorus*, *Potos flavus*, *Felis yagouaroundi*, and *Coendou prehensilis* (Table 6).

The density for *Sylvilagus brasiliensis* was based on the assumption of an equal probability for trapping *Sylvilagus* and *Didelphis*. This figure should be treated with caution. The estimate for *Tayassu tajacu* is based on recording tracks during transect walking and that for *Tapirus terrestris* on an extrapolation from known densities and track frequencies on Barro Colorado. The estimates for the density of *Panthera onca* and *Priodontes gigantea* are guesses based on an estimate of probable home range.

Road Kill Analysis

For the purposes of road kill analyses, we have taken as the standard the data gathered by Eisenberg but have cross-checked the relative abundance of road kills from the data compiled by O'Connell. From January 1975 to September 1977, Eisenberg traveled approximately 12,105 km. If we divide the travel time into four seasons based on the alternation of wet and dry periods, we may consider December through February as season I (the early dry season); March through May as season II (the height of the dry season, June through August as season III (the early rainy season) and September through November as season IV (the transition from the rainy season and the beginning of the dry season. Taking the accumulated distance traveled, a correction factor must be developed for the seasonality so that the under-representation of travel during season IV (915 km) could be corrected by a factor of 4.2.

The distance traveled by Eisenberg was further broken down into 4 habitat types: (1) The coast range, wherein approximately 4,555 km were traveled; (2) the dry forest (not seasonally inundated) where approximately 4,240 km were traversed; (3) the transition between the dry forest and wet savanna comprising 960 km of travel; and finally, (4) the wet llanos itself, where 2,350 km were traveled. Referring to the four habitat types as A, B, C, and D, respectively, then a correction factor of 4.7 was required for habitat type C, and 1.9 for habitat type D.

The four species showing the highest frequency of road kills were *Didelphis marsupialis*, *Cerdocyon thous*, *Conepatus semistriatus*, and *Tamandua tetradactyla* (Table 15). *Didelphis* tends to show a peak in frequency of road kills during the height of the dry season. *Tamandua* shows a similar trend as does *Cerdocyon*. The frequency with which *Cerdocyon* is struck, however, may be seasonally elevated, not because of an increased abundance during this time, but because it is scavanging carcasses along the highway. *Tamandua* probably has an increased frequency of being struck because it moves greater distances when water sources become concentrated. The *Didelphis* peak appears to be correlated with the season of pouch young and juvenile dispersal. *Conepatus* shows an increased frequency of being struck during the beginning of the dry season which gradually declines in the wet season.

Conepatus appears to show its maximum density in the dry, deciduous tropical forest. *Cerdocyon* shows its greatest density typically in the wet llanos. *Tamandua* shows a high density in both the dry forest and llanos. *Didelphis* has a high density in the coast range, a low density in the dry forest and its maximum density in

Table 15. Road kill analysis according to area

Species	*Area*	*Actual value*	*Corrected value*
Didelphis marsupialis	A	15	15.0
	B	7	7.5
	C	3	14.2
	D	10	19.0
Cerdocyon thous	A	2	2.0
	B	7	7.5
	C	2	9.5
	D	14	26.7
Tamandua tetradactyla	A	3	3.0
	B	8	8.6
	C	0	0
	D	2	3.8
Conepatus semistriatus	A	2	2.0
	B	7	7.5
	C	1	4.7
	D	1	1.9

Table 16. Road kill analysis according to season

Species	*Season*	*Actual value*	*Corrected value*
Didelphis marsupialis	I	7	7.8
	II	19	19.6
	III	6	6.0
	IV	3	12.8
Cerdocyon thous	I	5	5.6
	II	12	12.4
	III	8	8.0
	IV	0	0
Tamandua tetradactyla	I	4	4.5
	II	5	5.2
	III	4	4.0
	IV	0	0
Conepatus semistriatus	I	5	5.6
	II	3	3.1
	III	2	2.0
	IV	1	4.3

the llanos (see Table 16). These road kill figures are not in disagreement with the data derived from collecting (see Handley, 1976).

The Calculation of Productivity Values

The productivity calculations in Table 8 (previous section) were calculated in the following manner. We made the productivity calculations based on ecological densities. Given the density estimate based on results of trap, mark and release and on the assumption of near unity for the sex ratio, the number of adult females per square kilometer could be calculated. Since the mean number of young per adult female per year is known in every case, either from trapping data or from data determined from zoological parks, the mean number of young produced per adult female per year could be calculated. With species such as *Dasypus novemcinctus*, *Tamandua tetradactyla*, *Alouatta seniculus*, *Hydrochaeris hydrochaeris* and *Odocoileus virginianus*, it was assumed that the female produced only one litter per year. For those species that produce more than one litter per year, the assumption of a second litter was included in the calculation.

To calculate productivity, an estimate was made on the number of young surviving for a period of a year. For uniparous species, the value was corrected down, assuming a fixed probability of mortality. For multiparous species and especially for those for which trap, mark and release data were available, certain assumptions were made concerning the number of young surviving at given time intervals within a hypothetical year. For example, in *Didelphis marsupialis*, it was assumed that 7 young survive to 90 days. By 180 days, 3 young survive; by 270 days, 1.5 young survive, and by 360 days, only 1 young survived. The incremental growth for an individual during this time period was then multiplied by the surviving number and added to the previous total. When, as in the case of *Didelphis marsupialis*, the possibility exists for a second litter, a second set of calculations were performed and corrected down, if a lowered probability for a second litter exists based on the trap, mark and release data. For those species where the possibility existed for descendents of a breeding female to breed in the year of their birth, such as *Marmosa robinsoni*, *Rhipidomys* sp., and *Zygodontomys brevicauda*, the additional productivity by descendents was also calculated, based on the survivorship data from the trap, mark and release studies. The net result was the calculation of the cumulative mean weight of young produced per female per year with suitable corrections for the assumption of constant mortality rate in those multiparous species for which data were available. For those species where data were not available, average assumptions were made.

Finally, the productivity in terms of kilograms per square kilometer per year was obtained by taking the previous value and multiplying it through by the number of breeding females on the plot. In addition, those species, such as *Tamandua tetradactyla*, *Alouatta seniculus*, *Hydrochaeris hydrochaeris* and *Odocoileus virginianus*, which have young surviving from a given female into the following year, an assumption was made concerning growth of these descendents in parallel with the growth and production of new young. This latter value was not plotted in the figures, but is duly entered in Table 8 as the final column.

Literature Cited

Brady, C. A.
1979. Observations on the behavior and ecology of the crab-eating fox (*Cerdocyon thous*). Pages 161–171 in *Vertebrate Ecology in the Northern Neotropics*, edited by John F. Eisenberg. Washington, D. C.: Smithsonian Institution Press

Brokx, P.
In press White-tailed deer in South America in *White-tailed Deer Ecology and Management*, edited by L. K. Halls. Harrisburg, Pa.: Stackpole Co.

Caughley, G.
1977. *Analysis of Vertebrate Populations*. London/New York: John Wiley and Sons.

Chivers, D. J.
1969. On the daily behaviour and spacing of howling monkey groups. *Folia Primatologica*, 10:48–102.

Eisenberg, J. F. and R. W. Thorington, Jr.
1973. A preliminary analysis of a neotropical mammal fauna. *Biotropica*, 5:150–161.

Fleming, T. H.
1971. Population ecology of three species of neotropical rodents. *Misc. Pubs. Mus. Zool., Univ. Mich.*, no. 143.
1973. Numbers of mammal species in North and Central American forest communities. *Ecology*, 54:555–563.

Glanz, W.
1978. Population ecology of the neotropical red-tailed squirrel, *Sciurus granatensis*, in relation to food availability. Abst. 184, in Abstracts of Technical Papers 58th Annual Meeting of the American Society of Mammalogists, Athens, Georgia.

Handley, C. O., Jr.
1976. Mammals of the Smithsonian Venezuelan Project. *Brigham Young Univ. Sci. Bull., Biol. Ser.*, 29(5):1–91.

Heaney, L. R. and R. W. Thorington, Jr.
1978. Ecology of neotropical red-tailed squirrels, *Sciurus granatensis*, in the Panama Canal Zone. *J. Mammal.*, 59(4):846–851.

Hendrichs, H.
1978. Untersuchungen zur Säugetierfauna in einem paläotropischen und einem neotropischen Bergregenwaldgebiet. *Säugetierk. Mitteil., BLV Verlagsgesselschaft mbH München* 40, 25(3):214–225.

Kleiman, D. G., J. F. Eisenberg and E. Maliniak.
1979. Reproductive parameters of caviomorph rodents. Pages 173–183 in *Vertebrate Ecology in the Northern Neotropics*, edited by John F. Eisenberg. Washington, D. C.: Smithsonian Institution Press

Montgomery, G. G. and Y. D. Lubin.
1977. Prey influences on movements of neotropical anteaters. Pp. 103–131 in *Proceedings of the 1975 Predator Symposium*, edited by R. Phillips and C. Jonkel. Missoula: Montana Forest and Conservation Experiment Station.
1978. Movements of *Coendou prehensilis* in the Venezuelan llanos. *J. Mammal.*, 59(4):887–888.

Montgomery, G. G. and M. E. Sunquist.
1975. Impact of sloths on neotropical forest energy flow and nutrient cycling. Pages 69–111 in: *Tropical Ecological Systems: Trends in Terrestrial and Aquatic Research*, edited by F. B. Golley and E. Medina. Ecological Studies 11. New York: Springer-Verlag.
1978. Habitat selection and use by two-toed and three-toed sloths. Pages 329–359 in: *The Ecology of Arboreal Folivores*, edited by G. G. Montgomery. Washington, D.C.: Smithsonian Institution Press.

O'Connell, M. A.
1979. Ecology of didelphid marsupials from northern Venezuela. Pages 73–87 in *Vertebrate Ecology in the Northern Neotropics*, edited by John F. Eisenberg. Washington, D. C.: Smithsonian Institution Press.

Ojasti, J.
1973. *Estudio Biologico del Chiguire o Capibara*. Caracas: Fundo Nacional de Investigaciones Agropecuarias.

Oppenheimer, J. R.
1968. Behavior and ecology of the white-faced monkey, *Cebus capucinus*, on Barro Colorado Island, C. Z. Ph.D. Dissertation, University of Illinois, Urbana.

Smythe, N.
1970. Ecology and behavior of the agouti (*Dasyprocta punctata*) and related species on Barro Colorado Island, Panama. Ph.D. Thesis, University of Maryland, College Park.
1978. The natural history of the Central American agouti (*Dasyprocta punctata*). Smithsonian Contributions to Zoology, No. 257.

Troth, R. G.
1979. Vegetational types on a ranch in the central llanos of Venezuela. Pages 17–30 in *Vertebrate Ecology in the Northern Neotropics*, edited by John F. Eisenberg. Washington, D. C.: Smithsonian Institution Press.

Walsh, J. and R. Gannon.
1967. *Time Is Short and the Water Rises*. Camden, New Jersey: Thomas Nelson and Sons.

SECTION 6:

Avian Studies

The rufous and white wren (*Thryothorus rufalbus*) nest represents a balance of selective factors, a trade-off typical of many tropical birds. While the domed bit of detritus might in itself not appear as potential prey to an inexperienced predator, the stinging ants living in the thorns of the acacia help to insure its anonymity: a perfect solution? But the nest is easily seen in the sparsely-leaved tree and a new enemy, the parasitic striped cuckoo, receives a benefit from the wren's use of ants. If the nest were more stealthily placed, it would be harder for the cuckoo to find. What will be the future nest? A lifetime could be spent studying this.

Introduction

The distribution and abundance of birds in northern South America has been vastly influenced by climatic changes. It is apparent that the north coastal range of Venezuela served as a refugium during the Pleistocene and has all the attributes of an "ecological island." The north coast range is a separate geological formation from that of the Andes and, given the proposed history of aridity in northern Venezuela during the Pleistocene, one could envision the tops of the north coast range as being islands of forest surrounded by extreme xeric-adapted vegetation. Morton points out that certain characteristics of the avifauna in Guatopo National Park suggest a long-term isolation and the conspicuous absence of certain bird taxa attests to this fact.

The avifauna of the llanos is derivative with no endemic species. One can envision the rivers taking their origins from the north coast range with their attendant gallery forests serving as conduits for the colonization of the llanos at the end of tht Pleistocene xeric period. Indeed, the forest in the vicinity of the Rio Guarico which forms the eastern boundary of Hato Masaguaral contains faunal elements similar to Guatopo, but as one proceeds away from this forest cover toward the open llanos, species more adapted to seasonally deciduous forests predominate.

The section on avian ecology opens with an analysis of the avifauna of Masaguaral by Betsy Thomas. She provides a basic list of birds found on and near the ranch. In addition, she has estimated the relative abundance of species and indicates the season of the year in which they use Masaguaral. Her data concerning breeding or nesting records are especially useful. Once again, the pronounced seasonality imposed by the rainfall pattern is reflected in the temporal patterning of breeding by the llanos avifauna. Different patterns emerge, however, when the different taxa are considered separately. For example, the timing of nesting in the smaller hawks in the dry season may, in fact, reflect an adaptation to the relative abundance of small rodents that would serve as useful prey items for feeding nestlings. On the other hand, aquatic birds tend to breed with the onset of rains when the lagoons will increase in their breadth and depth, stimulating breeding activity by fishes and amphibians. Thus, although the timing of reproduction by different bird taxa may contrast sharply in the llanos, the seasonal flux in rainfall ultimately regulates the season of breeding for many of the species. This regulation is imposed in terms of the relative abundance of the preferred food items for feeding the nestlings.

Eugene Morton analyzes the avifauna of Guatopo and Masaguaral, thereby allowing a comparison between the two habitat types. His comments concerning the formation of mixed feeding assemblages in these two diverse habitats are especially instructive. J.F.E.

BETSY TRENT THOMAS
Tomas Enterprises, Apatado, 80844
Caracas 108, Venezuela

The Birds of a Ranch in the Venezuelan Llanos

ABSTRACT

A list of the 243 species of birds is summarized from six consecutive years of observation. Monthly occurrence and breeding notations are correlated with the annual rainfall profile. Brief descriptions of usual numbers and habitat preferences of each species are included. Fourteen additional species seen in the area, but not observed on the ranch, are listed separately.

RESÚMEN

Se resume seis años de estudio de una lista de 243 especies de aves. Presencia y crianza mensual son correlaciondas con el nivel de lluvia caída por año. Se incluye una descripción de los numeros de aves y preferencias de morada para cada especie. Se da una lista separado de 14 especies adicionales vistas en el area pero no observadas en el Hato.

Introduction

Most of the center of Venezuela, an area of about 1,000 by 250 km, is covered by savannas called llanos. They are low altitude, nearly flat grasslands interspersed by small groups of trees and occasional gallery forests, the latter being found only along permanent water courses. The area is classified as dry tropical forest (Ewel and Madriz, 1968) based on the Holdridge system.

The llanos are sedimentary. They are bounded on the south by the very old Guayana Shield (Haffer, 1974) which begins near the Orinoco River. They are terminated on the west by the northern Andes and on the north by the coastal cordillera; both are more recent uplifts (Haffer 1974:129–130). The eastern llanos end at the delta of the Orinoco River.

The avifauna of the llanos is poorly known except for general lists by Berlepsch and Hartert (1902), Cherrie (1916), Friedmann and Smith (1950, 1955), Phelps and Phelps (1958, 1963) and Meyer de Schauensee and Phelps (1978). Shorter papers about single species are by Schwartz (1975), Thomas (1978) and Wiley and Wiley (1977). During six years, May 1972 through November 1978, I kept a list by month of the species of birds that I observed on a cattle ranch in the llanos while I made another avian study. My study area, the ranch Fundo Pecuario Masaguaral, is situated about halfway between Calabozo, Estado Guarico and San Fernando de Apure, Estado Apure, 08°31′N 67°35′W. It is near the middle of the Venezuelan llanos.

The ranch, approximately 3,400 ha, has had the unusual history of being preserved as a private faunal and floral reserve for over 30 years. The conservation-dedicated owner, Sr. Tomás Blohm, has encouraged and assisted many scientific studies on his property.

About 76 percent of Masaguaral is covered by grassland with scattered trees, the palm *Copernicia tectorum* and small tree and palm clumps called *matas*. This part is lightly grazed by beef cattle. The other 24 percent, about 800 ha, is gallery forest. The average altitude of the ranch is c. 63 m and the annual temperature variation of 24° to 33° C is small. There are two seasons, wet and dry, each of about six months duration.

Deep water wells equipped with diesel-driven pumps keep three small lagoons and adjacent marshes wet during the dry season. These lagoons serve as feeding and roosting areas for large numbers of birds from January to June. This artificial source of water is an important contribution to wildlife conservation for an area well beyond the ranch boundaries. The lagoons and marshes are lost in the general flooding of the ranch during the wet season. Nine windmills supplying water for cattle are a minor source of water on the ranch.

Methods and Procedures

The list of birds in Table 1 is a summary of the species observed on Masaguaral, their usual numbers, breeding months and customary habitats. I include Venezuelan vernacular names, many of which are onomatapoeic. Nomenclature generally follows Meyer de Schauensee and Phelps (1978). Table 2 is a list of birds I have observed within 40 kilometers of the ranch, but not encountered there. Reports of birds seen on the ranch by well qualified observors, which I was unable to confirm during the study period, are not included.

No birds were collected, although some small flycatchers were identified from mist net catches. My records of the Tyrannidae, notoriously difficult in field identification, and nocturnal species are no doubt too few. During the annual flooding of the ranch it was not possible to visit all areas of the property every month, thus probably some uncommon species have not been recorded.

The observations began during what proved to be a period of exceptionally dry years as the cumulative rainfall data indicate. Figure 1 shows the averages of the monthly rainfall from Corozo Pando, 6 km south of the ranch, as reported by the Division de Hidrologia, Ministerio de Obras Publicas of Venezuela. Monthly bird occurence and breeding can be correlated with the rainfall profile by matching Figure 1 to each page of Table 1.

In Table 1 an 'x' indicates that the bird was observed during the month; a 'B' indicates breeding as determined by nest building in species not known to maintain nests at other times of year, copulation, a nest with eggs or young, or counted back from observations of adults feeding fledglings; a '·' indicates the bird was not observed during that month in any of the six years.

Abundance is a subjective term depending on the observor's experience, the season, the time of day, and the habitat. For a discussion of the problem see Ridgely (1976:19–21). My notations are based on vocal as well as visual records. Abundant means the bird can always be found on the ranch, sometimes in large numbers. Common means it is likely to be encountered on 75 percent or more of days of field observation; fairly common on 50 percent; uncommon on 25 percent and occasional on less than 25 percent of field days. Rare indicates I have found the bird less

Figure 1. Average monthly rainfall from Corozo Pondo (6 km south of Masaguaral), as reported by the Division de Hidrologia, Ministerio de Obras Publicas of Venezuela.

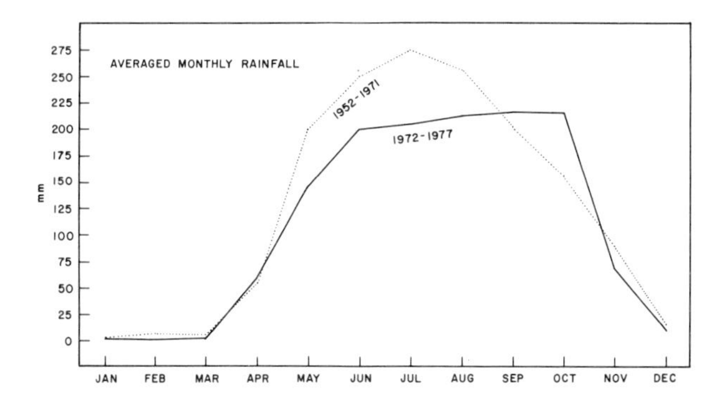

than five times. Irregular refers to species of sporadic occurrence.

When birds are listed as found in more than one habitat I have placed the more frequent habitat first. Lagoons are permanent bodies of water of more than 1 ha and maintained artificially during the dry season. Ponds are usually seasonal and generally of less than 1 ha. Marshes are flooded areas with standing vegetation. The Rio Guárico and the Caño Caracol are small rivers, both of which became a series of water pools during April and May of the driest years.

The gallery forest, surrounding the Rio Guárico and the Caño Caracol, consists of closed canopy medium to tall trees, predominantly deciduous in the dry season

Table 1. Birds observed on Masaguaral by months.[1]

	J	*F*	*M*	*A*	*M*	*J*	*J*	*A*	*S*	*O*	*N*	*D*	
Red-legged tinamou *Crypturellus erythropus* Soisola pata roja	·	x	·	·	B	x	x	x	x	x	x	x	Fairly common, singly; gallery forest; fresh egg shells 28 May 1972.
Least Grebe *Podiceps dominicus* Patico zambullidor	·	·	·	x	·	·	·	·	·	·	·	·	Rare, two birds in Apr. 1973 on a newly created lagoon.
Neotropic cormorant *Phalacrocroax olivaceus* Cotúa olivácea	x	x	x	·	·	x	x	x	x	x	·	·	Irregular, usually 1–10 when seen, never over 50; marshes and lagoons.
Anhinga *Anhinga anhinga* Cotúa agujita	x	x	x	x	x	x	x	x	x	x	x	x	Fairly common, singly rarely pairs; Caño Caracol, lagoons and ponds.
White-necked heron *Ardea cocoi* Garza morena	x	x	x	x	x	x	x	x	x	x	x	x	Common, singly; aquatic areas; probably breeds on ranch but data lacking.
Great egret *Casmerodius albus* Garza blanca real	x	x	x	x	x	x	B	B	B	B	B	x	Common, 1–20 birds in non-breeding months; aquatic areas; maximum number of nests in heronry 140.
Snowy egret *Egretta thula* Garcita blanca	x	x	·	x	x	x	x	B	B	B	x	x	Fairly common, singly; aquatic areas; maximum number of nests in heronry 4.
Little blue heron *Florida caerulea* Garcita azul	x	x	x	x	x	x	B	B	B	B	x	x	Fairly common, singly; aquatic areas; maximum number of nests in heronry 5.

[1] Species are identified by common, scientific, and Venezuelan vernacular names. x = observed, B = breeding, · = not observed. (See text, p. 214 for definitions of terms of relative abundance.)

Species	J	F	M	A	M	J	J	A	S	O	N	D	Notes
Striated heron *Butorides striatus* Chicuaco cuello gris	x	x	x	x	x	x	B	B	B	B	x	x	Common, singly; aquatic areas; during breeding months perhaps 100 pairs; usually nests alone.
Cattle egret *Bubulcus ibis* Garcita reznera	x	x	x	x	x	x	B	B	B	B	x	x	Common, small flocks; savanna and open grassland near cattle; maximum nests in heronry 80.
Whistling heron *Syrigma sibilatrix* Garza silbadora	x	x	x	x	x	x	x	x	x	x	x	x	Fairly common, usually pairs; open grassland, marshes and near ponds.
Capped heron *Pilherodius pileatus* Garciola real	x	x	·	x	x	·	·	·	·	·	x	x	Uncommon, singly and pairs; small ponds in gallery forest, occasionally savanna ponds.
Black-crowned night-heron *Nycticorax nycticorax* Guaco	x	x	x	·	x	·	B	B	B	B	x	x	Uncommon; near lagoons and rivers; 50 immatures roosted together Nov 1973; maximum number of nests in heronry 6.
Yellow-crowned night-heron *Nyctanassa violacea* Chicuaco enmascarado	·	·	·	x	x	x	x	x	x	B	·	·	Uncommon; near lagoons and rivers; maximum number of nests in heronry 2.
Rufescent tiger-heron *Tigrisoma lineatum* Pájaro vaco	x	x	x	x	x	x	B	B	B	x	x	x	Common, singly; aquatic areas; nests alone in *matas* and in gallery forest near ponds.
Boat-billed heron *Cochlearius cochlearius* Pato cuchara	x	·	·	·	·	x	·	B	B	x	·	x	Occasional, singly; ponds and lagoons; 13 nests together, not in heronry, Aug 1976.
Wood stork *Mycteria americana* Gabán	x	x	x	x	x	x	x	x	x	x	x	x	Irregular, numbers vary from 1–100; mostly immatures; in trees by marshes and lagoons.
Maguari stork *Ciconia maguari* Cigüeña or gabán pionío	x	x	x	x	x	B	B	B	B	B	B	x	Common, but few birds Feb–May, singly and groups; edges of marshes and lagoons; maximum number of nests 29.
Jabirú *Jabiru mycteria* Garzón soldado	x	x	x	x	x	x	B	B	B	B	B	B	Common, but few birds; pairs and singly; lagoons and marshes; maximum number of nests alone and near heronry 6.
Buff-necked ibis *Theristicus caudatus* Tautaco	x	x	x	x	x	x	x	x	x	x	x	x	Fairly common, usually in pairs; short dry grass of open savanna.
Sharp-tailed ibis *Cercibis oxycerca* Tarotaro	x	x	x	x	x	x	x	x	·	x	x	·	Uncommon, usually pairs or small groups; wet savanna and marshes near lagoons.
Green ibis *Mesembrinibis cayennensis* Corocoro negro	x	x	x	x	x	x	B	x	·	·	x	x	Uncommon, singly and pairs; ponds in gallery forest and lagoon edges.

Species	J	F	M	A	M	J	J	A	S	O	N	D	Notes
Whispering (Bare-faced) ibis *Phimosus infuscatus* Tara	x	x	x	x	x	x	B	B	B	B	x	x	Occasional Jan–Jun, common during breeding months; marshes and lagoon edges; nests in heronry near *C. maguari*.
White ibis *Eudocimus albus* Corocoro blanco	·	·	x	·	x	x	·	·	x	·	·	·	Rare, 2 maximum number seen at one time; marshes with *E. ruber*.
Scarlet ibis *Eudocimus ruber* Corocoro colorado	x	x	x	x	x	x	·	x	x	·	x	x	Uncommon, usually 1–10 birds but over 50 in Aug. 1975 half were in immature plumage; marshes.
Glossy ibis *Plegadis falcinellus* Corocora castaño	·	·	·	x	x	·	·	x	x	·	·	·	Irregular, 45 birds Apr 1973, 15 birds Aug–Sept 1973, 10 birds Apr–May 1977; marshes.
Roseate spoonbill *Ajaia ajaja* Garza paleta	x	x	x	x	x	x	x	x	x	x	x	x	Fairly common, usually 1–10 in non-breeding plumage, but 40 birds Sept 1978, some in breeding colors; ponds, marshes and lagoons.
Horned screamer *Anhima cornuta* Aruco	·	·	·	·	·	x	·	x	x	x	x	·	Rare, one pair; marshes near lagoons; not seen in some years.
Fulvous whistling-duck *Dendrocygna bicolor* Yaguaso colorado	x	x	x	x	x	x	·	x	·	·	x	x	Common Jan–Jun 20–400 diurnal roosting at lagoons with other *Dendrocygna*, rare other months.
White-faced whistling-duck *Dendrocygna viduata* Yaguaso cariblanco	x	x	x	x	x	x	B	x	x	x	x	x	Abundant Jan-Jun 50–15,000 diurnal roosting at lagoons, other months only 10–50 birds.
Black-bellied whistling-duck *Dendrocygna autumnalis* Guirirí	x	x	x	x	x	x	x	x	B	B	x	x	Abundant Jan–June 100–10,000 diurnal roosting at lagoons, other months only 20–70 birds.
Blue-winged teal *Anas discors* Barraquete aliazul	x	·	x	·	·	·	·	·	·	·	·	·	Rare migrant, 1–2 birds with *Dendrocygna*; lagoons.
Comb duck *Sarkidiornis melanotos* Pato de monte	x	x	x	x	x	x	·	·	·	·	x	x	Common Jan–Jun 1–250 diurnal roosting with *Dendrocygna*; lagoons.
Moscovy duck *Cairina moschata* Pato real	x	x	x	x	x	x	x	x	·	·	x	x	Uncommon, singly or small groups but 50 birds Mar. 1973; trees near ponds and lagoons; juveniles seen in Apr. and Dec.
King vulture *Sarcoramphus papa* Rey zamuro	x	x	x	·	x	x	x	x	x	x	x	x	Fairly common, 1–5 birds; near and over gallery forest and in savanna feeding; juveniles seen in Jul. and Nov.
Black vulture *Coragyps atratus* Zamuro	x	x	x	x	x	x	x	x	x	x	x	x	Abundant, flocks of up to 60 birds; savanna.

	J	F	M	A	M	J	J	A	S	O	N	D	
Turkey vulture *Cathartes aura* Oripopo	x	x	x	x	x	x	x	x	x	x	x	x	Abundant; savanna; migrant race Nov–Dec, resident race all year; one juvenile in Aug.
Lesser yellow-headed vulture *Cathartes burrovianus* Oripopo cabeza amarilla menor	x	x	x	x	x	x	x	x	x	x	x	x	Common, usually only 1–5 birds; savanna.
White-tailed kite *Elanus leucurus* Gavilán maromero	x	x	x	x	x	x	x	x	x	x	x	x	Common, singly or pairs; savanna; a juvenile seen in Feb.
Pearl kite *Gampsonyx swainsonii* Cernícalo	x	·	·	x	x	·	x	x	x	·	·	·	Occasional, singly and pairs; large trees in savanna and in *matas*.
Gray-headed kite *Leptodon cayanensis* Gavilán palomero	·	·	·	·	·	·	·	·	·	x	x	·	Rare, single birds; gallery forest.
Hook-billed kite *Chondrohierax uncinatus* Gavilán pico ganchudo	·	·	·	·	·	x	·	x	x	·	·	·	Rare, single birds; gallery forest.
Everglade kite *Rostrhamus sociabilis* Gavilan caracolero	·	x	·	x	·	x	x	x	x	·	x	x	Uncommon, usually singly or pairs; Caño Caracol and marshes; common in rice fields near ranch.
Slender-billed kite *Helicolestes hamatus* Gavilán pico de hoz	·	·	·	·	·	x	B	B	B	B	x	·	Fairly common, 2–3 breeding pairs; gallery forest and *matas*.
Bicolored hawk *Accipiter bicolor* Gavilán pantalón	·	·	·	x	·	·	·	·	·	·	x	·	Rare, single birds; gallery forest.
White-tailed hawk *Buteo albicaudatus* Gavilán tejé	x	x	x	x	x	x	x	x	x	x	x	x	Fairly common, singly and pairs; savanna; a juvenile seen in Jul.
Zone-tailed hawk *Buteo albonotatus* Gavilán negro	x	·	x	·	·	x	·	x	x	·	·	·	Occasional, single birds; savanna; one Jun. bird in primary molt.
Roadside hawk *Buteo magnirostris* Gavilán habado	x	x	x	x	B	B	B	x	x	x	x	x	Common, singly; *matas* and gallery forest.
Gray hawk *Buteo nitidus* Gavilán gris	x	·	x	x	x	x	x	x	x	·	x	x	Occasional, singly; gallery forest; not seen since Jul. 1976.
Bay-winged hawk *Parabuteo unicinctus* Gavilán andapié	x	x	·	·	·	x	x	x	·	x	·	·	Occasional, singly or pairs; savanna.
Black-collared hawk *Busarellus nigricollis* Gavilán colorado	x	x	x	x	x	x	x	x	x	·	x	x	Fairly common, usually singly; Caño Caracol and near lagoons; 2 adults with 2 juveniles Apr.

Species	J	F	M	A	M	J	J	A	S	O	N	D	Notes
Savanna hawk *Hetrospizias meridionalis* Gavilán pita venado	x	x	x	x	B	B	B	B	B	x	x	x	Abundant; savanna.
Great black-hawk *Buteogallus urubitinga* Aguila negra	x	x	x	x	x	x	x	x	x	x	x	x	Fairly common, singly; large trees and palms in savanna; juveniles seen Jul–Nov.
Ornate hawk-eagle *Spizaetus ornatus* Aguila de penacho	·	·	·	·	x	·	·	·	x	x	x	·	Rare, single birds; in and over gallery forest.
Crane Hawk *Geranospiza caerulescens* Gavilán zancón	x	x	x	x	x	x	x	x	x	x	x	x	Common, singly rarely pairs; savanna and gallery forest; immature plumaged bird in Nov.
Osprey *Pandion haliaetus* Aguila pescadora	·	·	·	·	·	·	·	x	x	x	·	·	Rare, single birds; lagoons and Caño Caracol.
Laughing falcon *Herpetotheres cachinnans* Halcón macagua	x	x	x	x	x	x	x	x	x	x	x	x	Fairly common, singly more often pairs; *matas* and gallery forest.
Collared forest-falcon *Micrastur semitorquatus* Halcón semiacollarado	·	·	·	·	·	x	·	·	x	·	x	·	Rare, singly; gallery forest.
Yellow-headed caracara *Milvago chimachima* Caricare sabanero	x	x	x	x	x	x	x	x	x	x	x	x	Common, singly; open savanna often near cattle; a juvenile seen in Dec.
Crested caracara *Polyborus plancus* Caricare encrestado	x	x	x	x	x	x	x	x	B	B	B	B	Abundant, singly, pairs and groups, 65 adults and immatures together Jul. 1973; *matas* and savannas.
Bat falcon *Falco rufigularis* Halcón golondrina	·	·	·	·	·	x	·	·	x	·	·	x	Rare, singly; gallery forest and *matas*.
Aplomado falcon *Falco femoralis* Halcón aplomado	x	x	x	x	·	x	x	x	·	·	x	·	Uncommon, singly or pairs; savanna.
American kestrel *Falco sparverius* Halcón primito	B	B	x	x	x	x	x	x	x	x	x	x	Fairly common, singly or pairs; open savanna.
Rufous-vented chachalaca *Ortalis ruficauda* Guacharaca del norte	x	x	x	x	x	x	x	x	x	x	x	x	Abundant, groups of 5–10 birds; *matas* and gallery forest.
Yellow-knobbed curassow *Crax daubentoni* Paují de copete	x	·	·	·	x	x	x	x	x	·	x	·	Occasional, groups of 2–5 birds; gallery forest.
Crested bobwhite *Colinus cristatus* Perdiz encrestada	x	x	x	x	x	x	x	x	x	B	B	x	Abundant, flocks of 5–10 birds; savanna.

Species	J	F	M	A	M	J	J	A	S	O	N	D	Notes
Limpkin *Aramus guarauna* Carrao	x	x	x	x	x	x	x	x	B	B	x	x	Common Jun–Oct, fairly common other months; marshes, ponds and Caño Caracol:
Gray-necked wood-rail *Aramides cajanea* Cotara caracolera	x	x	x	x	x	x	B	B	B	x	x	x	Common, singly; gallery forest and *matas* bordering marshes and lagoons.
Paint-billed crake *Neocrex erythrops* Polla pico rojo	·	·	·	·	·	·	·	·	x	x	x	x	Rare, single birds; marshes.
Purple gallinule *Porphyrula martinica* Gallito azul	x	·	·	·	·	x	B	B	B	x	x	x	Common Jun–Nov, uncommon other months; marshes.
Azure gallinule *Porphyrula flavirostris* Gallito claro	·	·	·	·	·	·	·	x	x	x	·	x	Rare, probably more during rainy season than indicated; marshes.
Sunbittern *Eurypyga helias* Tigana	x	x	x	x	x	·	x	x	x	x	x	x	Fairly common, singly and pairs; ponds in gallery forest; juveniles seen Oct–Dec.
Wattled jacana *Jacana jacana* Gallito de laguna	x	x	x	x	x	x	x	B	x	x	B	x	Common; all aquatic areas.
Southern lapwing *Vanellus chilensis* Alcaraván	x	x	x	x	B	B	x	x	x	x	x	x	Common, usually pairs; open grassland.
Solitary sandpiper *Tringa solitaria* Playero solitario	x	x	·	x	x	·	x	x	x	x	x	x	Common, singly; savanna ponds rarely gallery forest ponds; earliest seasonal record 23 Jul 1977, latest 4 May 1977.
Greater yellowlegs *Tringa melanoleuca* Tigüi-tigüe grande	x	·	·	·	·	·	·	x	·	x	x	·	Occasional, single birds; savanna ponds.
Spotted sandpiper *Actitis macularia* Playero coleador	x	x	·	·	·	·	·	x	x	x	x	x	Occasional, single birds; savanna ponds.
Least sandpiper *Calidris minutilla* Playerito menudo	·	·	·	·	·	·	·	x	x	x	·	·	Rare, 1–4 birds; Caño Caracol and savanna ponds.
Common snipe *Capella gallinago* Becasina chillona	x	·	·	·	x	x	x	·	·	·	·	·	Rare, 1–4 birds; wet savanna; May–Jul birds may have been *G. paraguaiae*; all birds appeared to be transients.
Common stilt *Himantopus himantopus* Viuda patilarga	x	x	x	x	x	x	·	·	·	·	·	·	Occasional, 1–5 birds; ponds and lagoon edges.
Double-striped thick-knee *Burhinus bistriatus* Dara	x	x	B	B	x	x	x	x	x	x	x	x	Fairly common, usually a pair occasionally 3–4; open grassland.

Species	J	F	M	A	M	J	J	A	S	O	N	D	Notes
Black skimmer *Rynchops nigra* Pico de tijera	·	x	·	x	x	·	·	·	·	x	x	·	Occasional, usually singly, but 20 birds Feb 1974; lagoons.
Bare-eyed pigeon *Columba corensis* Paloma ala blanca	·	·	·	·	·	·	·	·	·	·	x	·	One 1978, with large flock of *C. cayennensis* feeding on first sorghum planted on ranch.
Pale-vented pigeon *Columba cayennensis* Paloma colorada	x	x	B	x	x	x	x	x	x	x	x	x	Fairly common, singly, pairs or flocks up to 30 from Aug–Nov; savanna, marshes, gallery forest.
Eared dove *Zenaida auriculata* Paloma sabanera	x	x	x	x	x	x	x	x	B	x	B	x	Abundant, flocks; savanna.
Plain-breasted ground-dove *Columbina minuta* Tortolita sabanera	x	x	x	x	x	x	x	x	x	B	B	x	Common, flocks; savanna; often with *C. talpacoti*.
Ruddy ground-dove *Columbina talpacoti* Tortolita rojiza	x	x	x	x	x	x	x	x	x	x	B	x	Abundant, flocks; savanna.
Blue ground-dove *Claravis pretiosa* Palomita azul	x	x	x	x	·	·	·	·	x	x	B	x	Uncommon, usually pairs; gallery forest and savanna.
Scaled dove *Scardafella squammata* Palomita maraquita	x	x	x	x	x	x	x	x	B	B	B	x	Abundant, flocks; savanna.
White-tipped dove *Leptotila verreauxi* Paloma turca	x	x	x	x	x	x	B	x	B	B	B	x	Common, singly or pairs; *matas* and gallery forest.
Scarlet macaw *Ara macao* Guacamayo bandera	x	x	x	x	x	·	x	x	x	x	x	x	Uncommon, pairs and small flocks; usually over and in gallery forest, occasionally in savanna trees.
Brown-throated parakeet *Aratinga pertinax* Perico cara sucia	x	x	B	x	x	x	x	x	x	x	x	x	Abundant, pairs and flocks up to 40; savanna, *matas* and gallery forest.
Green-rumped parrotlet *Forpus passerinus* Periquito	x	x	x	x	B	x	B	B	B	B	B	x	Abundant, pairs and flocks up to 80 in the dry season; savanna.
Orange-chinned parakeet *Brotogeris jugularis* Churica	x	B	x	x	x	x	x	x	x	x	x	x	Common, pairs and flocks up to 30 birds; savanna trees and *matas*.
Yellow-headed parrot *Amazona ochrocephala* Loro real	x	x	x	x	x	x	x	x	x	x	x	x	Fairly common, pairs flying in early morning and late afternoon; savanna and gallery forest.
Hoatzin *Opisthocomus hoazin* Guacharaca de agua	x	x	x	x	B	B	B	B	B	B	B	B	Common, small groups; Caño Caracol and Rio Guárico; occasionally *matas* near ponds.

Species	J	F	M	A	M	J	J	A	S	O	N	D	Remarks
Dwarf Cuckoo *Coccyzus pumilus* Cuclillo gusanero	·	·	·	·	x	x	B	B	x	x	x	·	Uncommon, singly; bushes and trees of the wet savanna.
Yellow-billed cuckoo *Coccyzus americanus* Cuclillo pico amarillo	·	·	·	x	x	·	·	·	·	·	·	·	Rare, singly; savanna *matas* and bushes.
Dark-billed cuckoo *Coccyzus melacoryphus* Cuclillo grisáceo	·	·	·	·	·	x	x	B	x	x	·	·	Occasional, singly; bushes of wet savanna; seen only in 1973 and 1977.
Gray-capped cuckoo *Coccyzus lansbergi* Cuclillo acanelado	·	·	·	·	·	x	x	x	·	·	·	·	Rare, singly; bushes of wet savanna.
Squirrel cuckoo *Piaya cayana* Piscua	x	x	x	x	x	x	x	x	x	x	x	x	Common, singly rarely pairs; large trees and vines of *matas* and gallery forest.
Greater ani *Crotophaga major* Garrapatero hervidor	·	·	·	·	x	x	x	x	B	B	x	x	Seasonally common, flocks of 4–10 birds; *matas* and gallery forest.
Smooth-billed ani *Crotophaga ani* Garrapartero común	x	x	x	·	x	x	x	B	x	x	x	x	Fairly common, small flocks; savanna bushes and thickets.
Groove-billed ani *Crotophaga sulcirostris* Garrapatero curtidor	x	x	x	x	x	x	B	x	B	B	B	x	Common, flocks of 5–15; savanna bushes and thickets.
Striped cuckoo *Tapera naevia* Saucé	x	x	x	x	x	x	x	x	B	x	x	x	Common, singly; savanna.
Great horned owl *Bubo virginianus* Lechuzón orejudo	x	·	·	x	x	·	x	x	x	x	x	x	Fairly common, singly; savanna trees and *matas*; juveniles seen in Jul–Aug.
Ferruginous pygmy-owl *Glaucidium brasilianum* Pavita ferruginea	x	x	x	x	B	x	x	x	x	x	x	x	Fairly common, singly and pairs; savanna trees and *matas*.
Burrowing owl *Athene cunicularia* Mochuelo de hoyo	x	·	·	·	·	x	x	·	·	·	·	·	One pair, not seen since 1973; open grassland.
Great potoo *Nyctibius grandis* Nictibio grande	·	x	x	·	·	x	·	x	x	x	x	·	Uncommon, singly; roosting in gallery forest and hunting in savanna *matas*.
Common potoo *Nyctibius griseus* Nictibio grisáceo	·	·	·	·	x	x	·	·	x	·	·	·	Rare, single birds heard at night; savanna *matas*.
Lesser nighthawk *Chordeiles acutipennis* Aguaitacamino chiquito	·	·	·	·	x	·	x	x	·	·	x	x	Occasional; open grassland; probably more common than data indicates.

	J	F	M	A	M	J	J	A	S	O	N	D	
Nacunda nighthawk *Podager nacunda* Aguaitacamino barriga blanca	·	·	·	·	x	·	·	x	·	·	·	·	Rare, May 1972, Aug 1975; open grassland.
Pauraque *Nyctidromus albicollis* Aguaitacamino común	x	x	x	x	x	x	x	x	x	x	B	x	Common; open grassland at night and *matas* during the day.
White-tailed nightjar *Caprimulgus cayennensis* Aguaitacamino rastrojero	x	x	·	·	·	·	·	x	x	x	x	·	Uncommon; grasslands and bushes near lagoons.
Spot-tailed nightjar *Caprimulgus maculicaudus* Aguaitacamino cola pintada	·	·	·	·	·	·	·	·	·	·	x	·	Rare, one bird stunned by jeep; open grassland.
Fork-tailed palm-swift *Reinarda squamata* Vencejo tijereta	x	x	x	x	x	x	x	x	x	x	x	x	Common, 1–10 birds; flying over open savanna.
Black-throated mango *Anthracothorax nigricollis* Mango pechinegro	·	x	·	·	·	·	·	·	·	·	·	·	Rare, a single female; gallery forest.
Ruby-topaz hummingbird *Chrysolampis mosquitus* Tucusito rubí	·	·	·	·	·	·	·	x	x	·	x	·	Occasional, usually singly; savanna bushes.
Blue-chinned sapphire *Chlorestes notatus* Colibrí verdecito	·	·	·	·	x	·	·	·	x	x	·	·	Rare, singly; gallery forest.
Blue-tailed emerald *Chlorostilbon mellisugus* Esmeralda coliazul	x	x	x	x	x	x	·	x	x	x	x	x	Uncommon, singly; gallery forest and savanna bushes and trees.
White-tailed goldenthroat *Polytmus guainumbi* Colibrí gargantidorado	x	·	x	·	·	·	x	x	x	B	B	·	Occasional, usually singly; wet savanna and marshes.
Glittering-throated emerald *Amazilia fimbriata* Diamante gargantiverde	x	x	x	x	x	x	B	x	x	x	x	x	Fairly common, singly; savanna bushes, *matas* and gallery forest.
Copper-rumped hummingbird *Amazilia tobaci* Amazilia bronceada coliazul	x	x	·	x	·	·	·	x	x	·	x	x	Uncommon, singly; gallery forest.
Amethyst woodstar *Calliphlox amethystina* Tucusito amatista	x	·	·	·	·	·	·	x	·	x	·	x	Rare, singly; feeds low in gallery forest edge.
Ringed kingfisher *Ceryle torquata* Martín pescador grande	x	·	·	·	·	·	x	x	x	x	x	x	Fairly common, singly and pairs; Caño Caracol and Rio Guárico.
Amazon kingfisher *Chloroceryle amazona* Martín pescador matraquero	x	x	·	·	·	·	x	·	x	·	x	·	Uncommon, singly and pairs; Caño Caracol.

Species	J	F	M	A	M	J	J	A	S	O	N	D	Notes
Green kingfisher *Chloroceryle americana* Martín pescador pequeño	x	·	x	·	·	x	x	x	x	·	x	·	Uncommon, singly; Caño Caracol.
Pygmy kingfisher *Chloroceryle aenea* Martín pescador pigmeo	·	x	·	·	x	·	x	·	x	x	x	x	Uncommon, usually pairs; ponds in gallery forest and ponds of wet savanna.
Rufous-tailed jacamar *Galbula ruficauda* Tucuso barranquero	x	x	x	x	B	B	x	x	x	x	x	x	Common, singly and pairs; gallery forest and *matas*.
Russet-throated puffbird *Hypnelus ruficollis* Bobito	x	x	x	x	x	x	x	B	B	x	x	x	Common, singly and pairs; *matas* and occasionally in gallery forest.
Scaled piculet *Picumnus squamulatus* Telegrafista escamado	x	x	x	x	B	·	·	x	B	x	x	·	Uncommon, usually singly; gallery forest and *matas*.
Spot-breasted woodpecker *Chrysoptilus punctigula* Carpintero pechipunteado	x	x	x	x	B	B	B	B	x	x	x	x	Fairly common, singly; *matas*.
Lineated woodpecker *Dryocopus lineatus* Carpintero real barbirrayado	x	x	x	x	x	x	x	x	x	x	x	x	Fairly common, usually singly; large savanna trees occasionally gallery forest.
Red-crowned woodpecker *Melanerpes rubricapillus* Carpintero habado	x	x	B	B	B	B	x	x	x	x	x	x	Common, singly and pairs; savanna trees, *matas* and gallery forest.
Red-rumped woodpecker *Veniliornis kirkii* Carpintero rabadilla roja	x	x	x	x	x	·	x	·	x	x	·	·	Uncommon, pairs and small groups in trees of *matas*.
Crimson-crested woodpecker *Campephilus melanoleucos* Carpintero real pico amarillo	x	·	·	x	x	x	x	x	x	x	x	x	Fairly common, singly rarely pairs; gallery forest occasionally large trees in *matas*.
Straight-billed woodcreeper *Xiphorhynchus picus* Trepador subesube	x	x	x	x	B	x	x	B	x	x	x	x	Common, singly; *matas* and gallery forest.
Streak-headed woodcreeper *Lepidocolaptes souleyetii* Trepadorcito listado	·	x	·	x	x	B	x	x	·	·	·	·	Occasional, singly; large trees in *matas*.
Red-billed scythebill *Campylorhamphus trochilirostris* Trepador pico de garfio	x	x	x	x	x	x	x	x	x	x	x	x	Fairly common, singly; large trees in *matas* and gallery forest.
Pale-breated spinetail *Synallaxis albescens* Güitío gargantiblanco	x	·	x	·	x	x	x	x	x	x	x	x	Fairly common, pairs; dense savanna thickets.
Yellow-throated spinetail *Certhiaxis cinnamomea* Güitío de agua	x	x	x	B	B	B	B	B	B	x	x	x	Common, pairs; bushes and thickets at all aquatic areas.

Species	*J*	*F*	*M*	*A*	*M*	*J*	*J*	*A*	*S*	*O*	*N*	*D*	Notes
Plain-fronted thornbird *Phacellodomus rufifrons* Güaití	x	x	x	x	x	x	B	B	B	B	B	x	Common, small groups; savanna bushes, trees and *matas*.
Streaked xenops *Xenops rutilans* Pico lezna rayado	x	x	·	·	x	·	·	·	·	·	x	·	Occasional, singly; gallery forest.
Black-crested antshrike *Sakesphorus canadensis* Hormiguero copetón	x	x	x	x	x	x	B	x	x	x	x	x	Common, pairs; understory of gallery forest and wet *matas*.
Barred antshrike *Thamnophilus doliatus* Pavita hormiguera común	x	x	x	x	x	x	x	x	x	x	x	x	Fairly common, pairs; in and near edges of gallery forest and *matas*.
White-fringed antwren *Formicivora grisea* Coicorita	x	x	x	x	x	x	x	x	x	x	x	x	Fairly common, pairs; gallery forest and *matas*.
Cinereous becard *Pachyramphus rufus* Cabezón cinéreo	·	x	·	·	·	·	·	x	x	x	·	·	Rare, singly; savanna trees and bushes; birds in juvenal plumage Aug–Sept.
White-winged becard *Pachyramphus polychopterus* Cabezón aliblanco	·	·	·	·	x	x	x	x	x	x	x	·	Occasional, pairs; savanna trees and *matas*; probably breeding but data lacking.
Black-crowned tityra *Tityra inquisitor* Bacaco pequeño	·	·	·	·	B	x	B	B	x	·	·	·	Uncommon, pairs; large savanna trees and gallery forest.
Lance-tailed manakin *Chiroxiphia lanceolata* Saltarín cola de lanza	x	x	x	x	x	x	x	x	x	x	x	x	Fairly common, small groups; appears confined to one area of gallery forest.
Pied water-tyrant *Fluvicola pica* Viudita acuática	x	x	x	B	B	B	B	B	B	x	x	x	Common, pairs; savanna ponds and lagoons.
White-headed marsh-tyrant *Arundinicola leucocephala* Atrapamoscas duende	x	x	x	x	x	x	x	B	x	x	B	x	Common, pairs; savanna ponds.
Vermilion flycatcher *Pyrocephalus rubinus* Atrapamoscas sangre de toro	x	x	x	x	x	x	x	x	x	x	B	x	Common, pairs; wet areas of the savanna.
Cattle tyrant *Machetornis rixosus* Atrapamoscas jinete	x	x	x	x	x	B	B	x	x	B	x	x	Common, singly; open savanna; perches on domestic and wild animals.
Fork-tailed flycatcher *Muscivora tyrannus* Atrapamoscas tijereta	x	x	·	x	x	x	x	x	x	x	·	·	Abundant May–Jul, flocks of 100–500, occasional in other months; open grassland.
Tropical kingbird *Tyrannus melancholicus* Pitirre chicharrero	x	x	x	x	B	B	x	x	x	x	x	x	Abundant, singly and pairs; savanna trees and bushes.

Species	J	F	M	A	M	J	J	A	S	O	N	D	Status
Gray kingbird *Tyrannus dominicensis* Pitirre gris	x	x	x	B	x	x	x	x	x	x	x	x	Fairly common, singly; open savanna and low bushes; resident.
Variegated flycatcher *Empidonomus varius* Atrapamoscas veteado	x	·	·	·	·	·	x	·	·	·	·	·	Rare, singly; gallery forest.
Piratic flycatcher *Legatus leucophaius* Atrapamoscas ladrón	·	·	B	x	·	·	·	B	·	·	·	·	Rare, pairs; near Caño Caracol.
White-bearded flycatcher *Conopias inornata* Atrapamoscas barbiblanco	x	x	B	B	B	B	B	B	x	x	x	x	Common, pairs and small family groups; savanna trees and *matas*.
Boat-billed flycatcher *Megarhynchus pitangua* Atrapamoscas picón	x	x	x	·	x	x	x	x	x	x	x	x	Fairly common, pairs; usually high in trees of *matas* and gallery forest.
Streaked flycatcher *Myiodynastes maculatus* Gran atrapamoscas listado	x	·	B	B	x	x	x	x	·	x	x	·	Uncommon, singly and pairs; usually in gallery forest, rarely in *matas*.
Rusty-margined flycatcher *Myiozetetes cayanensis* Atrapamoscas pecho amarillo	x	x	B	B	B	B	B	B	x	x	x	x	Common, pairs; savanna trees, bushes and *matas*, gallery forest edge.
Social flycatcher *Myiozetetes similis* Pitirre copete rojo	x	x	x	x	B	B	B	x	x	x	x	x	Common, singly and pairs; savanna bushes, *matas* and gallery forest.
Great kiskadee *Pitangus sulphuratus* Cristofué	x	x	B	B	B	B	B	B	B	x	x	x	Abundant, singly; savanna trees, *matas* and gallery forest edge.
Lesser kiskadee *Pitangus lictor* Pecho amarillo orillero	x	x	x	x	x	x	x	x	x	x	x	x	Uncommon, pairs; low, near water edged by bushes and trees in gallery forest and savanna.
Short-crested flycatcher *Myiarchus ferox* Atrapamoscas garrochero chico	x	·	·	x	x	x	x	x	x	x	x	·	Uncommon, singly; *matas*.
Brown-crested flycatcher *Myiarchus tyrannulus* Atrapamoscas garrochero colirufo	x	x	x	x	x	x	x	x	x	x	x	x	Fairly common, singly; savanna bushes and *matas*.
Dusky-capped flycatcher *Myiarchus tuberculifer* Atrapamoscas cresta negra	·	x	x	·	x	x	·	x	·	·	·	·	Uncommon, singly; gallery forest.
Fuscous flycatcher *Cnemotriccus fuscatus* Atrapamoscas fusco	·	·	·	·	x	x	·	·	·	x	·	·	Rare, singly; gallery forest.
Bran-colored flycatcher *Myiophobus fasciatus* Atrapamoscas pechirrayado	·	x	·	x	·	x	·	x	·	x	x	·	Occasional, singly; gallery forest edge, rarely savanna.

Species	J	F	M	A	M	J	J	A	S	O	N	D	Notes
Yellow-breasted flycatcher *Tolmomyias flaviventris* Pico Chato amarillento	x	x	x	x	B	B	B	B	B	x	x	x	Fairly common, singly; gallery forest and *matas.*
Common Tody-flycatcher *Todirostrum cinereum* Titirijí lomicenizo	x	x	x	x	B	B	B	B	B	x	x	x	Common, pairs; savanna bushes, trees and *matas.*
Slate-headed tody-flycatcher *Todirostrum sylvia* Titirijí cabecicenizo	x	·	·	·	x	x	·	x	·	x	·	·	Fairly common, singly; gallery forest.
Pale-eyed pygmy-tyrant *Atalotriccus pilaris* Atrapamoscas pigmeo ojiblanco	x	x	x	·	x	·	x	x	·	·	x	·	Fairly common, singly; gallery forest and *matas.*
Yellow tyrannulet *Capsiempis flaveola* Atrapamoscas amarillo	·	·	·	x	·	·	·	x	x	·	x	·	Occasional, singly; savanna *matas* and gallery forest.
Pale-tipped tyrannulet *Inezia subflava* Inezia de vientre amarillo	x	x	x	x	x	x	x	x	x	x	x	x	Fairly common, singly; gallery forest, savanna bushes.
Yellow-bellied elaenia *Elaenia flavogaster* Bobito copetón vientre amarillo	x	·	x	x	x	x	x	x	·	x	x	x	Fairly common, singly; savanna bushes, trees and *matas.*
Small-billed elaenia *Elaenia parvirostris* Bobito copetón pico corto	·	·	·	·	x	x	x	x	x	·	·	·	Uncommon, austral migrant, singly; savanna bushes and *matas.*
Plain-crested elaenia *Elaenia cristata* Bobito crestiapagado	x	·	·	x	·	x	·	·	x	·	x	x	Uncommon, singly; savanna bushes and *matas.*
Lesser elaenia *Elaenia chiriquensis* Bobito copetón mono blanco	·	x	·	x	·	x	x	x	·	·	·	·	Uncommon, singly; savanna bushes and *matas.*
Forest elaenia *Myiopagis gaimardii* Bobito de selva	·	x	·	·	x	·	·	·	·	x	·	·	Occasional, singly; in mixed flocks of gallery forest.
Mouse-colored tyrannulet *Phaeomyias murina* Atrapamoscas color ratón	x	x	x	x	x	x	x	x	x	·	x	·	Fairly common; savanna bushes and *matas*; a Sept. bird was in primary molt.
Southern beardless tyrannulet *Camptostoma obsoletum* Atrapamoscas lampiño	B	B	x	x	x	x	x	x	x	·	x	B	Fairly common, singly; savanna bushes, thickets and *matas.*
White-winged swallow *Tachycineta albiventer* Golondrina de agua	B	x	x	x	x	x	x	x	x	x	x	x	Several resident pairs; Caño Caracol, lagoons and wet savanna.
Gray-breasted martin *Progne chalybea* Golondrina urbana	x	x	·	·	·	·	x	·	·	x	x	·	Irregular; savannas; immature plumaged birds in Oct.

Species	J	F	M	A	M	J	J	A	S	O	N	D	Notes
Blue-and-white swallow *Notiochelidon cyanoleuca* Golondrina azul y blanco	x	x	·	·	·	·	x	x	x	·	x	·	Irregular; savanna; a bird in immature plumage in Jul.
Rough-winged swallow *Stelgidopteryx ruficollis* Golondrina ala de sierra	·	x	·	x	x	·	x	x	·	x	x	·	Occasional; savanna.
Bank swallow *Riparia riparia* Golondrina parda	x	x	·	·	·	·	·	·	·	·	·	x	Fairly common, in flocks with *Hirundo rustica*; savanna.
Barn swallow *Hirundo rustica* Golondrina de horquilla	x	x	x	x	x	·	·	·	x	x	x	x	Common, but irregular numbers; usually 10–100 birds but over 5,000 at nearby roost Jan–Feb 1976; savanna.
Bicolored wren *Campylorhynchus griseus* Cucarachero currucuchú	x	x	x	x	B	x	B	B	x	x	x	x	Common, pairs and small groups; savanna bushes, trees and *matas*.
Stripe-backed wren *Campylorhynchus nuchalis* Cucarachero chocorocoy	x	x	x	B	B	B	B	B	B	x	x	x	Common, groups of 4–10 birds; savanna bushes, trees and *matas*.
Buff-breasted wren *Thryothorus leucotis* Cucarachero flanquileonado	x	x	·	·	x	·	·	·	x	·	·	·	Uncommon, pairs; bushes of gallery forest edge and *matas*.
House wren *Troglodytes aedon* Cucarachero común	x	x	x	x	x	x	B	x	x	B	x	x	Fairly common; *matas*.
Tropical mockingbird *Mimus gilvus* Paraulata llanera	x	x	B	B	x	x	x	x	x	x	x	x	Abundant; savanna bushes and trees; juveniles also seen in Sept.
Veery *Catharus fuscescens* Paraulata cachetona	·	·	·	·	·	·	·	·	·	x	x	·	Rare, only in 1978; in gallery forest and in *matas*.
Pale-breasted thrush *Turdus leucomelas* Paraulata montañera	x	x	·	x	x	x	B	x	x	x	·	·	Uncommon, singly; *matas*.
Bare-eyed thrush *Turdus nudigenis* Paraulata ojo de candil	x	x	x	x	x	x	B	B	B	x	x	x	Fairly common, singly; *matas*.
Tropical gnatcatcher *Polioptila plumbea* Chirito de chaparrales	x	x	x	x	x	x	x	x	x	x	x	x	Common, pairs; *matas* and gallery forest.
Rufous-browed peppershrike *Cyclarhis gujanensis* Sirirí	x	x	x	x	x	x	x	x	x	x	x	x	Common, singly; *matas* and gallery forest.
Red-eyed vireo *Vireo olivaceus* Julián chiví ojirrojo	·	·	x	x	x	x	x	x	x	x	x	·	Fairly common, singly; often in mixed flocks; gallery forest.

Species	J	F	M	A	M	J	J	A	S	O	N	D	Notes
Golden-fronted greenlet *Hylophilus aurantiifrons* Verderón luisucho	x	x	x	x	x	x	x	x	x	x	x	x	Common, pairs; gallery forest
Scrub greenlet *Hylophilus flavipes* Verderón patipálido	x	x	·	x	x	x	x	·	x	x	x	·	Uncommon, singly; gallery forest.
Bananaquit *Coereba flaveola* Reinita común	x	x	x	B	B	x	B	B	x	x	x	x	Common, singly; savanna bushes, trees, *matas* and gallery forest.
Chestnut-vented conebill *Conirostrum speciosum* Mielerito azul	x	x	·	·	x	x	·	x	·	·	x	·	Uncommon, singly and pairs; gallery forest.
Tropical parula *Parula pitiayumi* Reinita montañera	·	x	·	·	x	·	·	·	·	x	x	·	Occasional, pairs; gallery forest.
Yellow warbler *Dendroica petechia* Canario de mangle	x	x	x	·	·	·	·	·	·	x	x	·	Occasional, singly; savanna trees, bushes and *matas*.
Northern waterthrush *Seiurus noveboracensis* Reinita de charcos	x	·	·	x	·	·	·	·	x	x	x	·	Occasional, singly; near Caño Caracol and, rarely, wet *matas*.
American redstart *Setophaga ruticilla* Candelita migratoria	·	x	·	x	·	·	·	·	x	x	·	·	Rare, singly; *matas* and gallery forest.
Shiny cowbird *Molothrus bonariensis* Tordo mirlo	x	x	x	x	B	B	B	B	B	B	x	x	Common; savanna bushes and trees.
Crested oropendola *Psarcolius decumanus* Conoto negro	·	·	x	x	·	·	·	·	·	·	·	·	Rare, one and two birds observed only in 1977; gallery forest.
Yellow-rumped cacique *Cacicus cela* Arrendajo común	x	x	B	B	B	B	B	B	B	x	x	x	Common, groups; small numbers in nonbreeding season; *matas* and gallery forest.
Carib grackle *Quiscalus lugubris* Tordo negro	x	x	x	x	B	B	B	B	B	B	B	x	Abundant; savanna bushes and trees.
Yellow-hooded blackbird *Agelaius icterocephalus* Turpial de agua	x	·	·	x	·	x	x	B	B	B	B	x	Seasonally common, flocks up to 600; transient flock of 45 Apr 1977; wet savanna and marshes.
Orange-crowned oriole *Icterus auricapillus* Gonzalito real	·	x	·	·	·	·	x	x	·	·	x	·	Occasional, singly; near gallery forest edge.
Troupial *Icterus icterus* Turpial común	x	x	x	x	x	B	B	B	B	x	x	x	Common, singly; savanna trees and *matas*.

Species	J	F	M	A	M	J	J	A	S	O	N	D	Status and habitat
Yellow oriole *Icterus nigrogularis* Gonzalito	x	x	x	x	B	B	B	B	B	x	x	x	Common, singly; savanna trees and *matas*.
Oriole blackbird *Gymnomystax mexicanus* Tordo maicero	x	x	x	x	x	B	x	x	x	x	x	x	Fairly common, pairs or small groups, but one flock of 80 in Apr 1977; savanna bushes and trees.
Red-breasted blackbird *Leistes militaris* Tordo pechirrojo	x	x	x	x	x	·	x	x	x	x	x	·	Irregular, 2–300 birds, large numbers Mar–May 1977 with 80% in immature plumage; open grassland.
Eastern meadowlark *Sturnella magna* Perdigón	x	x	x	x	x	·	x	x	x	x	x	x	Fairly common; open grassland.
Bobolink *Dolichonyx oryzivorus* Tordo arrocero	·	·	·	·	·	·	·	·	·	x	·	·	Irregular, 50 birds in 1978; low bushes in open marsh.
Trinidad euphonia *Euphonia trinitatis* Curruñatá saucito	x	x	x	x	x	x	x	x	x	x	x	x	Common, pairs; high in large trees of gallery forest and *matas*.
Thick-billed euphonia *Euphonia laniirostris* Curruñatá piquigordo	x	x	·	·	·	·	·	B	·	·	·	·	Occasional, pairs; gallery forest.
Burnished-buff tanager *Tangara cayana* Tangara monjita	·	·	·	·	·	·	·	x	·	·	·	·	Rare, a single bird 1978; trees in savanna.
Blue-gray tanager *Thraupis episcopus* Azulejo de jardín	x	x	x	x	B	x	B	x	x	x	x	x	Abundant, pairs and small flocks in Oct; *matas* and gallery forest edge.
Glaucous tanager *Thraupis glaucocolpa* Azulejo verdeviche	x	x	x	x	x	x	x	B	x	x	x	x	Fairly common, pairs; savanna bushes, trees and *matas*.
Palm tanager *Thraupis palmarum* Azulejo de palmeras	x	·	·	·	·	·	·	·	·	·	·	·	Rare; savanna.
Hooded tanager *Nemosia pileata* Frutero de coronita	x	x	x	x	x	·	x	x	·	x	·	·	Uncommon, pairs; high in trees of gallery forest and *matas*.
Guira tanager *Hemithraupis guira* Pintasilgo buchinegro	·	x	·	·	·	·	·	x	·	·	·	·	Rare, pairs; gallery forest.
Grayish saltator *Saltator coerulescens* Lechosero ajicero	x	x	x	x	B	x	x	x	x	x	x	x	Abundant, pairs and small groups; savanna bushes, trees and *matas*.
Orinocan saltator *Saltator orenocensis* Lechosero pechiblanco	x	x	x	x	B	x	x	x	x	x	x	x	Fairly common, pairs and small groups; savanna bushes and trees; juveniles seen May–Sept.

Species	J	F	M	A	M	J	J	A	S	O	N	D	Status
Red-capped cardinal *Paroaria gularis* Cardenal bandera alemana	x	x	x	x	x	B	B	B	B	B	B	x	Common, pairs and small groups; bushes of wet areas of savanna and Caño Caracol.
Blue-black grosbeak *Cyanocompsa cyanoides* Picogordo azul	·	·	·	·	·	·	·	x	·	·	·	·	Rare, a single male in Aug 1976.
Dickcissel *Spiza americana* Arrocero americano	x	x	x	·	·	·	·	·	·	·	x	·	Occasional; savanna; flocks up to 10,000 in nearby rice fields and on migration.
Blue-black grassquit *Volatinia jacarina* Semillero chirrí	x	x	x	x	x	x	B	x	x	x	B	x	Common; savanna bushes and thickets.
Gray seedeater *Sporophila intermedia* Espiguero pico de plata	x	x	x	x	x	x	x	x	x	x	x	x	Fairly common; savanna bushes.
Lined seedeater *Sporophila lineola* Espiguero bigotudo	·	·	·	·	·	·	x	x	x	x	x	x	Seasonally common, flocks up to 100 birds; savanna bushes; this and *S. bouvronides* may be races of a single species (Schwartz 1976).
Lesson's seedeater *Sporophila bouvronides* Espiguero bengalí	·	·	·	·	·	·	·	x	x	x	x	·	Rare, single males; gallery forest.
Ruddy-breasted seedeater *Sporophila minuta* Espiguero canelillo	x	·	·	·	x	x	x	x	x	x	x	·	Fairly comon; savanna bushes.
Large-billed seed-finch *Oryzoborus crassirostris* Semillero picón	·	·	·	·	·	·	x	x	x	B	x	·	Occasional, singly and pairs; savanna bushes and gallery forest edge.
Saffron finch *Sicalis flaveola* Canario de tejado	x	x	x	x	x	x	B	B	B	B	B	x	Abundant; savanna.
Pileated finch *Coryphospingus pileatus* Granero cabecita de fósforo	·	x	x	x	x	·	x	x	x	x	x	·	Uncommon, pairs; savanna bushes.
Black-striped sparrow *Arremonops conirostris* Curtío	·	·	·	x	·	·	·	·	·	·	·	·	Rare; gallery forest.
Grassland sparrow *Ammodramus humeralis* Sabanerito de pajonales	x	x	x	x	x	x	x	x	x	x	x	x	Common; open grassland.

Table 2. Birds seen within 40 kilometers of Masaguaral but not found on the ranch during the study period

Amazonetta brasiliensis	*Tringa flavipes*
Ictinea plumbea	*Phaetusa simplex*
Buteo swainsoni	*Sterna superciliaris*
Buteo brachyurus	*Tyto alba*
Circus buffoni	*Rhinoptynx clamator*
Hoploxypterus cayanus	*Brachygalba goeringi*
Pluvialis dominica	*Satrapa icterophrys*

and generally with an open understory. Savanna is grassland with scattered trees and palms or bushy thickets. Open grassland refers to savanna with only grass and some forbs. Interrupting the savanna are associations of five or more trees with a closed canopy, often with a bushy understory; these are indicated by the vernacular name *matas*.

Less direct evidence of the annual avian cycle of some species is indicated by notes of observations of juvenal and immature plumaged birds or, in a few instances when all other data are lacking, adult molt.

In summary, the 243 species of birds in Table 1 represent about 18 percent of Venezuelan birds. However the 54 families observed on the ranch represent more than 66 percent of the families found in Venezuela (Meyer de Schauensee and Phelps 1978). Approximately 6 percent (n = 15) of the birds in Table 1. are long-distant migrants from the northern hemisphere, while only 2 percent (n = 5) are known to be austral migrants.

Acknowledgments

It is a pleasure to acknowledge my indebtedness to Tomás Blohm for permission to use his ranch as my study area. I want to thank William H. Phelps, Jr. for permission to consult, on many occasions, his ornithological collection. I also thank Tomás Blohm, R. Haven Wiley, Eugene Morton, Paul Schwartz, Charles T. Collins and many others for companionship in the field, observations of Masaguaral birds, and discussions about them. Haven Wiley and Charles Collins made helpful suggestions on a first draft of this paper.

Literature Cited

von Berlepsch, H., and E. Hartert
1902. On the Birds of the Orinoco Region. *Novitates Zoologicae*, 9:1–134.

Cherrie, G. K.
1916. A Contribution to the Ornithology of the Orinoco Region. *Bulletin of the Brooklyn Institute of Arts and Sciences*, 2(6):133–374.

Ewel, J. J., and A. Madriz
1968. *Zonas de Vida de Venezuela.* Caracas: Ministerio de Agricultura y Cria.

Friedmann, H., and F. D. Smith, Jr.
1968. A Contribution to the Ornithology of Northeastern Venezuela. *Proceedings of the United States National Museum*, 100(3268):411–538.
1955. A Further Contribution to the Ornithology of Northeastern Venezuela. *Proceedings of the United States National Museum*, 104(3345):463–524.

Haffer, J.
1974. *Avian Speciation in Tropical South America.* 390 pages. The Nuttall Ornithological Club, number 14.

Myeer de Schauensee, R., and W. H. Phelps, Jr.
1978. *A Guide to the Birds of Venezuela.* Princeton: Princeton University Press.

Phelps, W. H.; and W. H. Phelps, Jr.
1958. Lista de las Aves de Venezuela con su Distribución. Volume 2, pt. 1, Passeriformes. *Boletin de la Sociedad Venezolana de Ciencias Naturales*, 19(90):1–317.
1963. Lista de las Aves de Venezuela con su Distribución. Volume 1, pt. 2, Passeriformes. *Boletin de la Sociedad Venezolana de Ciencias Naturales*, 24(104, 105):1–479.

Ridgely, R. S.
1976. *A Guide to the Birds of Panama.* Princeton: Princeton University Press.

Schwartz, P.
1975. Solved and unsolved problems in the *Sporophila Lineola/bouvronides* complex (Aves:Emberizidae). *Annals of Carnegie Museum*, 45(14):277–285.

Thomas, B. T.
1978. The dwarf cuckoo in Venezuela. *Condor*, 80:105–106.
In press. Plumage succession of nestling Maguari storks. *Boletin de la Sociedad Venezolana de Ciencias Naturales.*
In press. Introduction to the behavior and breeding of the White-bearded Flycatcher.

Wiley, R. H. and M. S. Wiley
1977. Recognition of neighbors' duets by stripe-backed wrens *Campylorhynchus nuchalis. Behavior*, 62(1–2):10–34.
In press. Spacing and timing in the nesting ecology of tropical Blackbird.

EUGENE S. MORTON
National Zoological Park
Washington, D. C. 20008

A Comparative Survey of Avian Social Systems in Northern Venezuelan Habitats

RESÚMEN

Se discute la influencia de la distribución de alimentos en sistemas sociales de aves en un bosque húmedo, en el parque Nacional de Guatopo, en los llanos en Masaguaral, en el bosque de galería sobre que limita con el río Guarico. En Guatopo y Masaguaral, las especies se unen con conespecies cuando comen alimentos aglutinados, pero son solitarias cuando los alimentos están dispersos. La proporción de especies de la misma clase que son gregarias de especies de la misma clase que son solitarias son es igual en Guatopo y Masaguaral, pero en Guatopo estos grupos son prequeños, fracuentemente dos aves, mientras que en Masaguaral se encuentran grande grupos. En Guatopo, grupos de la misma especie, sólo usan alimentos aglutinados, pero, en Masaguaral existen grande grupos de especies de la misma clase independientemente de la distribución del alimento. Los datos de Guatopo están comparados con los datos de Buskirk en Costa Rica. Se concluye que hay algunas diferencias entre los dos estudios que no apoyan la idea de que la predacion es más o menos importante que la distribución de alimentos cuando se predicen sistemas sociales.

Se discuten los numeros de especies comun es entre los tres lugares. Guarico es muy similar a Guatopo en varios aspectos. La presencia de especies en ambas areas y la existencia de especies errantes adaptades a la vida de bosque en el bosque de galería apoya la idea de que éste bosque es la ruta principal de colonización para Guatopo. Los números de oscine y nonoscine en Guatopo con respecto a los números de colonos potencial en cada grupos difieren.

ABSTRACT

The influence of food distribution on avian social systems in a moist forest, Guatopo National Park, llanos at Masaguaral, and the gallery forest bordering the Guarico River is discussed. At Guatopo and Masaguaral species join with conspecifics when exploiting clumped foods but are solitary if the food is dispersed. The proportions of intraspecificially gregarious to intraspecifically solitary species is the same in both areas but at Guatopo these groups are small, usually only two birds, while at Masaguaral large groups occur. At Guatopo conspecific groups exploit only clumped food but at Masaguaral large conspecific flocks occur independent of the distribution of food. Reasons to explain this are discussed.

The Guatopo data are compared with Buskirk's data from Costa Rica. It is concluded that several major differences exist that do not support the idea that predation is of less importance than food dispersion in predicting social systems.

The number of species in common between Guatopo, Masaguaral, and Guarico is discussed. Guarico is similar to Guatopo on many respects. The presence of species in both areas and the occasional occurence of wandering forest-adapted species in the gallery forest supports the idea that the gallery forest is a main route of colonization for Guatopo. The number of oscine and non-oscine passerines in Guatopo differs with respect to the number of potential colonizers in each group. It is suggested that omnivorous food habits correlate with the ability to colonize new forest patches and that oscines are superior colonizers.

Introduction

The several hundred species of birds that live in the northern Venezuelan region exhibit social behavior that is not qualitatively different from species of other tropical regions. Some species join others of the same or different species in foraging flocks for varying portions of the day or year. However, a quantitative analysis of this behavior may provide insight into the selection pressures that favor their characteristic ways of obtaining food efficiently and at the same time avoiding predators (Moynihan 1962; Morse 1970; Buskirk 1976).

Buskirk (1976) has recently discussed a unified approach to studying social systems on the community level. He assigned each species in his 8.8 ha study area of evergreen lower montane rain forest near Monteverde, Costa Rica, to social categories and food dispersion type and assessed their predation vulnerability. He concluded that flocking serves primarily an antipredation function, flocking composition (intra- versus interspecific composition) being primarily the product of competition patterns related to food dispersion. This is in accord with Willis (1972) and others who note that flocking disappears in areas without avian predators. However, Morton (in press) notes changes in social behavior correlated with habitat structure in some species, and that the tendency to join flocks may be more labile than usually assumed. The need for more testing of the correlations between food dispersion, flocking, and predation pressure was pointed out by Buskirk (1976). Here I will compare Buskirk's results with an analysis of the social behavior of the 227 bird species found thus far in Guatopo National Park (see Eisenberg, et al., 1979, for a description of the study area), an area about 600–700 m lower in elevation than Buskirk's 1550-m plot but still characterized as lower montane evergreen forest, although mainly second growth. Secondly, I will contrast and compare avian social systems in Guatopo with an adjacent but quite different area, the llanos near Masaguaral. Mrs. Betsy Trent Thomas (see p. 213) has compiled an annotated list of the avifauna of Masaguaral and the adjacent gallery forest bordering the Guarico River. Lastly, I will return to Guatopo to discuss the relative abilities of oscine and non-oscine passerines as "competitors" for niche space in tropical forest.

The Avifauna of Guatopo National Park

The Guatopo list (Appendix 1) represents species found in a relatively small area of the park (see map, Eisenberg, et. al., 1979) and excludes the dry fringes of llanos habitat characterized by the ubiquitous nests of Plain-fronted Thornbirds (*Phacellodomus rufifrons*). The habitat consists of second growth forest 15–20 m in height with occasional emergent trees, edge and heliconia thickets. One trail (Trail 1) winds through more mature forest. Most of these areas were from 550 to 900 m elevation. A brief survey was made of cloud forest habitat on Morro de Apa at 1550 m (see map, Eisenberg, et. al., 1979). The habitats are described in more depth in O'Connell, 1979). My trips to Guatopo took place only during the dry season (November, January, and February) within a four-year period. Little evidence of breeding was found during these trips, in contrast to Panama where many species are dry season breeders (Morton 1973, 1977).

The social organization of each species is categorized under intra- and interspecific tendencies. A species that forages alone is termed "N" in both these categories in the Appendix. One that joins mixed species flocks but is the sole representative of its species in them would be termed "N" intraspecifically but "F" interspecifically. Thus, intraspecific social grouping is symbolized N for solitary species, P for those occurring in pairs (generally permanently pairbonded individuals), S if 3–5 conspecifics co-occur, and L if larger flocks of conspecifics are characteristic. Under interspecific social grouping, only N for non-mixed species flock-joiners and F for those that do join other species are used. Food dispersion categories are D for dispersed food such as insects, C for clumped food such as abundant fruits on a tree or many flowers producing nectar or I for Intermediate. These relative categories follow Buskirk (1976). It should be mentioned that species having permanent monogamous pair bonds are still listed as N intraspecifically if they do not forage together, for example, *Thryothorus rufalbus*.

Intraspecific Social Organizations at Guatopo

There is a close concordance in results of statistical treatment of the data in Appendix 1 between Buskirk's Costa Rican forest community and Guatopo (Table 1). Species feeding on dispersed foods tend not to join with conspecifics (Table 1, I). At the same time species feeding on clumped resources tend to join conspecifics. This relationship is even more impressive when it is realized that 74 percent of the 46 species listed as gregarious are, in fact, only marginally gregarious, being found in pair bonds as opposed to small or large groups of conspecifics and many of these feed on dispersed foods (Table 1, II). Since dispersed food resources and joint defense of a permanent territory by a pair are correlated (pers. obs.) it is possible that the preponderance of pairs in the gregarious category has less to do with the dispersed food per se than with the defense of the territory in which it is found. So, the highly significant relationship between food distribution and social systems seems real given the bias against it I made by lumping pair bonded species into the gregarious category.

More species than would be expected by chance exhibit a gregarious intraspecific tendency if their food is in the clumped or intermediate food dispersion categories (Table 1, I). Group sizes for those feeding on intermediate and clumped foods are larger than those feeding on dispersed foods (Table 1, II).

In contrast to Buskirk's (1976) findings, Guatopo birds do not show any statistical tendency to join into interspecific groups based on food distribution categories (Table 1, III). This difference is based on two factors. In Guatopo there are 18 hummingbird species as opposed to 6 at Monteverde. The hummingbirds make up most of the species in the solitary, clumped food group. The group contributing the most to the significant X^2 in his data is the meagre two species representing interspecifically gregarious clumped resource feeders (two toucans, 2 percent of his avifauna). In contrast, Guatopo has 11 such species (5 percent of the avifauna). Also, Guatopo has nearly three times the number of interspecifically solitary species that feed on dispersed items than gregarious species that feed on dispersed items. Buskirk's species were divided nearly equally in these two groups. I believe the difference is due to differences in habitat diversity within our respective study sites. Buskirk's uniform forest habitat contrasts with my second growth and edge habitat. At Guatopo, large assemblages of insectivores that join mixed species flocks are not as prevalent as assemblages of omnivores that often join insectivores and exhibit both intra- and interspecific gregariousness (Table 2). This masked the statistical importance of strictly insectivorous interspecifically gregarious species that Buskirk reported and that are characteristic of the interior of mature tropical forest elsewhere (Morton 1973).

Another difference was the highly significant difference in overall flocking behavior (Table 1, IV) that I obtained whereas Buskirk found no significance. Since "overall flocking behavior" means that any species that is either intra-and/or interspecifically gregarious is classified in the gregarious heading, it is not intuitively obvious why his overall flocking category would not show significance when both intra- and interspecific categories do when analyzed separately. I am not willing to make the assumption, as Buskirk did, that terrestrial foragers, hummingbirds, and "sentinel" (alias sit and wait) foragers are less vulnerable to predators than his "active arboreal" foragers. Differences in food dispersion on the two-dimensional terrestrial substrate contrasted to the three-dimensional arboreal one may cause characteristic competitive differences. However, there may be a significant difference in the class of predators influencing the two substrates. Several genera of snakes (*Boa, Bothrops, Oxybelus*, etc.) wait in canopy foliage for unaware birds. Since ground foraging birds are not forced to fly from point to point this predator technique is not evident for them. Thus, instead of trying to fit solitary terrestrially foraging birds into the idea that these are less vulnerable to predators, therefore they do not flock, may be unrealistic. Hummingbirds are also vulnerable to predation (see references in Groves, 1978). This assumption is not needed to explain flocking/food-distribution correlations in the Venezuelan data. As Buskirk pointed out, flocking behavior is not derived from a unitary selective force and to argue that either predation or enhanced food finding is responsible for gregarious behavior is like arguing that behavior is either innate or learned.

Table 1. Food distribution and social organizations of birds at Guatopo National Park

	Food distribution			X^2	*P*<
	Dispersed	*Inter-mediate*	*Clumped*		
Intraspecific organization					
Solitary	73[1] (57.83)	15 (23.33)	16 (22.84)	17.51	.001
Gregarious	46 (61.17)	33 (24.67)	31 (24.16)		
Intraspecific group size					
2	34 (23.20)	17 (17.65)	7 (17.15)		
3–5	9 (17.00)	13 (12.93)	20.5 (12.57)	22.81	.001
6+	3 (5.80)	5 (4.41)	6.5 (4.29)		
Interspecific organization					
Solitary	89.5	29.5	38	3.81	N.S.
Gregarious	30.5	18.5	11		
Overall flocking organization					
Solitary	60.5 (47.88)	10 (19.31)	16 (19.31)	14.03	.001
Gregarious	58.5 (71.12)	38 (28.69)	32 (28.69)		

[1] Number of species.

Numbers in parentheses equal expected value.

Table 2. Species associated in Guatopo mixed species flocks

Species	% Occurrence	Diet[2]
Insectivorous/Omnivorous flocks (N = 9)		
Hylophilus aurantiifrons	100	I/F
Myiopagis gaimardii[1]	78	O
Leptopogon superciliaris[1]	66	I
Setophaga americana	56	I
Picumnus squamulatus	44	I
Xenops minutus[1]	44	I
Hemithraupis guira[1]	44	F/I
Sittasomus griseicapillus	33	I
Campylorhamphus trochilirostris	33	I
Tolmomyias flaviventris[1]	33	I/F
Ornithion semiflavum	33	O
Ramphocaenus melanurus	33	I
Cyclarhis gujanensis	33	F/I
Parula pitiayumi	33	I
Basileuterus culicivorous[1]	33	I
Tangara gyrola[1]	33	F/I
Tachyphonus rufus	33	F/I
Ramphocelus carbo	33	F/I
Lepidocolaptes souleyetii	22	I
Herpsilochmus rufimarginatus	22	I?
Pachyramphus polychopterus	22	I/F
Myiarchus tuberculifer	22	I
Todirostrum cinereum	22	I
Polioptila plumbea	22	I
Coereba flaveola	22	N/O
Myioborus minatus[1]	22	I
Dendrocincla fuliginosa	11	I
Dysithamnus mentalis[1]	11	I
Mionectes olivaceus[1]	11	F
Tyrannulus elatus	11	F/I
Icterus nigrogularis	11	F/I
Dendroica cerulea[1]	11	I
Dendroica striata[1]	11	I
Vermivora peregrina[1]	11	N/O
Euphonia xanthogaster	11	F
Piranga rubra	11	O
Thraupis episcopus	11	O
Thraupis palmarium	11	O
Chlorophanes spiza[1]	11	F/N
Tangara arthus[1]	11	F/I
Saltator maximus	11	O
Saltator coerulescens	11	O
Frugivorous/Nectarivorous flocks (N = 3)		
Tachyphonus rufus	100	F/O
Ramphocelus carbo	100	F/O
Thraupis episcopus	100	O
Dacnis cayana[1]	66	F/N
Chlorophanes spiza[1]	66	F/N
Cyanerpes caeruleus[1]	66	F/N
Turdus leucomelas	33	F/I
Icterus nigrogularis	33	F/I
Euphonia laniirostris	33	F
Euphonia xanthogaster	33	F
Tangara gyrola[1]	33	F/I
Tangara cyanoptera[1]	33	F/I
Saltator maximus	33	O
Saltator coerulescens	33	O

[1] Observed only in mixed species flocks.

[2] F = frugivorous; N = nectarivorous; I = insectivorous; O = omnivorous

Guatopo Avian Social Organization Compared with Masaguaral and Guarico

The llanos of Masaguaral and the nearby gallery forest share disproportionate numbers of species with Guatopo. The habitats are described in detail in Troth, 1979. The llanos are characterized by extremely wet conditions from June through October and dry, almost desert conditions the remainder of the year. Breeding seems nearly confined to the beginning of the wet season in May and June for most species. *Camptostoma obsoletum* is an outstanding exception for its ball-like nest is hidden in the dried heads of a tall (1–2 m) mint (*Hyptis* sp.). The nest contains pieces of the *Hyptis* dried flowers embedded in cotton-like material and would be neither cryptic nor waterproof if breeding took place in the wet season.

The bird species list (Appendix 2) is taken from Betsy Thomas's preceding chapter but includes only those species she lists as uncommon through abundant. Gallery forest birds are listed in Appendix 3, which includes rare to abundant species. Masaguaral shares 49 of its 142 common species with Guatopo (18 percent). Even though the Guarico gallery forest is contiguous with Masaguaral, bird species restricted to the gallery forest are more like those in Guatopo than those in the llanos proper. Forty-five (49 percent) of the 92 gallery species are shared with Guatopo.

Masaguaral shows the same trend in intraspecific social organization as Guatopo. Species that feed on clumped food tend to be gregarious intraspecifically

(Table 3, I) while those feeding on dispersed foods are nearly equally split between intraspecifically gregarious and solitary species. Guatopo has about twice as many solitary species as intraspecifically gregarious species feeding on dispersed food. As in Guatopo, there is no relation between interspecific organization and food distribution (Table 3, III). One hundred twenty-six (92 percent) of the species are interspecifically solitary, only 11 species (8 percent) are prone to join other species.

It is significant that no species using clumped resources join other species in foraging for them. The relationship between food distribution and social organization is again present when overall flocking behavior is considered (Table 3, IV).

The gallery forest (Guarico) again shows a significant correlation between food distribution and intraspecific social organization but there is no correlation in interspecific or overall flocking organization with food distribution (Table 4). More species (22 percent vs. 8 percent) at Guarico join mixed flocks than at Masaguaral. Guarico and Guatopo are about the same with 28 percent of the Guatopo species joining others in organized mixed species flocks. Even with the 45 species shared in common removed from the comparison, the remaining Guarico species do not differ from the Guatopo assemblage (X^2 = .30, N.S.) in intraspecific organization. However, with these species removed from the Guarico list the two sites differ significantly in interspecific organization ($X^2 = 8.13$, $p < .01$). The species shared, therefore, are those that participate in mixed species flocks in both sites and are chiefly responsible for the similarity in mixed species flocks in both areas.

If we now compare the proportions of species in the three habitats that are either gregarious or solitary without regard to food distribution, more significant differences arise. While there is no difference in the tendency for conspecifics to join one another (Table 5, I), the size of these intraspecific groups is quite different (Table 5, II). Masaguaral has significantly more species that join mixed species flocks than has Masaguaral (Table 5, III).

Comparing Guatopo with Guarico we find no differences in either intraspecific, interspecific, or group size comparisons (Table 6). Comparing Masaguaral with Guarico we find the same relation as the Guatopo/Masaguaral comparison: no intraspecific difference, but more large flocks of conspecifics and fewer mixed species flock joiners at Masaguaral (Table 7).

I will carry this descriptive analysis one step further and compare the now-familar categories for flocking tendencies between Buskirk's Monteverde forest birds and Guatopo forest species (Table 8). In all comparisons except intraspecific flock-size distribution, the two sites differ significantly. Guatopo has relatively more intraspecific flocking species, more overall flocking behavior in its species, and less mixed species flock joiners. Of the 98 Monteverde species, 22 percent are also listed among the 227 species at Guatopo and 61 percent of the genera at Monteverde are also among the Guatopo genera. If I remove hummingbirds from consideration, since they represent the greatest taxonomic difference in the two faunas, Guatopo and Monteverde do not differ in intraspecifically gregarious and solitary species ($X^2 = 0.74$, N.S.). However, they do still differ significantly in interspecific social interactions ($X^2 = 7.80$, $p < .01$) and in overall flocking behavior ($X^2 = 6.59$, $p < .02$). Buskirk (1976) suggests that interspecific flocking occurs most often in species that feed on dispersed resources, have low intraspecific group sizes, and are relatively vulnerable to predation (if they did not join flocks). He reached this conclusion after removing hummingbirds, sit-and-wait predators, and terrestrial foragers from consideration. When this is done for the Guatopo avifauna, I

Table 3. Food distribution and social organization of birds at Masaguaral

	Food distribution			X^2	*P<*
	Dispersed	*Intermediate*	*Clumped*		
Intraspecific organization					
Solitary	53.5 (48.79)	12 (11.83)	3 (7.88)	6.85	.05
Gregarious	45.5 (50.21)	12 (12.17)	13 (8.12)		
Intraspecific group size[1]					
2	13.5	6	3		
3–5	13.5	3	1	1.25	N.S.
6+	18.5	4	6		
Interspecific organization[1]					
Solitary	87	20	19	.14	N.S.
Gregarious	7	4	0		
Overall flocking organization					
Solitary	55 (47.45)	10 (11.25)	3 (4.29)	10.95	.01
Gregarious	42 (49.55)	13 (11.75)	16 (9.71)		

[1] Intermediates combined equally with Dispersed and Clumped for X^2.
Numbers in parentheses equal expected value.

Table 4. Food distribution and social organization of birds at Guarico (Gallery Forest)

	Food distribution			X^2	$P<$
	Dispersed	*Intermediate*	*Clumped*		
Intraspecific organization					
Solitary	35.5 (30.71)	6 (6.83)	4 (7.96)	6.54	.05
Gregarious	18.5 (23.29)	6 (5.18)	10 (6.04)		
Intraspecific group size[1]					
2	15	3	5		
3–5	2.5	1	4	3.49	N.S.
6+	1	0	2		
Interspecific organization					
Solitary	43.5	6	13		N.S.
Gregarious	9.5	7	1		
Overall flocking organization					
Solitary	29	4	4	4.70	N.S.
Gregarious	25	10	10		

[1] Intermediates combined with Dispersed and Clumped for X^2.
Numbers in parentheses equal expected value.

Table 6. Comparisons of social interactions among species in Guatopo and Guarico

	Solitary		*Gregarious*	X^2	$P<$
Intraspecific social interactions					
Guatopo	104		110	1.59	N.S.
Guarico	45.5		34.5		
Intraspecific group size	2	3–5	6+		
Guatopo	58	42.5	14.5	3.52	N.S.
Guarico	23	7.5	3		
Interspecific social interactions					
Guatopo	157		60	1.01	N.S.
Guarico	62.5		17.5		

Table 5. Comparisons of social interactions among species in Guatopo and Masaguaral

	Solitary	*Gregarious*		X^2	$P<$
Intraspecific social interactions					
Guatopo	104	110		.014	W.S.
Masaguaral	68.5	70.5			
Intraspecific group size	2	3–5	6+		
Guatopo	58 (50.45)	42.5 (37.60)	14.5 (26.95)	20.15	.001
Masaguaral	22.5 (30.05)	17.5 (22.40)	28.5 (16.05)		
Interspecific social interactions					
Guatopo	157 (173.48)	60 (43.52)		20.17	.001
Masaguaral	126 (109.52)	11 (27.48)			

Numbers in parentheses equal expected value.

Table 7. Comparisons of social interactions among species in Guarico and Masaguaral

	Solitary		*Gregarious*	X^2	P<
Intraspecific social interactions					
Guarico	45.5		34.5	1.18	N.S.
Masaguaral	68.5		70.5		
Intraspecific group size	2	3–5	6+		
Guarico	23 (14.94)	7.5 (8.21)	3 (10.35)	14.34	.001
Masaguaral	22.5 (30.56)	17.5 (16.79)	28.5 (21.15)		
Interspecific social interactions					
Guarico	62.5 (69.49)		17.5 (10.51)	8.48	.010
Masaguaral	126 (119.01)		11 (17.99)		

Numbers in parentheses equal expected value.

Table 8. Comparison of social interactions among species in Guatopo and Monteverde, Costa Rica[1]

	Solitary		*Gregarious*	X^2	P<
Intraspecific social interactions					
Monteverde	64 (52.4)		33 (44.60)	8.12	.01
Guatopo	104 (115.6)		110 (98.40)		
Intraspecific flock size	2	3–5	6+		
Monteverde	13	17	8	3.44	N.S.
Guatopo	58	42.5	14.5		
Interspecific social interactions					
Monteverde	53 (63.94)		42 (31.06)	8.23	.01
Guatopo	157 (146.06)		60 (70.94)		
Overall flocking behavior					
Monteverde	39 (30.16)		29 (37.84)	6.13	.02
Guatopo	86.5 (95.34)		128.5 (119.66)		

[1] Data for Monteverde from Buskirk (1976).
Numbers in parentheses equal expected value.

still get a significant relationship between food distribution and social tendency on the intraspecific level ($X^2 = 12.19$, $p < .01$) but still no significant relationship on the interspecific level ($X^2 = 2.16$, $p < .50$). That is, my data agree completely on the relation between intraspecific flocking, dispersed resources, and low intraspecific group sizes, but do not show a relation as well with "low predation vulnerability". Unlike Buskirk's avifauna, most of the interspecifically gregarious species are intraspecifically gregarious as well. This may be due to relatively more rare species in Buskirk's mature forest study area or the mixed-edge, second-growth forest habitat at Guatopo. Since he lists 42 percent of his species as rare, many of them might be stragglers or dispersing individuals that are intraspecifically gregarious, at least in bonded pairs, where they are more common.

Summary and Significance of Social Structure and Food Distribution

Guatopo and Masaguaral

Both habitats show a significant relationship between food distribution and intraspecific gregariousness. More species join with conspecifics when exploiting clumped food while more are solitary if exploiting dispersed food. This matches previous arguments (Morton 1973) that the finding of clumped foods is facilitated by flocking and its abundance once found reduces competition for it. Intraspecific group sizes also were larger when clumped food was used in Guatopo but there was no such correlation at Masaguaral. Surprisingly, overall proportions of intraspecifically gregarious and solitary species are identical in both places (51 percent are gregarious). In Guatopo, most of these intraspecifically "gregarious" species are found in groups of two, whereas in Masaguaral most occur in large groups (Table 5, II). It is this last difference which is so obvious to the eye, with large flocks of doves, finches, blackbirds, and parrotlets so conspicuous there.

I found no significant relationship between interspecific gregariousness and food distribution within either habitat. The proportion of species that join mixed flocks does differ between habitats, with 28 percent of the Guatopos species joining others while only 8 percent of Masaguaral species join others. Significantly, at Masaguaral there are no species using clumped foods that are interspecifically gregarious while at Guatopo there are 11.

Why do large conspecific flocks occur at Masaguaral independent of food distribution while Guatopo birds join mixed species flocks? Why in Guatopo are those that do form large conspecific flocks tied to clumped foods? How does this relate to selection pressures acting on flocks? The basic difference, I suggest, is related to food-type differences. In Guatopo, fruit and insects constitute the general food base for birds throughout the year while in the llanos, fruit and seeds constitute the food base during the non-breeding season while fruit and insects form the breeding season food base. By "base" I mean that these are the primary foods used by the avian biomass taken as a whole. Most of the seed crop at Masaguaral is taken by birds feeding on the ground in a visually open habitat. Seed crops are widely dispersed and the food is not directly visible. In this sense the substrate and visibility of the food at a distance is similar to that of terrestrial forest insectivores. Unlike insect food, however, the best predictor of where seed is is the presence of other birds feeding upon the ground. Seed crops are used up while insect populations replenish themselves, at least in forest ground litter (A. S. Rand and Sally Levings, pers. comm. on data from Barro Colorado Island, Canal Zone). Variable degrees of defense of the terrestrial forest substrate are characteristic, ranging from complete exclusion of conspecifics (e.g., *Formicarius analis*) to home range-related changes in dominance at ant swarms (e.g., *Gymnopithys bicolor*, Willis, 1967), to kin group joint defense (e.g., *Cyporhinus arada*, Morton 1978). These differences correlate with foraging techniques, population densities, and the use of ants as concentrators of insect food. The seedeating terrestrial llanos birds do not defend seed resources because of the ephemeral nature of seeds, the use of already-feeding conspecifics to locate rich feeding areas (which is facilitated by the visually unobstructed habitat), and the large biomass of each species supported by seeds(primary productivity). Seedeating llanos birds show little dominance interaction when in flocks, instead selection has favored coordinated group take-off and tight flight flocking as an anti-hawk behavior. They are classic "selfish herds" (Hamilton, 1971).

Avian Adaptations to Dry Season Conditions at Masaguaral

The contrast in avian flock structure and in daily activity between the llanos and a forest like Guatopo is striking. Large flocks of *Aratinga pertinax* and *Brotogeris jugularis* (arboreal fruit and seedeaters), and *Forpus passerinus* (a seedeater) are in nearly continuous evidence. *Forpus* is found in greater numbers with flocks of 100 not unusual. It is also the most general in its feeding substrate being common on the ground along with doves and *Sicalis* finches, and in fruiting trees. It is also the only species that feeds on the abundant seed

Figure 1. Map of Masaguaral indicating transect route and survey stations (see text).

heads of mints such as the 1–3 m *Hyptis* sp. although other species feed on these seeds after they are scattered to the ground.

A quantitative estimation of bird abundance is shown in a census conducted on 26 November 1975 at six locations ½ km apart (Figure 1). At the census date, only station one was near water, and this was artificially due to pumping from a well. The list in Table 9 shows the results with the average number of each species per station giving an index to abundance; the number of stations where a species was recorded gives an index to ubiquitousness. The importance of grass and other seeds to the llanos avifauna is apparent. Seedeaters occurred at an average 3.89 stations per species with an average 6.33 individuals seen per station; all other species averaged 2.92 stations and only 1.37 individuals per station. The discrepancy would be even greater if two *Sporophila* species, which are at low population sizes in the dry season, were eliminated from the seedeater category and the Barn Swallow, *Hirundo rustica*, a migrating aerial insect eater eliminated from the "other" category.

The seedeating species illustrate different adaptations for exploiting this resource, even though *Forpus*, *Sicalis*, and *Scardafella* often forage together on the ground although in single species groups. *Scardafella* forage singly but more often (84 percent of observations) in small groups of 2 to 12 (ave. 5.2, N = 15). *Scardafella* and *Zenaida* are the only truly llanos-restricted doves, the others being found near humid forest edge as well. *Scardafella* literally digs for seeds, throwing dry soil about 10 cm at a rate of .68 tosses per second. The bird thus is able to find surface seeds as well as those up to about 1 cm under the surface. There was no significant difference in pecking rate in birds feeding alone versus those in flocks. *Sicalis* concentrate on surface seeds and move frequently to new areas as these are exhausted whereas *Scardafella* rework the same areas. *Forpus* forage on the ground to a lesser extent than either, favoring the dried heads of various mints, especially *Hyptis* that grow to 2 m. *Hyptis* seeds are ca 1 mm-diameter winged seeds that remain persistently in the seed heads. *Forpus* climb about these heads and rip them open, dumping many more seeds onto the ground below than they consume. There, other *Forpus* and *Sicalis*, associating with crown-foraging *Forpus*, glean this concentration of seed.

Forpus show a decided schedule of activity. John

Table 9. **Results of a census taken 26 November 1975 at six stations as shown in Fig. 1**

Species	*Number stations (of 6)*	*Average number individuals per 6 stations*	*Food habits*[1]
Ardea cocoi	1	0.33	—
Casmerodius albus	4	3.50	—
Butorides striatus	1	0.33	—
Mycteria americana	3	1.83	—
Euxenura maguari	6	2.33	—
Jabiru mycteria	1	0.17	—
Cercibis oxycerca	1	0.50	—
Plegadis falcinellus	1	0.17	—
Dendrocygna viduata	1	2.50	—
D. autumnalis	1	4.17	—
Coragyps atratus	5	1.50	—
Cathartes aura	6	4.33	—
Elanus leucurus	3	0.83	—
Helicolestes hamatus	1	0.17	—
Heterospizias meridionalis	2	0.50	—
Geranospiza caerulescens	1	0.33	—
Milvago chimachima	1	0.17	—
Polyborus plancus	3	1.00	—
Tringa solitaria	2	0.50	—
Zenaida auriculata	6	9.83	S
Columbina minuta	1	0.33	S
Columbina talpacoti	6	6.83	S
Scardafella squammata	6	11.83	S
Leptotila verreauxi	5	2.50	S
Aratinga pertinax	4	2.17	F/S
Forpus passerinus	5	6.50	S/F
Brotogeris jugularis	4	1.33	F/S
Crotophaga ani	4	0.83	I/F
Amazilia fimbriata	1	0.33	Nectar
Hypnelus ruficollis	3	0.50	I[2]
Melanerpes rubricapillus	2	0.33	F/I
Xiphorhynchus picus	2	0.50	I
Phacellodomus rufifrons	6	2.67	I
Fluvicola pica	1	0.17	I
Pyrocephalus rubinus	1	0.33	I/F
Tyrannus melancholicus	2	0.33	F/I
Myiozetetes cayanensis	4	1.50	F/I
Pitangus sulphuratus	5	1.17	O
Myiarchus tyrannulus	2	0.33	I/F
Todirostrum cinereum	3	0.83	I
Hirundo rustica	5	10.67	I
Campylorhynchus griseus	5	1.50	I
Campylorhynchus nuchalis	4	2.17	I
Troglodytes aedon	1	0.17	I
Mimus gilvus	5	2.17	F/I
Cyclarhis gujanensis	1	0.17	F/I
Quiscalis lugubris	1	0.33	S/I
Agelaius icterocephalus	1	0.67	S/I
Icteris nigrogularis	4	1.00	F/I
Thraupis episcopis	6	2.50	F/I
Saltator coerulescens	3	1.00	O
Sporophila intermedia	1	0.17	S
Sporophila minuta	1	0.17	S
Sicalis flaveola	4	8.83	S

[1] Food habits for non-aquatic species only. S = seeds, I = insects, F = fruit, O = omnivorous.

[2] Includes small vertebrates.

Eisenberg and I made five censuses of Forpus activity at two-hour intervals beginning at 0845. The trail was ca 1 km in length, beginning at a waterhole at one end and ending at large *Hyptis* plots. Each transit took one hr. *Forpus* concentrated on *Hyptis* feeding from 0730 until 1100. From 1100 to 1500 hrs they concentrated at the water hole, bathing and drinking, and in large *Ficus* trees where some fruit was taken but most time was spent preening and resting. At 1530 many flocks of four to five Forpus flew back toward the direction of the *Hyptis* concentration, where ca 50 were already feeding. From 1100 to 1300 hrs. they were completely absent from this area. At 1810 many began roosting in a densely leafed tree, isolated from other trees, and at 1820 I flushed ca 200 from this tree. They returned and spent the night there.

The utilization of fruit by llanos birds is extensive with *Ficus* sp. predominating in importance during the dry season. Table 10 lists bird species actually observed feeding on *Ficus* fruit at three trees within a 15-minute observation time. Several of these species, notably *Crotophaga ani* and *Machetornis*, were not known as fruit eaters before these observations. Unlike many forest frugivores, who travel in mixed species flocks, the llanos fruit eaters moved independently as single birds or in small to medium sized groups of conspecifics.

Table 10. Avian species feeding on *Ficus trigonata* listed in order of abundance, during 15-minute observation periods (N = 5)

Species	*Sizes of groups*
Thraupis episcopis	5–7
Mimus gilvus	2–4
Pitangus sulphuratus	1–2
Turdus nudigenis	4–5
Forpus passerinus	4–20
Brotogeris jugularis	2–6
Aratinga pertinax	2–6
Cacicus cela	1–8
Myiozetetes cayanensis	2
M. similis	2
Tityra inquisitor	2
Icterus icterus	2
Saltator coerulescens	2–4
Megarhynchus pitangua	2
Saltator orenocensis	2–3
Melanerpes rubricapillus	1
Columba cayennensis	1–2
Tyrannus melancholicus	1
T. dominicensis	2
Crotophaga ani	3–6
Machetornis rixosus	1
Icterus chrysater	1–2

The llanos fig tree fruits were also morphologically different from bird dispersed forest *Ficus* that I am familiar with in Panama. The Panama bird-dispersed figs have synchronous ripening involving a distinctive color change from green to yellow to reddish. In contrast, it is difficult to identify ripe fruit in the llanos figs, the final color being greenish-yellow with numerous small brown-red blotches. This may be an adaptation to slow down the birds' foraging, keeping birds in the tree for a longer time. In this way, the birds themselves serve as attractants to others rather than the conspicuous color of the fruit. This difference in llanos *Ficus* "strategy" may be due to the density and concurrent competition between trees to attract seed dispersers and possibly to the lack of organized mixed species bird flocks searching for fruit in the llanos.

A final contrast between llanos and forest birds, and one that deserves further study, is the strength of the mobbing response to predators, particularly snakes. The frequency with which I encountered birds mobbing snakes was high, about once per day within hearing distance of the dormitory at Masaguaral. Several of these were observed from the beginning of the interaction. Once a pair of *Campylorhynchus griseus* discovered a *Boa* on the ground. They uttered harsh, rapid barking sounds while perching 1–2 m above the snake. Within five seconds, two *Mimus*, four to five *Sicalis*, three *Thraupis episcopis*, one *Melanerpes*, and one female *Molothrus* arrived, followed within one minute by a pair of *Myiozetetes cayennensis* and *Pitangus sulphuratus*. The mobbing persisted for ca ten minutes until the snake disappeared down a hole. All other mobbing occurrances were of snakes or the small owl, *Glaucidium brasilianum*, in tree crowns. Playbacks of recorded mobbing and "squeaks" mimicking a bird in distress evoke immediate and prolonged mobbing in many species, some also having forest-edge populations elsewhere which do not mob. This strong and rapid response may be due to the high density and even distribution of some species. However, I believe the isolation of each small group of trees from one another, characteristic of the llanos, makes the location of a snake more certain and the avoidance of a particular tree crown more effective. The birds are able to avoid predators here better than in areas with continuous canopy. Thus, mobbing is more "effective" in the llanos and more species respond.

The Avifaunal Composition of Guatopo

One of the most interesting and significant aspects of the avifauna of Guatopo is the lack of certain groups of birds there. Guatopo is isolated from the species pool of what is generally termed "amazonas" by the seasonally arid llanos to the south and west, by the Andes to the west, and narrow forest strips between mountains and the Rio Orinoco to the east (now no forest exists). In absolute numbers of species, Guatopo is depauperate compared with a truly amazonian area. Guatopo's 227 resident species can be compared with 269 at La Selva, Costa Rica (Slud 1960) and 259 on Barro Colorado Island (Eisenmann 1952). In addition, the proportions of birds differ. At Guatopo oscine passerine species outnumber suboscines (proportion of suboscines to oscines equal 0.85) whereas at La Selva and Barro Colorado Island the ratio of suboscines to oscines is higher (1.22 and 1.27, respectively). There are no motmots (Momotidae) in Guatopo, nor are there barbets (Capitonidae) and the ubiquitous forest genera, *Thamnophilus* and *Myrmeciza*, are conspicuously absent. At Guatopo only the scrub-inhabiting *Thamnophilus doliatus* and *Myrmeciza longipes* are found, not their usually ubiquitous forest counterparts such as *T. punctatus* and *M. exsul.* On the other hand, there are ample illustrations that the oscines have taken on the niches usually occupied by these forest suboscines.

There are no ant-following antbirds in Guatopo, yet there are swarms of *Eciton* which are followed by birds. At two swarms I was able to study intensively, *Turdis fumigatus* (6), *Turdus leucomelas* (2), *Cyanocompsa cyanoides* (4) had taken on antbird-like roles at the swarm. Larger than normal numbers of a normal ant follower (ca. 15), the ant-tanger *Eucometis penicillata*, were there as well as the woodcreeper *Dendrocincla fuliginosa*, also a common ant follower. The wren, *Microcerculus marginatus* (2), was observed along with these species at another ant swarm along trail 1, in mature forest. Other woodcreepers such as *Xiphorhynchus guttatus* and *Sittasomus griseus* and possibly *Lepidocolaptes souleyetii* associated with these ant swarms. In the absence of "professional" ant following birds (Willis 1966a), the niche has been filled by oscine passerines and woodcreepers.

Table 11. Passerine families showing the number of widely distributed resident venezuelan species with habitat in Guatopo that are or could be there (see text).

Family	*No. species with habitat in Guatopo*	*No. species found in Guatopo*	*General food type*[1]
Suboscines			
Dendrocolaptidae	20	6 (30%)	I
Furnariidae	24	6 (25%)	I
Formicariidae	65	10 (15%)	I
Cotingidae	26	8 (31%)	O
Pipridae	22	4 (18%)	F
Tyrannidae	86	30 (35%)	O
Oscines[2]			
Troglodytidae	10	6 (60%)	I
Turdidae	13	5 (38%)	O
Vireonidae	15	4 (27%)	O
Icteridae	14	7 (50%)	O
Parulidae[2]	10	6 (60%)	I
Thraupidae	52	23 (44%)	O
Fringillidae	30	16 (53%)	O

[1] I = insectivorous, O = omnivorous, F = frugivorous.

[2] Corvidae and Hirundinidae, with only 5 and 4 possible species, are not included.

Numbers in parentheses = percent of possible species actually occurring in Guatopo.

The prevalence of oscines in Guatopo is highlighted by a comparison of the ability, of potential colonists to reach Guatopo. *A Guide to the Birds of Venezuela* (de Schauensee and Phelps 1978) was surveyed for species that have habitat in Guatopo and are widely distributed elsewhere in Venezuela. A comparison of the percentage of potential species versus those actually present in Guatopo is found in Table 11. This table shows the significant difference ($X^2 = 27.67$, $p < .001$) between oscine and non-oscine passerines in potential versus actual occurrence in Guatopo. Whether this represents a difference in dispersal ability or the ability to persist after colonization is not known, but I believe it is likely to be dispersal ability.

The types of food eaten and the social consequences of this may be a large factor in this difference. The avian families, whether suboscine or oscine, that contain the most omnivorous species, are those that are more numerous in the Guatopo "island" of forest ($X^2 = 4.23$, $p < .05$). Omnivorous species may be better able to traverse areas, such as the llanos, that have abundant fruit supplies at certain times even though these forest omnivores are not able to reproduce in non-forest areas. Additionally, once they reach Guatopo omnivores have an advantageous breeding energy budget strategy: they are able to spend more time than strict insectivores looking for insect food for their young since as adults they may feed themselves quickly on easily obtained fruit (Morton 1973).

In contrast, insectivores characteristically have permanent territories while omnivores are less territorial or are territorial only while breeding. This in itself may reduce the mobility of insectivores. There is some evidence of this in the presence of some high cloud forest species in Guatopo such as *Henicorhina leucophrys* occurring without *H. leucosticta*, its lowland counterpart, and *Chamaeza campanisona*. Perhaps high-elevation species are more able to reach mountain tops because they do not encounter territorial conspecifics blocking them along the way. I agree with Amadon (1973) that the oscines, or more specifically the omnivorous oscines, may be replacing sub-oscines in the neotropics but not from the standpoint that sub-oscines are "inferior" in competition with oscines. The argument by Slud (1960) and Willis (1966b) that the sub-oscines are equal competitors with oscines in the forest interior is probably true but perhaps not really cogent to the question of the gradual replacement of them by oscines. With the expansion and contraction of forests during climatic changes (Haffer 1974) it is the ability to disperse that may ultimately predict what groups remain. The oscines "have it."

Acknowledgments

Dr. Chris Parrish, Departamento de Mathematicas, Universidad Simon Bolivar, enthusiastically developed a list of birds species for Guatopo National Park based on several years' worth of trips there. This field work verified the species I observed during the meagre 64 field days I spent in Venezuela, only half of which were in Guatopo, and allowed me to concentrate on behavioral observations. Mrs. Betsy Trent Thomas has provided an excellent annotated list of the birds at Masaguaral and the gallery forest as a separate chapter in this volume. Mrs. Glenda Quintero began an extensive mist netting operation of one trail in Guatopo and added several species to the list. Dr. John Eisenberg provided the guiding force to initiate this and other studies of northern Venezuelan fauna together with Dr. Edgardo Mondolfi. Dr. Rudi Rudran provided continuous field coordination. Dr. Tomas Blohm kindly permitted the use of his ranch and facilities at Masaguaral. Mr. Luis Escalona provided quarters at Guatopo. I thank all of them for providing the opportunity for this study. To them and to Peggy O'Connell, Guy Greenwell, Peter and Lynn August, Ranji Rudran, Pedro Quintero, and Ken Green I add my thanks for companionship in the field.

Literature Cited

Amadon, D.
1973. Birds of the Congo and Amazon forests: A comparison. Pages 267–278 in *Tropical Forest Ecosystems in Africa and South America: A Comparative Review*, edited by B. J. Meggers, E. S. Ayensu, and W. D. Duckworth. Washington, D. C.: Smithsonian Institution Press.

Buskirk, W. H.
1976. Social systems in a tropical forest avifauna. *Am. Nat.*, 110:293–310.

Eisenmann, E.
1952. Annotated list of birds of Barro Colorado Island, Panama Canal Zone *Smithsonian Misc. Colls.*, 117:1–62.

Eisenberg, J. F., M. A. O'Connell, and R. V. August
1979. Density, productivity, and distribution of mammals in two Venezuelan habitats. Pages 187–207 in *Vertebrate Ecology in the Northern Neotropics*, edited by John F. Eisenberg, Washington, D.C.: Smithsonian Institution Press.

Graves, G. R.
1978. Predation on hummingbird by Oropendola. *Condor*, 80:251.

Haffer, J.
1974. Avian speciation in tropical South America. *Publ. Nutall Ornithol. Club*, 14:390.

Hamilton, W. D.
1971. Geometry for the selfish herd. *J. Theor. Biol.*, 31:295–311.

Morse, D. H.
1977. Feeding behavior and predator avoidance in heterospecific groups. *Bioscience*, 27:332–339.

Morton, E. S.
1977. Intratropical migration in the yellow-green vireo and piratic flycatcher. *Auk*, 94:97–106.
1978. Reintroducing recently extirpated birds into a tropical forest preserve. Pages 379–384 in *Endangered Birds: Management Techniques for Preserving Threatened Species*, edited by S. A. Temple. Madison: University of Wisconsin Press.

Moynihan, M. H.
1962. The organization and probable evolution of some mixed species flocks of neotropical birds. *Smithsonian Misc. Colls.*, 143:1–140.

O'Connell, M. A.
1979. Ecology of didelphid marsupials in Northern Venezuela. Pages 73–87 in *Vertebrate Ecology* in the Northern Neotropics, edited by John F. Eisenberg. Washington, D.C.: Smithsonian Institution Press.

Schauensee, R. M. de, and W. H. Phelps, Jr.
1978. *A Guide to the Birds of Venezuela*. Princeton: Princeton University Press.

Slud, P.
1960. The birds of Finca "La Selva," Costa Rica: A tropical wet forest locality. *Bull. Amer. Mus. Nat. Hist.*, 121:48–148.

Thomas, B. T.
1979. Birds of a ranch in the llanos of Venezuela. Pages 213–232 in *Vertebrate Ecology in the Northern Neotropics*, edited by John F. Eisenberg. Washington, D.C.: Smithsonian Institution Press.

Troth, R.
1979. Vegetational types on a ranch in the central llanos of Venezuela. Pages 17–30 in *Vertebrate Ecology in the Northern Neotropics*, edited by John F. Eisenberg. Washington, D.C.: Smithsonian Institution Press.

Willis, E. O.
1966a. Interspecific competition and the foraging behavior of plain-brown woodcreepers. *Ecology*, 47:667–672.
1966b. The role of migrant birds at swarms of army ants. *The Living Bird*, 5:187–231.
1967. The behavior of bicolored antbirds. Univ. Cal. Publ. Zool., 79:1–132.
1972. Do birds flock in Hawaii, a land without predators? *California Birds*, 3:1–8.

Appendix 1. Guatopo Bird Species

	Species	Social Organization		Food distribution[3]
		Inter-specific[1]	Intra-specific[2]	
Tinamidae				
Gray tinamou	*Tinamus tao*	N	N	D
Little tinamou	*Crypturellus soui*	N	N	D
Cathartidae				
King vulture	*Sarcoramphus papa*	—	—	—
Black vulture	*Coragyps atratus*	—	—	—
Turkey vulture	*Cathartes aura*	—	—	—
Accipitridae				
Swallow-tailed kite	*Elanoides forficatus*	N	L	D
Gray-headed kite	*Leptodon cayanensis*	N	N	D
Hook-billed kite	*Chondrohierax uncinatus*	N	N	D
Double-toothed kite	*Harpagus bidentatus*	N	N	D
Plumbeous kite	*Ictinia plumbea*	N/F	S	D
Broad-winged hawk	*Buteo platypterus*	N	N	D
Roadside hawk	*B. magnirostris*	N	N	D
Zone-tailed hawk	*B. albonotatus*	N	N	D
Swainson's hawk	*B. swainsoni*	—	—	—
Short-tailed hawk	*B. brachyurus*	N	N	D
Gray hawk	*B. nitidus*	N	N	D
White hawk	*Leucopternis albicollis*	N	N	D
Common black hawk	*Buteogallus anthracinus*	N	N	D
Great black hawk	*B. urubitinga*	N	N	D
Black hawk-eagle	*Spizaetus tyrannus*	N	N	D
Solitary eagle	*Harpyhaliaetus solitarius*	N	N	D
Falconidae				
Yellow-headed caracara	*Milvago chimachima*	—	—	—
Collared forest-falcon	*Micrastur semitorquatus*	N	N	D
Bat falcon	*Falco rufigularis*	N	N	D
Cracidae				
Rufous-vented chachalaca	*Ortalis ruficauda*	N	L	C
Crested guan	*Penelope purpurascens*	N	S	C
Rallidae				
Gray-necked wood-rail	*Aramides cajanea*	N	N	D
Columbidae				
Scaled pigeon	*Columba speciosa*	N	S	C
Ruddy pigeon	*C. subvinacea*	N	S	C
Blue ground-dove	*Claravis pretiosa*	N	P	?
Common ground-dove	*Columbigallina passerina*	N	S	D
Ruddy ground-dove	*C. talpacoti*	N	S	D

[1] Interspecific organization (N = none, solitary; F = joins mixed species flocks).
[2] Intraspecific organization (N = none, solitary; P = pair, 2 birds; S = small, 3–5 birds; L = large, 6 + birds).
[3] Food distribution (D = dispersed; C = clumped; I = intermediate).

	Species	Social Organization		Food distribution[3]
		Inter-specific[1]	Intra-specific[2]	
Gray-fronted dove	*Leptotila rufaxilla*	N	N	D
White-tipped dove	*L. verreauxi*	N	N	D
Ruddy quail-dove	*Geotrygon montana*	N	N	I
Psittacidae				
Military macaw	*Ara militaris*	N	L	C
Red and green macaw	*Ara chloroptera*	N	S	C
Brown-throated parakeet	*Aratinga pertinax*	N	S/L	C
Maroon-faced parakeet	*Pyrrhura leucotis*	N	L	C
Orange-chinned parakeet	*Brotogeris jugularis*	N	L/P	C
Blue-headed parrot	*Pionus menstruus*	N	L	C
Cuculidae				
Squirrel cuckoo	*Piaya cayana*	N	—	D
Pheasant cuckoo	*Dromococcyx phasianellus*	N	N	D
Groove-billed ani	*Crotophaga sulcirostris*	N	S	D
Strigidae				
Tropical screech owl	*Otus choliba*	N	N	D
Black and white owl	*Ciccaba nigrolineata*	N	N	D
Mottled owl	*C. virgata*	N	N	D
Caprimulgidae				
Common potoo	*Nyctibius griseus*	N	N	D
Pauraque	*Nyctidromus albicollis*	N	N	D
Apodidae				
White-collared swift	*Streptoprocne zonaris*	—	L	—
Gray-rumped swift	*Chaetura cinereiventris*	—	—	—
Vaux's swift	*Chaetura vauxi*	—	—	—
Short-tailed swift	*Chaetura brachyura*	—	L	—
White-tipped swift	*Aeronautes montivagus*	—	L	—
Lesser swallow-tailed swift	*Panyptila cayennensis*	N	L	D
Trochilidae				
Rufous-breasted hermit	*Glaucis hirsuta*	N	N	D
Long-tailed hermit	*Phaethornis superciliosus*	N	N	D
Sooty-capped hermit	*P. augusti*	N	N	D
Little hermit	*P. longuemareus*	N	N	D
White-necked jacobin	*Florisuga mellivora*	N	N	C
Brown violetear	*Colibri delphinae*	N	N	C
Sparkling violetear	*C. corascans*	N	N	C
Black-throated mango	*Anthracothorax nigricollis*	N	N	C
Ruby-topaz hummingbird	*Chrysolampis mosquitus*	N	N	C
Violet-headed hummingbird	*Klais guimeti*	N	N	C
Spangled coquette	*Lophornis stictolopha*	N	N	C
White-chested emerald	*Amazilia chionopectus* (?)	N	N	C
Glittering-throated emerald	*A. fimbriata*	N	N	C
Copper-rumped emerald	*A. tobaci*	N	N	C
Golden-tailed sapphire	*Chrysuronia oenone*	N	N	C

	Species	Social Organization		Food distribution[3]
		Inter-specific[1]	Intra-specific[2]	
White-vented plumeleteer	*Chalybura buffonii*	N	N	D
Violet-fronted brilliant	*Heliodoxa rubinoides*	N	N	D
Lazuline sabrewing	*Campylopterus falcatus*	N	N	D
Blue-chinned sapphire	*Chlorestes notatus*	N	N	C
Trogonidae				
Collared trogon	*Trogon collaris*	N	N	I
White-tailed trogon	*Trogon viridis*	N	N	I
Alcedinidae				
Green kingfisher	*Chloroceryle americana*	N	N	D
Galbulidae				
Rufous-tailed jacamar	*Galbula ruficauda*	N	N	D
Bucconidae				
White-necked puffbird	*Notharchus macrorhynchus*	N	N	D
Moustached puffbird	*Malacoptila mystacalis*	N	N	D
Ramphastidae				
Groove-billed toucanet	*Aulacorhynchus sulcatus*	N	S	C
Black-mandibled toucan	*Ramphastos ambiguus*	N	P	I
Picidae				
Scaled piculet	*Picumnus squamulatus*	F	N	D
Lineated woodpecker	*Dryocopus lineatus*	N	N	D
Red-crowned woodpecker	*Melanerpes rubricapillus*	N	N	D
Golden-olive woodpecker	*Piculus rubiginosus*	N	N	D
Red-rumped woodpecker	*Veniliornis kirkii*	F	P	D
Crimson-crested woodpecker	*Campephilus melanoleucos*	N	N	D
Dendrocolaptidae				
Olivaceous woodcreeper	*Sittasomus griseicapillus*	F	N	D
Plain-brown woodcreeper	*Dendrocincla fuliginosa*	F	N	D
Buff-throated woodcreeper	*Xiphorhynchus guttatus*	F	N	D
Stong-billed woodcreeper	*Xiphocolaptes promeropirhynchus*	F	P	D
Streaked-headed woodcreeper	*Lepidocolaptes souleyetii*	F	P	D
Red-billed scythebill	*Campylorhamphus trochilirostris*	F	N	D
Furnariidae				
Pale-breasted spinetail	*Synallaxis albescens*	N	N	D
Stripe-breasted spinetail	*S. cinnamomea*	N	P	D
Crested spinetail	*Cranioleuca subcristata*	F	P	D
Plain xenops	*Xenops minutus*	F	P	D
Gray-throated leafscraper	*Sclerurus albigularis*	N	P	D
Montane foliage-gleaner	*Anabacerthia striaticollis*	F	P	D
Formicariidae				
Great antshrike	*Taraba major*	N	P	D
Fasciated antshrike	*Cymbilaimus lineatus*	N	P	D
Barred antshrike	*Thamnophilus doliatus*	N	P	D
Plain antvireo	*Dysithamnus mentalis*	F	P	D
Slaty antwren	*Myrmotherula schisticolor*	F	P	D

	Species	Social Organization		Food distribution[3]
		Inter-specific[1]	Intra-specific[2]	
Rufous-winged antwren	*Herpsilochmus rufimarginatus*	F	P	D
Jet antbird	*Cercomacra nigricans*	N	P	D
White-bellied antbird	*Myrmeciza longipes*	N	P	D
Short-tailed antthrush	*Chamaeza campanisona*	N	N	D
Black-faced antthrush	*Formicarius analis*	N	N	D
Cotingidae				
Chestnut-crowned becard	*Pachyramphus castaneus*	F	N	D
Cinereous becard	*P. rufus*	F	N	D
White-winged becard	*P. polychopterus*	F	N	D
Black-tailed tityra	*Tityra cayana*	N	P	C
Masked tityra	*T. semifasciata*	N	P	C
Golden-breasted fruiteater	*Pipreola aureopectus*	F	N	C
Handsome fruiteater	*Pipreola formosa*	N	?	C
Pipridae				
Golden-headed manakin	*Pipra erythrocephala*	N	N	C
Lance-tailed manakin	*Chiroxiphia lanceolata*	N	S	C
Wire-tailed manakin	*Teleonema filicauda*	N	?	C
Thrush-like manakin	*Schiffornis turdinus*	N	N	I
Tyrannidae				
Black phoebe	*Sayornis nigricans*	N	N	D
Tropical kingbird	*Tyrannus melancholicus*	N	N	D
Variegated flycatcher	*Empidonomus varius*	F	N	I
Piratic flycatcher	*Legatus leucophaius*	N	N	C
Streaked flycatcher	*Myiodynastes maculatus*	N	N	I
Rusty-margined flycatcher	*Myiozetetes cayanensis*	N	P	I
Social flycatcher	*M. similis*	N	P	I
Great kiskadee	*Pitangus sulphuratus*	N	P	I
Boat-billed flycatcher	*Megarhynchus pitangua*	N	P	I
Olive-sided flycatcher	*Nuttallornis borealis*	N	N	D
Dusky-capped flycatcher	*Myiarchus tuberculifer*	N/F	N	D
Great-crested flycatcher	*M. crinitus*	N	N	I
Tropical pewee	*Contopus cinereus*	N	N	D
Cinnamon flycatcher	*Pyrrhomyias cinnamomea*	N	N	D
Bran-colored flycatcher	*Myiophobus fasciatus*	N	N	D
White-throated spadebill	*Platyrinchus mystaceus*	N	P	D
Yellow-olive flycatcher	*Tolmomyias sulphurescens*	F	N	D
Yellow-breasted flycatcher	*T. flaviventris*	F	N	D
Common tody-flycatcher	*Todirostrum cinereum*	N	P	D
Scale-crested pygmy-tyrant	*Lophotriccus pileatus*	N	N	D
Pale-eyed pygmy-tyrant	*Atalotriccus pilaris*	N	?	D
Paltry tyrannulet	*Tyranniscus vilissimus*	N	N	I
Yellow-bellied elaenia	*Elaenia flavogaster*	N	N	I
Forest elaenia	*Myiopagis gaimardii*	F	P	I
Scrub flycatcher	*Sublegatus modestus*	N	N	D
Yellow-crowned tyrannulet	*Tyrannulus elatus*	N	P	D
Yellow-bellied tyrannulet	*Ornithion semiflavum*	F	P	I
Slaty-capped flycatcher	*Leptogogon superciliaris*	F	P	D
Olive-striped flycatcher	*Mionectes olivaceus*	F	P	I

	Species	Social Organization		Food distribution[3]
		Inter-specific[1]	Intra-specific[2]	
Ochre-bellied flycatcher	*Pipromorpha oleaginea*	N	N	C
Southern beardless tyrannulet	*Camptostoma obsoletum*	F	P	I
Hirundinidae				
Gray-breasted martin	*Progne chalybea*	N	S	D
Blue and white swallow	*Notiochelidon cyanoleuca*	N	L	D
Rough-winged swallow	*Stelgidopteryx ruficollis*	N	S	I
Corvidae				
Green jay	*Cyanocorax yncas*	N	S	I
Troglodytidae				
Moustached wren	*Thryothorus genibarbis*	N	P	D
Rufous-breasted wren	*T. rutilus*	N	P	D
Rufous and white wren	*T. rufalbus*	N	N	D
House wren	*Troglodytes aedon*	N	N	D
Nightingale wren	*Microcerculus marginatus*	N	N	D
Mimidae				
Tropical mockingbird	*Mimus gilvus*	N	P	I
Turdidae				
Gray-cheeked thrush	*Catharus minimus*	N	S	D
Pale-breasted thrush	*Turdus leucomelas*	N	S	I
Cocoa thrush	*T. fumigatus*	N/F	S	I
Bare-eyed thrush	*T. nudigenis*	N	S	I
White-necked thrush	*T. albicollis*	N	S	I
Sylviidae				
Long-billed gnatwren	*Ramphocaenus melanurus*	N	P	D
Vireonidae				
Rufous-browed peppershrike	*Cyclarhis gujanensis*	F/N	N	I
Golden-fronted greenlet	*Hylophilus aurantiifrons*	F	S	D
Scrub greenlet	*H. flavipes*	N	P	D
Icteridae				
Giant cowbird	*Scaphidura oryzivora*	F	S	I
Crested oropendola	*Psarocolius decumanus*	N	L	I
Russet-backed oropendola	*P. angustifrons*	N	L	I
Yellow-rumped cacique	*Cacicus cela*	N	S	I
Carib grackle	*Quiscalus lugubris*	N	L	I
Yellow oriole	*Icterus nigrogularis*	N	N	I
Orange-crowned oriole	*I. auricapillus*	N	P	I
Baltimore oriole	*I. galbula*	N	S	C
Parulidae				
Black and white warbler	*Mniotilta varia*	F	N	D
Tropical parula	*Parula pitiayumi*	F	P	D
Cerulean warbler	*Dendroica cerulea*	F	S	D
Blackpoll warbler	*D. striata*	F	N	D
Northern waterthrush	*Seiurus noveboracensis*	N	N	D
American redstart	*Setophaga ruticilla*	N/F	N	D

	Species	Social Organization		Food distribution[3]
		Inter-specific[1]	*Intra-specific*[2]	
Slate-throated redstart	*Myioborus miniatus*	F	P	D
Golden-crowned warbler	*Basileuterus culicivorus*	F	P	D
Black-crested warbler	*B. nigrocristatus*	N	N	D
Flavescent warbler	*B. flaveolus*	N	N	D
Bananaquit	*Coereba flaveola*	F	N	I
Thraupidae				
Purple honeycreeper	*Cyanerpes caeruleus*	F	P	C
Red-legged honeycreeper	*C. cyaneus*	F	S	C
Green honeycreeper	*Chlorophanes spiza*	F	P	C
Blue dacnis	*Dacnis cayana*	F	S	C
Orange-bellied euphonia	*Euphonia xanthogaster*	F	S/P	C
Thick-billed euphonia	*E. laniirostris*	F	P	C
Bay-headed tanager	*Tangara gyrola*	F	S	C/I
Burnished-buff tanager	*T. cayana*	F	S	C/I
Black-headed tanager	*T. cyanoptera*	F	S	C/I
Golden tanager	*T. arthus*	F	S	C/I
Speckled tanager	*T. guttata*	F	S	C/I
Fawn-breasted tanager	*Pipraeida melanonota*	N	P	I
Blue-gray tanager	*Thraupis episcopus*	F	S	I
Palm tanager	*T. palmarum*	F	P	I
Silver-beaked tanager	*Ramphocelus carbo*	F	L/S	I
Hepatic tanager	*Piranga flava*	F/N	P	I
Summer tanager	*Piranga rubra*	F/N	N	I
White-lined tanager	*Tachyphonus rufus*	F	L	I
Guira tanager	*Hemithraupis guira*	F	S	I
Rose-breasted thrush-tanager	*Rhodinocicla rosea*	N	P	D
Gray-headed ant-tanager	*Eucometis penicillata*	F	S	C
Common bush-tanager	*Chlorospingus opthalmicus*	F	L	C/I
Fringillidae				
Buff-throated saltator	*Saltator maximus*	N	P	I
Grayish saltator	*S. coerulescens*	N	P	I
Rose-breasted grosbeak	*Pheucticus ludovicianus*	F/N	S	I
Blue-black grosbeak	*Cyanocompsa cyanoides*	F/N	P	I
Blue-black grassquit	*Volatinia jacarina*	N	S	C
Gray seedeater	*Sporophila intermedia*	N	S	C
Lined seedeater	*S. lineola*	N	S	C
Yellow-bellied seedeater	*S. nigricollis*	N	S	C
Ruddy-breasted seedeater	*S. minuta*	N	S	C
Lesser seed-finch	*Oryzoborus angolensis*	F	P	D
Large-billed seed-finch	*O. crassirostris*	N	P	D
Ochre-breasted brush-finch	*Atlapetes semirufus*	F	P	D
Black-striped sparrow	*Arremonops conirostris*	N	P	D
Golden-winged sparrow	*Arremon schlegeli*	N	P	D
Lesser goldfinch	*Spinus psaltria*	N	S	C

Appendix 2. Masaguaral Bird Species

	Species	Social Organization		Food distribution[3]
		Inter-specific[1]	Intra-specific[2]	
Anhingidae				
Anhinga	*Anhinga anhinga*	N	N	D
Ardeidae				
White-necked heron	*Ardea cocoi*	N	N	D
Great egret	*Casmerodius albus*	N	N	D
Snowy egret	*Egretta thula*	N	N	D
Little blue heron	*Florida caerulea*	N	N	D
Striated heron	*Butorides striatus*	N	N	D
Cattle egret	*Bubulcus ibis*	N	S	I
Whistling heron	*Syrigma sibilatrix*	N	P	D
Black-crowned night-heron	*Nycticorax nycticorax*	N	N	D
Yellow-crowned night-heron	*Nyctanassa violacea*	N	N	D
Rufescent tiger-heron	*Tigrisoma lineatum*	N	N	D
Ciconiidae				
Maguari stork	*Ciconia maguari*	N	N/S	D
Jabiru	*Jabiru mycteria*	N	N/P	D
Threskiornithidae				
Buff-necked ibis	*Theristicus caudatus*	N	P	D
Sharp-tailed ibis	*Cercibis oxycerca*	N	S	D
Green ibis	*Mesembrinibis cayennensis*	N	N	D
Scarlet ibis	*Eudocimus ruber*	N	S	D
Roseate spoonbill	*Ajaia ajaja*	N	S	D
Anatidae				
Fulvous whistling-duck	*Dendrocygna bicolor*	F	L	D
White-faced whistling-duck	*Dendrocygna viduata*	F	L	D
Black-bellied whistling-duck	*Dendrocygna autumnalis*	F	L	D
Comb duck	*Sarkidiornis melanotos*	F	L	D
Moscovy duck	*Cairina moschata*	N	S	D
Cathartidae				
King Vulture	*Sarcoramphus papa*	N	N	D
Black vulture	*Coragyps atratus*	?	L	D
Turkey vulture	*Cathartes aura*	?	L	D
Lesser yellow-headed vulture	*Cathartes burrovianus*	?	S	D
Accipitridae				
White-tailed kite	*Elanus leucurus*	N	N	D
Everglade kite	*Rostrhamus sociabilis*	N	N	D
Slender-billed kite	*Helicolestes hamatus*	N	N	D
White-tailed hawk	*Buteo albicaudatus*	N	N	D
Roadside hawk	*Buteo magnirostris*	N	N	D

[1] Interspecific organization (N = none, solitary; F = joins mixed species flocks).

[2] Intraspecific organization (N = none, solitary; P = pair, 2 birds; S = small, 3–5 birds; L = large, 6+ birds).

[3] Food distribution (D = dispersed; C = clumped; I = intermediate).

	Species	Social Organization		Food distribution[3]
		Inter-specific[1]	Intra-specific[2]	
Black-collared hawk	*Busarellus nigricollis*	N	N	D
Savanna hawk	*Hetrospizias meridionalis*	N	N	D
Great black-hawk	*Buteogallus urubitinga*	N	N	D
Crane hawk	*Geranospiza caerulescens*	N	N	D
Falconidae				
Laughing falcon	*Herpetotheres cachinnans*	N	P	D
Yellow-headed caracara	*Milvago chimachima*	N	N	D
Crested caracara	*Polyborus plancus*	N	S	D
Aplomado falcon	*Falco femoralis*	N	N	D
American kestrel	*Falco sparverius*	N	N	D
Cracidae				
Rufous-vented chachalaca	*Ortalis ruficauda*	N	L	C
Phasianidae				
Crested bobwhite	*Colinus cristatus*	N	L	D
Aramidae				
Limpkin	*Aramus guarauna*	N	N	D
Rallidae				
Gray-necked wood-rail	*Aramides cajanea*	N	N	D
Purple Gallinule	*Porphyrula martinica*	N	N	D
Jacanidae				
Wattled jacana	*Jacana jacana*	N	N	D
Charadriidae				
Southern Lapwing	*Vanellus chilensis*	N	P	D
Scolopacidae				
Solitary Sandpiper	*Tringa solitaria*	N	N	D
Burhinidae				
Dougle-striped thick-knee	*Burhinus bistriatus*	N	P	D
Columbidae				
Pale-vented pigeon	*Columba cayennensis*	N	S	C
Eared dove	*Zenaida auriculata*	N	L	D
Plain-breasted ground-dove	*Columbina minuta*	F	L	D
Ruddy ground-dove	*Columbina talpacoti*	F	L	D
Scaled dove	*Scardafella squammata*	N	L	D
White-tipped dove	*Leptotila verreauxi*	N	N	D
Psittacidae				
Brown-throated parakeet	*Aratinga pertinax*	N	L	C
Green-rumped parrotlet	*Forpus passerinus*	N	L	C
Orange-chinned parakeet	*Brotogeris jugularis*	N	L	C
Yellow-headed parrot	*Amazona ochrocephala*	N	P	C

	Species	*Social Organization*		*Food distribution*[3]
		Inter-specific[1]	*Intra-specific*[2]	
Cuculidae				
Dwarf cuckoo	*Coccyzus pumilus*	N	N	D
Squirrel Cuckoo	*Piaya cayana*	N	N	D
Smooth-billed ani	*Crotophaga ani*	N	S	D
Groove-billed ani	*Crotophaga sulcirostris*	N	S/L	D
Striped cuckoo	*Tapera naevia*	N	N	D
Strigidae				
Great horned owl	*Bubo virginianus*	N	N	D
Ferruginous pygmy-owl	*Glaucidium brasilianum*	N	N	D
Caprimulgidae				
Pauraque	*Nyctidromus albicollis*	N	N	D
White-tailed nightjar	*Caprimulgus cayennensis*	N	N	D
Apodidae				
Fork-tailed palm-swift	*Reinarda squamata*	N	S	D
Trochilidae				
Blue-tailed emerald	*Chlorostilbon mellisugus*	N	N	C
Glittering-throated emerald	*Amazilia fimbriata*	N	N	C
Bucconidae				
Russet-throated puffbird	*Hypnelus ruficollis*	N	N	D
Picidae				
Spot-breasted woodpecker	*Chrysoptilus punctigula*	N	N	D
Lineated woodpecker	*Dryocopus lineatus*	N	N	D
Red-crowned woodpecker	*Melanerpes rubricapillus*	N	N	D
Red-rumped woodpecker	*Veniliornis kirkii*	N	P	D
Dendrocolaptidae				
Straight-billed woodcreeper	*Xiphorhynchus picus*	N	N	D
Furnariidae				
Pale-breasted spinetail	*Synallaxis albescens*	N	N	D
Yellow-throated spinetail	*Certhiaxis cinnamomea*	N	S	D
Plain-fronted thornbird	*Phacellodomus rufifrons*	N	S	D
Formicariidae				
Barred antshrike	*Thamnophilus doliatus*	N	P	D
Cotingidae				
Black-crowned tityra	*Tityra inquisitor*	N	P	C
Tyrannidae				
Pied water-tyrant	*Fluvicola pica*	N	N	D
White-headed marsh-tyrant	*Arundinicola leucocephala*	N	N	D
Vermillion flycatcher	*Pyrocephalus rubinus*	N	N	D
Cattle tyrant	*Machetornis rixosus*	N	N	C
Fork-tailed flycatcher	*Muscivora tyrannus*	N	L	I
Tropical kingbird	*Tyrannus melancholicus*	N	N	D
Gray kingbird	*Tyrannus dominicensis*	N	N	D
White-bearded flycatcher	*Conopias inornata*	N	P	D

	Species	Social Organization		Food distribution[3]
		Inter-specific[1]	Intra-specific[2]	
Boat-billed flycatcher	*Mergarhynchus pitangua*	N	P	I
Rusty-margined flycatcher	*Myiozetetes cayanensis*	N	P	I
Social flycatcher	*Myiozetetes similis*	N	P	I
Great kiskadee	*Pitangus sulphuratus*	N	N	I
Lesser kiskadee	*Pitangus lictor*	N	P	D
Short-crested flycatcher	*Myiarchus ferox*	N	N	I
Brown-crested flycatcher	*Myiarchus tyrannulus*	N	N	D
Common tody-flycatcher	*Todirostrum sylvia*	N	P	D
Yellow tyrannulet	*Capsiempis flaveola*	N	S	D
Pale-tipped tyrannulet	*Inezia subflava*	N	N/P	D
Yellow-bellied elaenia	*Elaenia flavogaster*	N	N	I
Small-billed elaenia	*Elaenia parvirostris*	N	N	I
Plain-crested elaenia	*Elaenia cristata*	N	N	I
Lesser elaenia	*Elaenia chiriquensis*	N	N	I
Mouse-colored tyrannulet	*Phaeomyias murina*	N	P	D
Southern beardless tyrannulet	*Camptostoma obsoletum*	N	N	I
Hirundinidae				
White-winged swallow	*Tachycineta albiventer*	N	N	D
Bank swallow	*Riparia riparia*	N	L	D
Barn swallow	*Hirundo rustica*	N	L	D
Troglodytidae				
Bicolored wren	*Campylorhynchus griseus*	N	P/S	D
Stripe-backed wren	*Campylorhynchus nuchalis*	N	S	D
House wren	*Troglodytes aedon*	N	N	D
Mimidae				
Tropical mockingbird	*Mimus gilvus*	N	N	I
Turdidae				
Pale-breasted thrush	*Turdus leucomelas*	N	N	I
Bare-eyed thrush	*Turdus nudigenis*	N	S	I
Sylviidae				
Tropical gnatcatcher	*Polioptila plumbea*	F	P	D
Vireonidae				
Rufous-browed peppershrike	*Cyclarhis gujanensis*	F	N	I
Parulidae				
Bananaquit	*Coereba flaveola*	F	N	I
Yellow warbler	*Dendroica petechia*	N	N	D
Icteridae				
Shiny cowbird	*Molothrus bonariensis*	N	L	C
Yellow-rumped cacique	*Cacicus cela*	N	L	C
Carib grackle	*Quiscalus lugubris*	N	L	I
Yellow-hooded blackbird	*Agelaius icterocephalus*	N	L	C
Troupial	*Icterus icterus*	N	L	C
Yellow oriole	*Icterus nigrogularis*	N	N	I
Oriole blackbird	*Gymnomystax mexicanus*	N	L	I
Red-breasted blackbird	*Leistes militaris*	N	N	D
Eastern meadowlark	*Sturnella magna*	N	N	D

Species		Social Organization		Food distribution[3]
		Inter-specific[1]	Intra-specific[2]	
Thraupidae				
Trinidad euphonia	*Euphonia trinitatis*	N	P	C
Blue-gray tanager	*Thraupis episcopus*	F	P	I
Glaucous tanager	*Thraupis glaucocolpa*	F	P	I
Fringillidae				
Grayish saltator	*Saltator coerulescens*	N	P/S	I
Orinocan saltator	*Saltator orenocensis*	N	P/S	I
Red-capped cardinal	*Paroaria gularis*	N	P	?
Blue-black grassquit	*Volatinia jacarina*	N	L	C
Gray seedeater	*Sporophila intermedia*	N	L	C
Lined seedeater	*Sporophila lineola*	N	L	C
Ruddy-breasted seedeater	*Sporophila minuta*	N	L	C
Saffron finch	*Sicalis flaveola*	N	L	D
Pileated finch	*Coryphospingus pileatus*	?	P	?
Grassland sparrow	*Ammodramus humeralis*	N	N	D

Appendix 3. Guarico Bird Species (Gallery Forest)

Species		Social Organization		Food distribution[3]
		Inter-specific[1]	Intra-specific[2]	
Tinamidae				
Red-legged tinamou	*Crypturellus erythropus*	N	N	D
Ardeidae				
Capped heron	*Philherodius pileatus*	N	N	D
Anatidae				
Moscovy duck	*Cairina moschata*	N	S	D
Accipitridae				
Gray-headed kite	*Leptodon cayanensis*	N	N	D
Hook-billed kite	*Chondrohierax uncinatus*	N	N	D
Bicolored hawk	*Accipiter bicolor*	N	N	D
Gray hawk	*Buteo nitidus*	N	N	D
Ornate hawk-eagle	*Spizaetus ornatus*	N	N	D

[1] Interspecific organization (N = none, solitary; F = joins mixed species flocks).

[2] Intraspecific organization (N = none, solitary; P = pair, 2 birds; S = small, 3–5 birds; L = Large, 6+ birds.

[3] Food distribution (D = dispersed; C = clumped; I = intermediate).

	Species	Social Organization		Food distribution[3]
		Inter-specific[1]	Intra-specific[2]	
Falconidae				
Collared forest-falcon	*Micrastur semitroquatus*	N	N	D
Bat falcon	*Falco rufigularis*	N	N	D
Cracidae				
Rufous-vented chachalaca	*Ortalis ruficauda*	N	L	C
Yellow-knobbed curassow	*Crax daubentoni*	N	S	?
Rallidae				
Gray-necked wood-rail	*Aramides cajanea*	N	N	D
Eurypygidae				
Sunbittern	*Eurypyga helias*	N	N	D
Jacanidae				
Wattled jacana	*Jacana jacana*	N	N	D
Columbidae				
Pale-vented pigeon	*Columba cayennensis*	N	S	C
Blue ground-dove	*Claravis pretiosa*	N	P	D
White-tipped dove	*Leptotila verreauxi*	N	N	D
Psittacidae				
Scarlet macaw	*Ara macao*	N	P	C
Yellow-headed parrot	*Amazona ochrocephala*	N	P	C
Opisthocomidae				
Hoatzin	*Opisthocomus hoazin*	N	S	C
Cuculidae				
Squirrel cuckoo	*Piaya cayana*	N	N	D
Greater ani	*Crotophaga major*	N	L	D
Nyctibiidae				
Great potoo	*Nyctibius grandis*	N	N	D
Trochilidae				
Blue-chinned sapphire	*Chlorestes notatus*	N	N	C
Blue-tailed emerald	*Chlorostilbon mellisugus*	N	N	C
Alcedinidae				
Ringed kingfisher	*Ceryle torquata*	N	N	D
Amazon kingfisher	*Chloroceryle amazona*	N	N	D
Green kingfisher	*Chloroceryle americana*	N	N	D
Pygmy kingfisher	*Chloroceryle aenea*	N	P	D
Galbulidae				
Rufous-tailed jacamar	*Galbula ruficauda*	N	N	D
Bucconidae				
Russet-throated puffbird	*Hypnelus ruficollis*	N	N	D
Picidae				
Scaled puculet	*Picumnus squamulatus*	F	N	D
Lineated woodpecker	*Dryocopus lineatus*	N	N	D

	Species	Social Organization		Food distribution[3]
		Inter-specific[1]	Intra-specific[2]	
Red-crowned woodpecker	*Melanerpes rubricapillus*	N	N	D
Crimson-crested woodpecker	*Campephilus melanoleucos*	N	N	D
Dendrocolaptidae				
Red-billed scythebill	*Campylorhamphus trochilirostris*	N	N	D
Furnariidae				
Streaked xenops	*Xenops rutilans*	F	N	D
Formicariidae				
Black-crested antshrike	*Sakesphorus canadensis*	N	P	D
Barred antshrike	*Thamnophilus doliatus*	N	P	D
White-fringed antwren	*Formicivora grisea*	N	P	D
Cotingidae				
Cinereous becard	*Pachyramphus rufus*	F/N	P	D
Black-crowned tityra	*Tityra inquisitor*	N	P	C
Pipridae				
Lance-tailed manakin	*Chiroxiphia lanceolata*	N	S	C
Tyrannidae				
Variegated flycatcher	*Empidonomus varius*	N	N	I
Piratic flycatcher	*Legatus leucophaius*	N	N	C
Boat-billed flycatcher	*Megarhynchus pitangua*	N	P	I
Streaked flycatcher	*Myiodynastes maculatus*	N	N	I
Great kiskadee	*Pitangus sulphuratus*	N	N	I
Lesser kiskadee	*Pitangus lictor*	N	P	D
Dusky-capped flycatcher	*Myiarchus tuberculifer*	F	N	D
Bran-colored flycatcher	*Myiophobus fasciatus*	N	N	D
Yellow-breasted flycatcher	*Tolmomyias flaviventris*	F	N	D
Common tody-flycatcher	*Todirostrum cinereum*	N	P	D
Slate-headed tody-flycatcher	*Todirostrum sylvia*	N	P	D
Pale-eyed pygmy-tyrant	*Atalotriccus pilaris*	N	N	D
Yellow tyrannulet	*Capsiempis flaveola*	N	S	D
Pale-tipped tyrannulet	*Inezia subflava*	N	N/P	D
Forest elaenia	*Myiopagis gaimardii*	F	P	I
Troglodytidae				
Buff-breasted wren	*Thryothorus leucotis*	N	P	D
House wren	*Troglodytes aedon*	N	N	D
Sylviidae				
Tropical gnatcatcher	*Polioptila plumbea*	F	P	D
Vireonidae				
Rufous-browed peppershrike	*Cyclarhis gujanensis*	F	N	I
Red-eyed vireo	*Vireo olivaceus*	F	N	I
Scrub greenlet	*Hylophilus flavipes*	N	P	D
Golden-fronted greenlet	*H. aurantiifrons*	F	P/S	D
Parulidae				
Bananaquit	*Coereba flaveola*	F	N	I
Chestnut-vented conebill	*Conirostrm speciosum*	F	S	I

	Species	Social Organization		Food distribution[3]
		Inter-specific[1]	Intra-specific[2]	
Tropical parula	*Parula pitiayumi*	F	P	D
Yellow warbler	*Dendroica petechia*	N	N	D
Blackpoll warbler	*D. striata*	F	N	D
American redstart	*Setophaga ruticilla*	F	N	D
Icteridae				
Crested oropendola	*Psarocolius decumanus*	N	S	I
Yellow-rumped cacique	*Cacicus cela*	N	L	C
Orange-crowned oriole	*Icterus auricapillus*	N	N	I
Thraupidae				
Trinidad euphonia	*Euphonia trinitatis*	N	P	C
Thick-billed euphonia	*Euphonia laniirostris*	F	P	C
Blue-gray tanager	*Thraupis episcopus*	F	P	I
Hooded tanager	*Nemosia pileata*	?	P	D
Guira tanager	*Hemithraupis guira*	F	P	I
Fringillidae				
Black-striped sparrow	*Arremonops conirostris*	N	N	D

SECTION 7:

Reptile Studies

In a meeting of major biomass components of the llanos ecosystem, a caiman and a capybara eye each other.

DALE L. MARCELLINI
National Zoological Park
Smithsonian Institution
Washington, D. C. 20008

Activity Patterns and Densities of Venezuelan Caiman (*Caiman crocodilus*) and Pond Turtles (*Podocnemis vogli*)

ABSTRACT

Activity patterns for *Caiman* and *Podocnemis* were studied in the llanos of Venezuela. Basking by turtles appears to be controlled by light and is almost exclusively confined to daylight hours. *Caiman* does not show such rigid peaks in its basking activity. Densities for *Caiman* and *Podocnemis* were estimated. It is concluded that the larger reptiles in the llanos make a significant contribution to the total vertebrate biomass.

RESÚMEN

Se estudia la modalidad de actividad en *Caiman* y *Podocnemis* en los llanos de Venezuela. Las permanencia de las tortugas al sol parece estar controlada por la luz y está limitada casi exclusiramente a las horas de luz de día. La *Caiman* no muestra tanta rigidez de horario en está actividad. La densidad de *Caiman* y *Podocnemis* fué calculada. Sin duda, estos reptiles hacen un importante contribución a la biomasa total de vertebrados en los llanos.

Introduction

In 1975 a commitment was made to study reptiles as a part of the Smithsonian Venezuelan Research Project. Interest was directed at two of the most evident and abundant reptiles in the llanos—the pond turtle (*Podocnemis vogli*) and the caiman (*Caiman crocodilus*). Investigations were concentrated on activity cycles and thermal ecology with some work on behavior. Research was conducted on Masaguaral in the State of Guarico and at Rancho El Frio in the State of Apure. Over 13 months of field work was done between September 1975 and July 1977. This work was conducted in the dry season (November–April), but some wet season (May–October) data were collected.

Three investigators were involved with the reptile work: the author who participated in all aspects of the research; Dr. Jeffrey Lang, who concentrated on the thermal ecology and social behavior of the caiman; and Mr. Scott Maness, who worked on the thermal ecology of the pond turtle. The results of the studies of Lang and Maness will be reported elsewhere.

In this paper I report on two aspects of the reptile research. First, I describe and compare the 24-hour activity cycles of caiman and turtles. Second, I report on the density and biomass of llanos caiman and turtles.

PART I: Activity Patterns

Diel activity cycles have been described for a number of reptiles (Bustard, 1968, 1970; Cloudsley-Thompson, 1961; Marcellini, 1971). These studies have dealt largely with lizards, but some work has been done on diel activity cycles of aquatic turtles and crocodilians. Cott (1961), Cloudsley-Thompson (1964) and Modha (1968) have reported on the 24-hour activity cycles of the Nile crocodile; Lang (1976) discussed the activity cycle of the American alligator and Stanton and Dixon (1975) briefly reported on the activity cycle of *Caiman crocodilus*. Turtle diel activity cycles have been mentioned in numerous papers but numerical data are rare. The activity cycle of the yellow-bellied turtle has been quantified by Auth (1975) and some work was done by Moll and Legler (1971) on the basking activity of Panamanian *Chrysemys scripta*. Other than the work of Moll and Legler and Stanton and Dixon, no quantitative investigations have been done on diel activity cycles of tropical New World chelonians or crocodilians.

This paper describes and compares the daily activity cycles of the spectacled caiman (*Caiman crocodilus*) and the pond turtle (*Podocnemis vogli*). Possible environmental determinants for these cycles are discussed and the present work is compared with the results of other studies on activity cycles of crocodilians and turtles.

Methods and Materials

Data were gathered during February, March, and April of 1976 at Masaguaral in the State of Guarico and Rancho El Frio in the State of Apure. Ponds of these ranches contain large populations of caimans and turtles. These populations are concentrated during the dry season (November through April) making censuses possible. Six census sites were used: two on Masaguaral, and four at Rancho El Frio. Three of the six sites were on ponds of a size that allowed the entire pond to be censused (less than 200 m in the largest dimension). Three sites were on large ponds (greater than 200 m in the largest dimension) and only animals in a specified area were counted. Hourly counts of turtles and caiman were made from predetermined census spots at each pond. A hand-counter was used and binoculars aided the day counts. Night censuses were done with head lamps. Day and night caiman counts were relatively easy because the animals are large and their eyes shine when struck by the light of a head lamp. Turtle counts were more difficult because of their relatively small size and lack of eye shine. Fortunately night turtle counts proved to be of little importance because few turtles were abroad after dark. The number of animals in and out of the water was noted. Individuals with more than half their total length out of the water were counted as basking and vice versa. In this paper, the term basking is synonymous with out-of-the-water. Censuses required about 15 minutes to complete and were started on the hour.

Over 200 hourly counts covered a 24-hour census period, but more data were taken from 0600 to 2200 hours. Basking and in-water counts for each hour at each study site were averaged and expressed as a percentage of the highest number of caiman or turtles censused at that site. Basking and in-water percentages for each hour at each site were averaged for each species to produce the 24-hour activity cycle graphed in Figures 1 and 2. Thus, each point of the graph represents a composite of census days and sites.

Air and water temperatures were obtained with a Schulthesis thermometer. Air temperatures were taken one meter from the ground in the shade. Temperatures were not recorded for each census hour at each site, but sufficient air temperatures were obtained (101) to construct a "typical" 24-hour temperature cycle. Water temperatures were taken approximately 30 cm from shore and 5 cm deep. These temperatures were also taken irregularly (97), but it is felt that they give a good indication of water temperature variation during the study period.

Results

Air temperatures ranged from 23–37° C with a mean of 30° C. A representative 24-hour air temperature cycle was obtained by averaging hourly temperatures taken at the six sites (Figures 1 and 2). Water temperatures ranged from 21–31° C with a mean of 26° C. Daily water temperatures were lowest in the morning and highest in the afternoon, but the 24-hour variation never exceeded 6° C and was generally less than 3° C.

Podocnemis vogli activity is largely restricted to the daylight hours with only a few irregular sightings occurring from 2000 to 0400 hours (Figure 1). *Podocnemis* are first seen in the water before sunrise, and they increase in numbers over a six-hour period to a morning peak at 1000 hours. At about 0700 hours, some turtles begin to come out of the water to bask and at 0800 hours an explosive emergence occurs until a peak is reached at about 1300 hours. There appears to be a drop in both in-the-water and basking counts at midday. This is followed by the highest counts in both categories. Basking activity drops abruptly from this afternoon peak to an absence of basking by 1900 hours. Turtles are still evident in the water after they have ceased basking but these counts also diminish to very low numbers (less than 1 percent of highest number censused) by 2000 hours.

Caiman crocodilus are visible throughout the 24-hour period with higher counts from 1800 to 0700 hours (Figure 2). At 0600 hours very few caiman are basking but many are visible in the water. As the numbers in the water decrease to a low at about 1600 hours, the number basking increases to a morning high at about 0900 hours. From this high, a decrease occurs until a basking low is reached at midday (1300–1400 hours). A second and more pronounced afternoon basking period follows this low with a peak from 1600 to 1800 hours. At about this time the number of caiman visible in the water begins to increase rapidly as the number basking drops. By 2100 hours, basking is essentially finished and the number in the water has peaked.

Turtles in the water did not exceed 25 percent of the highest number censused, while basking turtles approached 60 percent of the highest number censused. Basking caiman did not exceed 36 percent of the highest number censused, while percentages for caiman in the water were over 90 percent of the highest number censused.

Discussion

Podocnemis diel activity appears to be governed by light. Turtles are not often seen at night and, when they are seen, they are only visible for a moment. As

Figure 1. Twenty-four hour activity cycle for *Podocnemis vogli* based on average hourly counts from six ponds. The numbers of animals in and out of the water are expressed as a percentage of the highest number censused and are compared to average hourly air temperatures. Percentages less than one are graphed as zero.

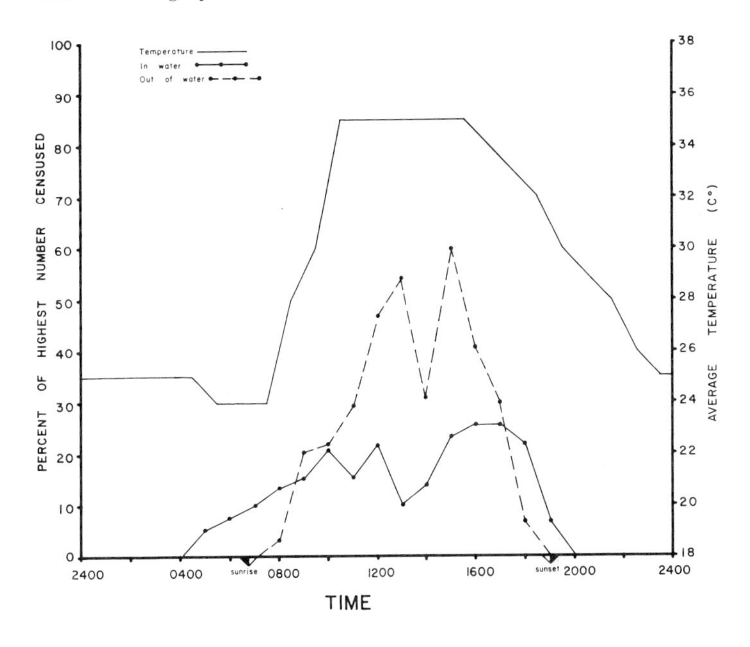

Figure 2. Twenty-four hour activity cycle for *Caiman crocodilus* based on average hourly counts from six ponds. The numbers of animals in and out of the water are expressed as a percentage of the highest number censused and are compared to average hourly air temperature. Percentages less than one are graphed as zero.

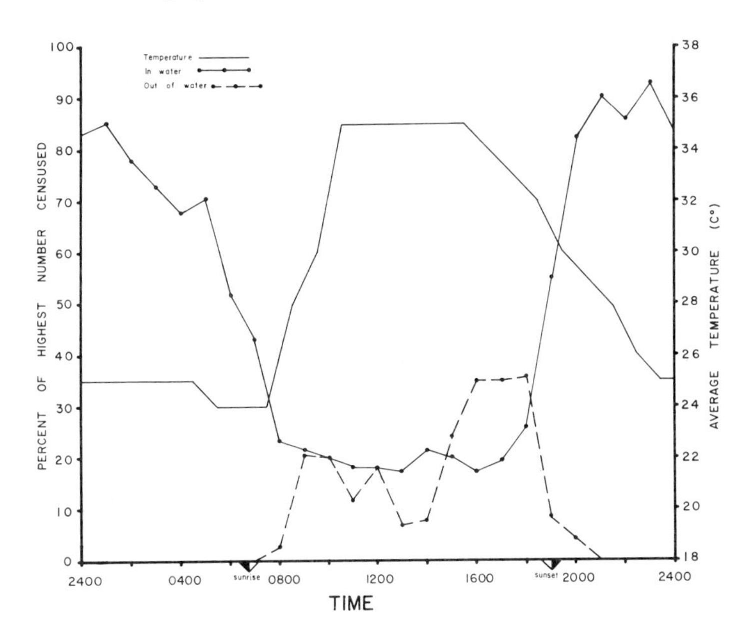

sunrise approaches the numbers of turtles visible in the water increases greatly and these numbers remain high throughout the day. As sunset nears turtle sightings decrease and few are visible after sunset. Turtle basking, on the other hand, seems to be governed in part by air temperature. *Podocnemis* do not come out of the water in numbers until air temperatures have nearly reached their daytime highs (Figure 1), but their retreat to the water begins while air temperatures are still high. The midday drop in basking turtles could be the result of overheating. This might also explain the sudden decrease in basking after 1500 hours while air temperatures are still high. Turtles may overheat in the afternoon and seek the water to cool off but the onset of darkness precludes a return to basking.

Caiman are visible throughout the 24-hour period, but light appears to govern if caiman are visible in the water or basking. Few caiman come out of the water before sunrise and few remain out after sunset. Air temperature seems to have little effect on basking emergence or retreat. Large numbers of caiman are out long before air temperatures have risen significantly from their night-time lows and most caiman return to the water while afternoon air temperatures are still high (Figure 2), but high air temperatures and concurrent overheating may be responsible for the reduction in midday basking activities.

Water temperature variations are slight (6° C maximum) and do not appear to effect the diel activity patterns of caiman or turtles.

Overcast skies and strong winds appeared to reduce basking activity of both caiman and turtles, but sufficient data are not available to adequately support this statement. The effect of rain was not determined because of a lack of rain during the study period.

Interspecific interactions did not appear to affect the activity patterns of the caiman and turtles. The animals seemed to ignore each other, even when basking together. No instances of caiman predation on turtles were observed, although animals were in close proximity at all hours.

The diel activity cycle determined for Nile crocodiles (Cloudsley-Thompson, 1964; Cott, 1961; Modha, 1968) is very similar to that of *Caiman crocodilus.* The crocodiles spend the night in the water and some begin basking before sunrise. Basking activity decreases before sunset and few animals are out after dark. Basking activity is bimodal with a peak from 0800 to 1000 hours, a midday low and another peak from 1600 to 1800 hours. In Modha's study no crocodiles were out basking at midday but this was thought to be a result of very high air temperatures.

Lang's work (1976) with captive juvenile American alligators presents a similar pattern to that described for crocodiles. The alligators began to move out of the water before sunrise and all were out by 0900 hours. In the evening, they moved back into the water (1700 to 1800 hours). A midday drop in basking activity was not quantified but it was mentioned.

Stanton and Dixon (1975), working at Masaguaral, showed basking patterns of *Caiman crocodilus* similar to those in the present paper. Differences are seen in higher morning basking peaks, a more pronounced depression of basking during midday hours, and a higher incidence of basking after sunset. These differences might be related to the fact that they censused only a small portion of one pond where local factors, such as exposure and water temperature, might affect basking.

The percentage of the population basking at peak hours in Nile crocodiles varies from 40 percent (Modha, 1968) to 70 percent (Cott, 1961), while peak in the water percentages are all over 90 percent. The data for *Caiman crocodilus* in the present study agrees fairly well with these results.

Diel activity patterns for turtles are infrequently reported in the literature and data are generally restricted to basking activity. *Chrysemys scripta* in Panama exhibit two basking peaks; one at 1000 hours and a second higher peak at 1300 hours (Moll and Legler, 1971). The peaks were separated by a low at 1100 hours. Basking was virtually nonexistent before sunrise and after sunset, and few turtles were seen after dark. Cagle (1950) mentions that *C. scripta* are active throughout the day with basking peaks at midmorning and mid-afternoon. Visible activity after dark was much reduced, but Cagle did see some turtles basking late in the evening. Auth (1975) reported that Florida *C. scripta* exhibit seasonally different diel activity cycles. In all cases, turtles were infrequently seen before sunrise and after sunset. The time of morning emergence to bask varied with the season, but return to the water was consistently completed by 1800 hours. In September, October, and August, turtles emerged in numbers at about the same time (0900 hours) but in November, morning emergence was delayed until 1000 hours. Basking curves for September, October and November were essentially unimodal with peak numbers being reached later in the day as the season progressed. The August basking curve was bimodal with a higher morning peak at 1000 hours, a depression at 1200 hours and a second much lower peak at 1400 hours. Ernst (1976) found that spotted turtles (*Clemmys guttata*) in Pennsylvania demonstrated a single basking peak with no turtles visible before sunrise or after sunset.

The diel activity of *Podocnemis vogli* is similar to those

reported above but differences in time of emergence and retreat and in midday basking pattern are found. These differences appear to be related to climatic and seasonal differences. Turtles emerge later and retreat earlier in colder climates and during the colder times of the year. Midday basking is unimodal when weather is cool and bimodal when the weather is warm.

Light and temperature have been mentioned as major factors controlling the diel cycle of Nile crocodiles (Cloudsley-Thompson, 1964; Cott, 1961; Modha, 1968) with light determining morning emergence and evening retreat while air temperature controls midday basking. Lang (1976) demonstrated that morning emergence to bask and evening retreat to the water by juvenile American alligators was cued by light.

The determinants for 24-hour activity cycles in turtles have been little studied but some work has been done with determinants for basking behavior. It is generally agreed that turtles do not bask and are generally less conspicuous after dark and that light controls visible activity (Auth, 1975; Ernst, 1976; Moll and Legler, 1971). Basking activity is apparently controlled by light intensity and air temperature (Auth, 1975; Boyer, 1965).

The literature on determinants for diel activity cycles in crocodilians and chelonians agrees well with the data in the present paper. It appears that light is important as a regulator of the type of activity engaged in. In turtles, light determines if they are visible in the water, basking or under water. Crocodilians are generally visible throughout the 24-hour period but light regulates basking emergence and retreat. Air temperature appears to be important to basking in both turtles and crocodilians. Air temperature seems to closely control basking in turtles. They do not emerge from the water until air temperatures are high and they appear to retreat to the water at midday to cool themselves, reemerging for another basking period in the afternoon. Crocodilians seem less dependent on air temperature than turtles, but midday basking patterns appear to be affected by air temperatures.

This paper has outlined the dry season diel activity cycles for Venezuelan *Caiman crocodilus* and *Podocnemis vogli.* It should be pointed out that the llanos environment changes abruptly in the wet season and that these changes greatly affect the turtle and caiman populations. More work is needed to determine if the diel activity cycles of these reptiles differ with the season.

PART II: Density and Biomass of *Caiman crocodilus* and *Podocnemis vogli* in the Llanos of Venezuela

Introduction

Investigations of reptile density and biomass have been primarily directed at northern hemisphere squamates. Studies of tropical crocodilians have been carried out (Cott, 1961; Modha, 1968; Stanton and Dixon,1975), but little quantative data on density and biomass are available. Work on tropical turtles has also been limited with few ecological studies (Moll and Legler, 1971). The large concentrations of caiman and turtles in the ponds of the llanos have resulted in an interest in demographic features of these populations.

This section reports on density and biomass of Venezuelan *Caiman crocodilus* and *Podocnemis vogli* with some discussion of seasonal changes in these parameters. The results of this study are compared to those of other investigations.

Materials and Methods

The data in this paper were collected during 1976 and 1977. Five ponds were utilized in this work; four on Masaguaral, and one on Rancho El Frio. The study ponds were measured and crude, depth profiles were made. Pond areas were calculated using average values derived from these measurements. The ponds differed in their size (Table 1) and permanence. Pond one dries yearly, pond two dries in low rainfall years, ponds three, four and the "figure eight pond" are permanent.

Numerical data on reptile populations were collected by censusing, using the methods described in the activity patterns portion of this paper. Biomass estimates were made using conservative weight figures of 1.5 kg for turtles (present study; Alarcon, 1969) and 15 kg for caiman (present study; Stanton and Dixon, 1975). Miscellaneous observations were made on caiman and turtle activity and behavior on both Rancho El Frio and Masaguaral.

Results

The number of caiman and turtles in a pond varied with season, pond type and, perhaps, year (Table 2). Temporary ponds have smaller populations of these reptiles than do permanent ponds. Caiman and turtles that frequent temporary ponds have been observed to move out of these ponds as the dry season progresses. Both caiman and turtles are known to disperse into the flooded llanos during the wet season. This fact is demonstrated by low pond counts in the wet season.

Table 1. Mean dimensions and calculated surface areas for five ponds

Location	*Pond*	*Mean dimensions*[1]			
		Length (m)	*Depth (m)*	*Width (m)*	*Area (m^2)*
Rancho Masaguaral	1	45	.87	10	450
	2	200	.82	30	6000
	3	120	.66	24	2880
	4	685	.34	92	63020
Rancho El Frio	"Figure 8"	29	1.00	20	580

[1] Dimensions for ponds 1 and 2 were obtained in the wet season while those for ponds 3, 4 and "Figure 8" are for the dry season.

Table 2. Numbers of caimans and turtles censused in five ponds during wet and dry seasons

Date	*Season*	*Ponds*									
		1 Temporary		*2 Temporary*		*3 Permanent*		*4 Permanent*		*"Figure 8" Permanent*	
		C	*T*	*C*	*T*	*C*	*T*	*C*	*T*	*C*	*T*
9/20/75	Wet	1	1	0	30	2	0	30	50		
3/4/76	Intermediate	0	0	0	0	31	5	60	93		
4/7/76	Dry					47	7	186	146		
1/16/76	Wet									28	17
4/13/76	Dry									189	314
4/20/77	Dry									230	600
5/2/77	Wet									9	23

C = caimans; T = turtles.

Table 3. Maximum numbers of caimans and turtles censused with calculated density and biomass for three ponds during the dry season

Locality	*Pond*	*Date*	*Turtles*			*Caimans*			*Combined*	
			Maximum Censused	*no/m^2*	*kg/m^2*	*Maximum Censused*	*no/m^2*	*kg/m^2*	*no/m^2*	*kg/m^2*
Rancho Masaguaral	3	4/7/76	6	.002	.003	46	.016	.240	.018	.243
	4	4/7/76	146	.002	.003	204	.003	.045	.005	.048
Rancho El Frio	"Figure 8"	4/18/76	314	.540	.810	189	.330	4.890	.680	5.420
		4/20/77	600	1.030	1.500	230	.400	5.900	1.400	7.400

Figure 3. Aggregation of *Caiman crocodilus* and *Podocnemis vogli* during the dry season on the llanos of Venezuela.

The severity of the dry season may vary yearly, resulting in differences in numbers of animals migrating to permanent ponds. Thus the "Figure 8" pond had a much higher population of reptiles in April 1977 than it had in April 1976 (Table 2).

The drying of temporary ponds combined with movement into permanent ponds results in large concentrations of caiman and turtles (Figure 3). At the height of the dry season, estimates of density and biomass were made for three permanent ponds (Table 3). The density of turtles ranged from .002 to 1.03 individuals per square meter, while caiman ranged from .003 to .40 per square meter. Biomass for turtles was from .003 to 1.5 kg/m^2 while caiman ranged from .045 to 5.90 kg/m^2. Combined biomass for the "Figure 8" pond in 1977 was 7.40 kg/m^2.

The 1977 research allowed the collection of some data on distances that caiman and turtles moved from the "Figure 8" pond during the wet seasons. Marked caiman were observed up to 5 km from the pond, while marked turtles were seen as far away as 1.5 km. In both cases, these animals were in the shallow water that flooded the llanos after the rains. Using these distances as radii of a circle with the pond at its center, areas of dispersion of 7.1 km^2 for turtles and 78.6 km^2 for caiman were calculated. These areas were used to estimate wet season density and biomass (Table 4) of 84.5 individuals/km^2 and 127 kg/km^2 for turtles, and 2.9 individuals/km^2 and 44 kg/km^2 for caiman. The combined values are 87.4 individual reptile/km^2 and 171 kg/km^2 biomass.

Table 4. Wet season density and biomass estimates for 230 caimans and 600 turtles in the area around the "Figure 8" pond

Turtles		*Caimans*		*Combined*	
no/km^2	*kg/km^2*	*no/km^2*	*kg/km^2*	*no/km^2*	*kg/km^2*
84.5	127.0	2.9	44.0	87.4	171.0
no/h^2	*kg/h^2*	*no/h^2*	*kg/h^2*	*no/h^2*	*kg/h^2*
.85	1.28	.03	.44	.88	1.72

[1] Calculated area of dispersion for turtles is 7.1 km^2 while for caimans it is 78.6 km.^{2}

Discussion

Wet season reptile density and biomass figures in this paper are comparable to estimates for mammals. Wet season turtle data are similar to estimates for mammalian foliage and fruit eaters, while caiman data are similar to estimates for mammalian omnivores (Odum, 1971).

Some of the dry season density figures in this paper are higher than previous estimates for turtles and crocodilians. Aquatic turtle densities of from $.014/m^2$ (Moll and Legler, 1971) to $.074/m^2$ (Auth, 1975) have been reported. The figures from Auth's work with Florida *Chrysemys* are higher than most figures in the present paper, but the "Figure 8" pond data greatly exceeds Auth's density estimates. If wet season turtle densities are considered, the present work shows lower densities than the 25 per hectare estimated for terrestrial *Terrapine* by Schwartz and Schwartz (1974).

Estimates for crocodilian densities in the literature are as high as $140/km^2$ in Nile crocodiles (Cott, 1961). Stanton and Dixon (1975), working at Masaguaral in some of the same ponds as the present study, calculated dry season densities for *Caiman crocodilus* of from .008 to $.089/m^2$. These figures are similar to data from most ponds in the present study, but the "Figure 8" pond exhibits caiman densities that are much higher than those reported by Stanton and Dixon.

The high densities and biomass values calculated for reptiles at and surrounding the "Figure 8" pond are probably not unusual for this part of the llanos. Many other ponds at Rancho El Frio had very large populations of reptiles in them. In some cases the ponds were close enough that there would be considerable overlap of reptile populations. In these cases, even higher wet season density and biomass estimates would result.

More data are needed to further elucidate density and biomass parameters for caiman and turtles in the llanos but it is apparent that these animals are significant members of the vertebrate community.

Acknowledgments

The author wishes to thank Sr. Tomás Blohm, owner of Fundo Pequario Masaguaral, and the Maldonado family, owners of Rancho El Frio. Without the hospitality, knowledge, and help of these people, the research would have been impossible. Dr. Edgardo Mondolfi offered vital support in the initial stages of the research. I thank Peggy O'Connell for her tireless aid with logistics. Mike Davenport, Robert Godshalk and Jeff Wyles were of invaluable assistance in the field work.

Literature Cited

Alarcon, P. H.
1969. Contribucion al conocimiento de la morphologia, ecologia, comportamiento y distribucion geografica de *Podocnemis vogli*, Testudinata, (Pelomedusidae). *Rev. Acad.Colombiana de Ciencias Exactas, Fisicas y Naturales,* 13(51):

Auth, D. L.
1975. Behavioral ecology of basking in the yellow-bellied turtle, *Chrysemys scripta* (Schoepff). *Bull. Fla. St. Mus.,* 20:1–45.

Boyer, D. R.
1965. Ecology of the basking habit in turtles. *Ecology,* 46: 100–118.

Bustard, H. R.
1968. Temperature dependent activity in the Australian gecko, *Diplodactylus vittatus. Copeia,* 3:606–612.
1970. Activity cycle of the tropical house gecko, *Hemidactylus frenatus. Copeia,* 1:173–176.

Cagle, F. R.
1950. The life history of the slider turtle, *Pseudemys scripta troostii* (Holbrook). *Ecol. Monogr.,* 20:31–54.

Cloudsley, Thompson, J. L.
1961. *Rhythmic Activity in Animal Physiology and Behaviour.* New York: Academic Press.
1964. Diurnal rhythm of activity in the Nile crocodile. *Anim. Behav.,* 12:98–100.

Cott, H. B.
1961. Scientific results of an inquiry into the ecology and economic status of the Nile crocodile (*Crocodilus niloticus*) in Uganda and Northern Rhodesia. *Trans. Zool. Soc. London,* 29:211–356.

Ernst, C. H.
1976. Ecology of the spotted turtle, *Clemmys guttata* (Reptilia, Testudines, Testudinidae), in southeastern Pennsylvania. *J. Herp.,* 10:25–33.

Lang, J. W.
1976. Amphibious behavior of *Alligator mississippiensis*: Roles of a circadian rhythm and light. *Science,* 191: 575–577.

Marcellini, D. L.
1971. Activity patterns of the gecko, *Hemidactylus frenatus. Copeia,* 4:631–635.

Modha, M. L.
1968. Basking behaviour of the Nile crocodile on Central Island, Lake Rudolf. *E. Afr. Wildl. J.,* 6:81–88.

Moll, E. O. and J. M. Legler.
1971. The life history of a neotropical slider turtle, *Pseudemys scripta* (Schoepff) in Panama. *Bull. L. A. Co. Mus.,* 11:1–102.

Odum, E. P.
1971. *Fundamentals of Ecology*. Philadelphia: W. B. Saunders Co.

Schwartz, C. W. and E. R. Schwartz.
1974. The three-toed box turtle in central Missouri. *Missouri Dept. Cons., Terrestrial Ser.*, 5:1–28.

Stanton, M. A. and J. R. Dixon.
1975. Studies on the dry season biology of *Caiman crocodilus crocodilus* from the Venezuelan llanos. *Memoria de la Sociedad de Ciencias Naturales La Salle*, 35:237–265.